Metal-Induced Crystallization

Metal-Induced Crystallization

Fundamentals and Applications

edited by
Zumin Wang
Lars P. H. Jeurgens
Eric J. Mittemeijer

Published by

Pan Stanford Publishing Pte. Ltd.
Penthouse Level, Suntec Tower 3
8 Temasek Boulevard
Singapore 038988

Email: editorial@panstanford.com
Web: www.panstanford.com

British Library Cataloguing-in-Publication Data
A catalogue record for this book is available from the British Library.

Metal-Induced Crystallization: Fundamentals and Applications

ISBN 978-981-4463-40-9 (Hardcover)
ISBN 978-981-4463-41-6 (eBook)

Printed in the USA

Contents

Preface

Crystalline semiconductors, often in the form of thin films, are crucial materials for many modern, advanced technologies in fields such as microelectronics, optoelectronics, display technology, and photovoltaic technology. Unfortunately, thin films of semiconductors produced by vapor deposition techniques are usually in a low-performance, amorphous form. The transformation of low-performance, amorphous semiconductors into high-performance, crystalline semiconductors is one of the most important and most challenging steps to advance the above-indicated technologies.

Metals in contact with amorphous semiconductors can induce the crystallization of these amorphous semiconductors at surprisingly low temperatures (as low as 120°C). This so-called metal-induced crystallization (MIC) process was firstly observed more than 40 years ago. In recent years, the MIC process has attracted great scientific and technological interest because it principally allows the production of crystalline semiconductor-based advanced devices at low temperatures directly on heat-sensitive components, such as plastics, which would otherwise have not been possible.

As a result of numerous investigations by different research groups all over the world, the fundamental aspects of the MIC process in various metal/amorphous semiconductor systems have now been disclosed, partly in great detail. MIC represents an intriguing solid–solid phase transformation phenomenon occurring/initiating particularly at surfaces and interfaces in thin-film (nano)systems. In such systems, the phase transformation is not simply controlled by bulk thermodynamics: the (excess) Gibbs energy associated with surfaces and interfaces in the systems may play a significant or even dominant role. Moreover, the kinetics of phase transformation in thin-film (nano)systems can be quite different from that in bulk systems, for example, because of the presence of a relatively large amount of defects (as grain boundaries and dislocations), possibly also providing fast diffusion paths. Fundamental research on the MIC process thus presents an attractive road to deeper insight into the role of interface energetics and kinetics in solid–solid phase transformations.

Advances in the fundamental understanding of the MIC process have resulted in pronounced progress in sophisticated applications of the MIC process in many technologies. This, for example, holds for the production of high-efficiency, low-cost, thin-film crystalline silicon solar cells of advanced flat-panel displays, and of Blu-ray data storage devices.

The present book, for the first time, summarizes the existing knowledge and broad range of applications of the MIC process of amorphous semiconductors. The book firstly addresses the current knowledge and achieved fundamental understanding of MIC processes (Chapters 1–4). Next, the book elucidates how to employ MIC processes in advanced technologies, for example, in novel, state-of-the-art solar cell and display technologies (Chapters 5–7). The aim is to give the reader a comprehensive perspective of the MIC process and thereby to stimulate the future development of novel crystalline semiconductor-based thin-film technologies.

Zumin Wang
Lars P. H. Jeurgens
Eric J. Mittemeijer
Stuttgart, Germany
Winter 2014

Chapter 1

Introduction to Metal-Induced Crystallization

Zumin Wang,[a] **Lars P. H. Jeurgens,**[a,*] **and Eric J. Mittemeijer**[a,b]
[a]*Max Planck Institute for Intelligent Systems, Heisenbergstraße 3, D-70569 Stuttgart, Germany*
[b]*Institute for Materials Science, University of Stuttgart, D-70569 Stuttgart, Germany*
z.wang@is.mpg.de

This chapter gives a historical overview of research performed on the metal-induced crystallization (MIC) process. The MIC temperatures and behaviors for a wide range of metal/amorphous semiconductor systems, as reported in the literature (data obtained using different experimental approaches), have been summarized and tabulated. The development of an understanding of the mechanisms controlling MIC, and related phenomena such as layer exchange, as well as the technological applications of these processes have been sketched as an introduction to later chapters of this book.

Present address: Swiss Federal Laboratories for Materials Science and Technology, Überlandstrasse 129, 8600 Dübendorf, Switzerland

Metal-Induced Crystallization: Fundamentals and Applications
Edited by Zumin Wang, Lars P. H. Jeurgens, and Eric J. Mittemeijer

ISBN 978-981-4463-40-9 (Hardcover), 978-981-4463-41-6 (eBook)
www.panstanford.com

1.1 A Brief History of Metal-Induced Crystallization

In 1969, Oki et al. [1] observed that amorphous Ge (a-Ge) crystallizes at surprisingly low temperatures when it is in contact with a metal such as Al, Ag, Au, Cu, or Sn. Shortly thereafter, Bosnell and Voisey reported that such decreased crystallization temperatures also occur for amorphous Si (a-Si) in contact with a metal [2]. In both studies, the amorphous semiconductors (and the metals) were prepared by vacuum evaporation and an electron diffraction technique was used to detect the occurrence of crystallization. Thereafter, more detailed electron microscopic investigations of this striking effect were carried out by Herd et al. [3] and Ottaviani et al. [4–6], and this phenomenon was named *metal-contact-induced crystallization* [3], nowadays usually referred to as *metal-induced crystallization* (MIC).

The MIC process was found to be associated with intermixing of the semiconductor and the metal, and small crystallites of Si or Ge could indeed be found to have formed in the metal [3–6]. On the basis of these observations, the MIC effect was interpreted as the result of initial dissolution of the semiconductor into the metal, followed by precipitation of the crystalline semiconductor out of the metal matrix [4–6]. An important role of fast atomic transport along the metal/semiconductor interface was indicated in these early studies [3–6]. A very different interpretation of the MIC effect was given by Brodsky and Turnbull [7], who instead suggested that MIC would be mediated by the formation of a low-temperature eutectic melt caused by lowering of the binary eutectic temperature when one of the two components (i.e., the semiconductor) is amorphous.

As compared to the above-described early interpretative efforts, understanding of the MIC process was greatly advanced in the early 1990s by the application of in situ heating transmission electron microscopy (TEM) techniques, which were developed in the late 1980s. By employing in situ heating high-resolution transmission electron microscopy (HRTEM), the MIC process in layered structures of simple eutectic metal semiconductor systems, such as crystalline Al (c-Al)/a-Si [8], crystalline Ag (c-Ag)/a-Ge [9], and c-Ag/a-Si [10], was investigated. It was shown that the MIC process does not involve the formation of any liquid phase: it is a fully solid-state process [8–12]. Furthermore, no formation of any (metastable) metal semiconductor compound(s) was detected during MIC in such systems [8–12]. The MIC process in a compound (silicide)-forming system such as Ni/a-Si was also investigated by in situ heating HRTEM [13, 14]. It was concluded in this work that in such a system MIC is mediated by the initial formation of a solid silicide phase ($NiSi_2$), which subsequently migrates into/within the a-Si film with formation of a crystalline Si (c-Si) phase in the wake of the $NiSi_2$ [13, 14].

In the late 1990s, (related) processes such as metal-induced layer exchange (MILE) [15–17] and metal-induced lateral crystallization (MILC) [18–20] were identified. It was found that upon MIC in, for example, a c-Al/a-Si bilayer, the Al and Si sublayers exchange their original locations [15]. As a result, a nearly continuous c-Si layer is formed at the original location of the metal (Al) sublayer at low temperatures (see Fig. 1.1D). Such a MILE process can be very useful for low-temperature production of thin-film crystalline semiconductors, in particular for thin-film photovoltaic applications (see Chapter 5). In MILC, a compound (e.g., silicide)-forming metal (e.g., Ni or Pd) is firstly grown in a patterned way on top of an amorphous semiconductor layer (e.g., a-Si). Upon annealing, it was found that the crystalline modification of the semiconductor not only grew at the metal-covered region but also grew laterally outside the metal coverage (see, e.g., Fig. 1.2D,E). Such MILC can be very useful for low-temperature preparation of polycrystalline Si thin films for applications in thin-film transistors in advanced flat-panel displays (see Chapter 6).

The thermodynamics and kinetics of MIC and MILE processes have been investigated systemically since 2003 in our research group led by Prof. Mittemeijer [21–37]. A unified, quantitative understanding of the MIC process occurring in various metal-semiconductor systems has been achieved on the basis of *interface thermodynamics*. On this basis the very different MIC temperatures and behaviors in a wide range of metal/amorphous semiconductor systems have been successfully predicted [29, 30, 33] (see Chapter 2). The *kinetics* of MIC and MILE were described on the basis of quantitative evaluation of the (inter)diffusion kinetics in metal/semiconductor layered systems by using Auger electron spectroscopy (AES) depth profile measurements [21, 23, 28] (see Chapter 3). Very recently, advanced

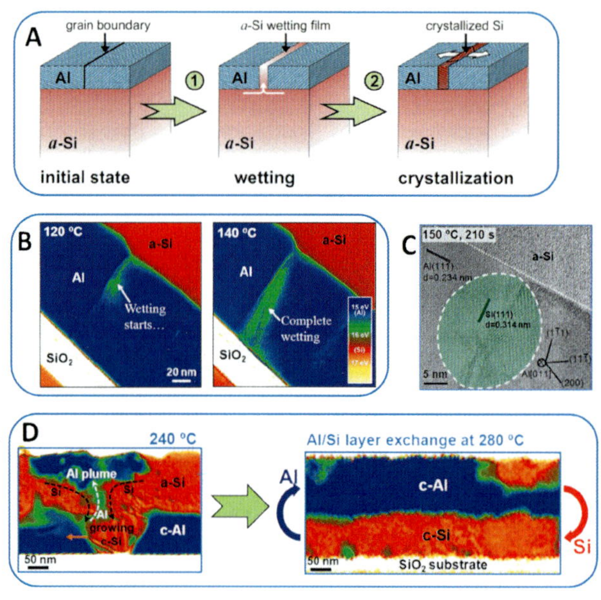

Figure 1.1 (A) Schematic illustration of the mechanism of Al-induced crystallization of a-Si in c-Al/a-Si bilayers. Initially, GBs in the Al overlayer get "wetted" by a-Si. Beyond a critical thickness of the "wetting" a-Si film at a temperature above 140°C, crystallization initiates at these "wetted" Al GBs and the formed c-Si grain subsequently grows laterally in the Al overlayer. Reprinted with permission from Ref. [33], Copyright 2009, John Wiley and Sons. (B) In situ heating valence energy–filtered TEM observation of "wetting" of an Al GB by a-Si during annealing of a 100 nm c-Al/150 nm a-Si bilayer. The increase in the plasmon loss energy at the location of the Al GB near the c-Al/a-Si interface observed at 120°C (left) demonstrates the initial "wetting" of the Al GB by Si. Complete "wetting" of the Al GB by Si has been realized at 140°C (right). Reprinted with permission from Ref. [36], Copyright 2011, John Wiley and Sons. (C) In situ heating HRTEM observation (cross-sectional view) of the nucleation of c-Si at a high-angle Al GB at 150°C. Reprinted with permission from Ref. [36], Copyright 2011, John Wiley and Sons. (D) In situ valence energy–filtered TEM observation (cross-sectional view) of a 150 nm a-Si/100 nm c-Al bilayer upon heating at 240°C, showing the growth of a c-Si nucleus in the Al bottom layer and, at the same time, the development of a mushroom-shaped Al "plume" of cloud-like morphology in the a-Si top layer. Upon heating at 280°C, Si and Al sublayers have practically exchanged their locations: layer exchange has occurred. Reprinted with permission from Ref. [37], Copyright 2012, American Chemical Society. *Abbreviation*: GB, grain boundary.

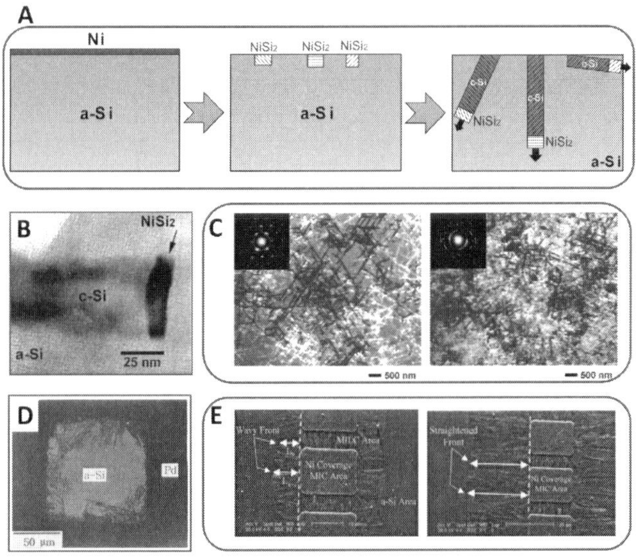

Figure 1.2 (A) Schematic illustration of the mechanism of Ni-induced crystallization of a-Si in c-Ni/a-Si bilayers. Upon heating, initially Ni reacts with a-Si to form $NiSi_2$ at the c-Ni/a-Si interface. c-Si then nucleates at the interface between $NiSi_2$ and a-Si. Continued crystallization of c-Si is realized by migration of $NiSi_2$ into/within the a-Si film, leaving c-Si in its wake. (B) TEM bright-field image of a migrating $NiSi_2$ precipitate with grown needle-like c-Si in its wake. Reprinted with permission from Ref. [59], Copyright 2001, Elsevier. (C) TEM images of a Ni-induced crystallized a-Si film at 400°C for 10 minutes (left) and 30 minutes (right). Continued crystallization of a-Si has occurred by the repeated $NiSi_2$-induced growth of needle-like Si crystallites in the initial a-Si film. Reprinted with permission from Ref. [59], Copyright 2001, Elsevier. (D) Pd (4 nm layer thickness)-induced lateral crystallization of a-Si film (150 nm layer thickness) upon annealing at 500°C for 5 hours (optical micrograph). Reprinted with permission from Ref. [18], Copyright 1995, AIP Publishing LLC. (E) Ni (5 nm layer thickness)-induced lateral crystallization of a-Si film (100 nm layer thickness) upon annealing at 500°C for 7 hours (left) and 21 hours (right) (orientation imaging microscopy image). Reprinted with permission from Ref. [20], Copyright 1998, AIP Publishing LLC.

real-time in situ analytic TEM techniques have been applied to disclose the detailed atomic-scale mechanisms of MIC and MILE, which were neither observed nor even recognized before [36, 37].

Research has been directed to (potential) applications of MIC, in particular since 2000. For example, thin-film c-Si solar cells with an efficiency of higher than 8% have been developed on the basis of an Al-induced crystallization approach [38] (for details of application of the MIC process in thin-film photovoltaic technologies, see Chapter 5). Furthermore, advanced display devices with exceptional performance have been developed by utilizing the low-temperature MILC process of a-Si (for details of application of the MILC process in advanced flat-panel displays, see Chapter 6). Very recently, MIC has also been applied for the production of Blu-ray data storage devices. In this case, because of the low crystallization temperature in metal/a-Si (or a-Ge) bilayer systems, the metal/a-Si (a-Ge) bilayer crystallizes upon low-power (i.e., low-cost) laser irradiation and can therefore serve as an excellent type of data storage medium (for details of application of the MIC process in Blu-ray data storage devices, see Chapter 7).

1.2 Experimental Methods for Investigating the Metal-Induced Crystallization Process

Many different experimental techniques have been employed to investigate the MIC process. Because of different sensitivities (to crystallization) of these techniques, the reported crystallization temperature(s) as determined by different techniques can be (even very) different for the same metal/amorphous semiconductor system (see Tables 1.1 and 1.2). The key features of major techniques used for MIC investigations are briefly summarized in the following sections.

1.2.1 X-Ray Diffraction

X-ray diffraction (XRD) is one the most powerful techniques for investigating the occurrence and kinetics of a phase transformation process, such as crystallization (see Chapter 11 in Ref. [39]). Crucial information on the (initial) nucleation and (subsequent) growth of a certain crystalline semiconductor phase can be extracted from the emergence and increase of the intensities of corresponding diffraction maxima. This method can be applied both to bulk specimens and to thin films and does not require any additional specimen preparation procedure. In combination with dedicated automated heating stages and chambers, which have been well developed in the last decade [39, 40], such in situ heating XRD experiments, usually under nonambient conditions (e.g., in vacuum or in a protective gas, like an argon atmosphere), have been performed to investigate the kinetics of MIC processes in various metal/amorphous semiconductor systems.

Besides monitoring the crystallization process, the XRD technique additionally allows detection of the possible formation of (any) compound phase(s) (e.g., silicides, germanides), which may accompany the MIC process.

Furthermore, the development of stresses in the metal and the crystallized semiconductor phase during MIC can be measured by XRD techniques (i.e., using the so-called d-$\sin^2 \psi$ method, where d = lattice spacing and ψ = specimen tilting angle [41]).

The XRD technique is not sensitive to the very initial (nucleation) stage of the crystallization process, because a minimal amount of (formed) crystalline phase is required to generate a detectable diffraction signal. As a result, the (metal-induced) crystallization temperatures determined by in situ XRD are often somewhat higher than those determined by other techniques (see, e.g., Tables 1.1 and 1.2) and thus can be taken as an upper limit for the (metal-induced) crystallization temperatures.

For example, in situ heating XRD has been applied to investigate MIC processes (including stress evolution) in c-Al/a-Si [30, 37] and c-Al/a-Ge [30, 32] bilayer systems. Recently, in situ heating XRD using synchrotron radiation has been applied to investigate the MIC of a-Si [42] and a-Ge [43] in contact with about 20 different metals.

1.2.2 Transmission Electron Microscopy

The TEM technique (often in combination with electron diffraction and analytical methods such as energy-dispersive X-ray analysis and electron energy loss spectroscopy) can provide direct and highly localized (diffraction contrast and/or high-resolution [HR]) images of the MIC process. Thus the initially formed crystalline semiconductor nuclei can be identified and their subsequent growth can be followed. As a result, the (metal-induced) crystallization temperatures as determined by in situ TEM/HRTEM are typically lower than those determined by other techniques and should thus be considered as more realistic estimates (see Tables 1.1 and 1.2).

In particular, HRTEM is able to resolve the atomic structures of the metal and the crystalline semiconductor and can thereby disclose the atomic-scale mechanism(s) of the MIC processes. By utilizing a heating TEM stage, the MIC process can be visualized in situ in real time by TEM and HRTEM.

One of the major drawbacks of the (HR)TEM technique is that the investigated specimen must be ultrathin (electron transparent). This requires complicated and time-consuming TEM specimen preparation procedures. Moreover, material microstructure and properties may change as a result of the TEM specimen preparation procedure and/or the dimensional constraints of TEM lamellae, thus obscuring or affecting MIC behaviors.

Investigations on MIC were among the first applications of the in situ heating TEM/HRTEM technique shortly after its development in the 1980s (e.g., Refs. [9, 13]). In situ TEM/HRTEM has been applied to investigate the MIC process in c-Al/a-Si [8], c-Ag/a-Si [10, 44], crystalline Au (c-Au)/a-Si [45], crystalline Ni (c-Ni)/a-Si [13, 14], c-Ag/a-Ge [9], and c-Au/a-Ge [46] systems. In very recent years, in situ heating, analytical HRTEM techniques, which can resolve the metal and the semiconductor at the subnanometer scale and which allow high time resolution, have been applied for investigations of the atomic mechanisms of MIC processes [36, 37].

1.2.3 Differential Scanning Calorimetry

Differential scanning calorimetry (DSC) measures the heat released (or consumed) upon phase transformation (e.g., in an MIC process) under either isochronal heating/cooling or isothermal conditions. It can detect sensitively the temperature for initiation of an MIC process. The occurrence (or not) of an intermediate step (e.g., by formation of a stable or metastable silicide phase) during MIC can also be exposed by DSC measurement. Furthermore, important thermodynamic parameters of MIC, such as the crystallization energy and activation energy for MIC, can be determined quantitatively by DSC.

As a major drawback, DSC measurements require a relatively large amount (gram scale) of metal/semiconductor test materials, a constraint that can be difficult to satisfy in investigations on thin-film systems (which is usually the case in MIC research). Consequently, almost always *multilayered* metal/amorphous semiconductor specimens, with more than tens of periods, are utilized for DSC measurements.

DSC analyses of MIC have been carried out for c-Al/a-Si [8], c-Al/a-Ge [32], c-Ag/a-Si [10], and c-Au/a-Si [45, 47] systems.

1.2.4 Spectroscopic Ellipsometry

Spectroscopic ellipsometry (SE) is widely applied to measure the optical properties of thin-film systems. The optical properties of a crystalline semiconductor are very different from those of its amorphous counterpart [48]. Real-time in situ spectroscopic ellipsometry (RISE) can therefore be used to monitor sensitively the initiation of (metal-induced) crystallization of amorphous semiconductors. Unlike the in situ heating (HR)TEM techniques, RISE is a nondestructive technique that does not require any additional specimen preparation step. RISE is also compatible with (ultra)high vacuum systems and is extremely sensitive to crystallization occurring particularly in very thin films.

As a drawback, the RISE technique requires that the investigated layer system be at least semitransparent in a certain wavelength range. Therefore, RISE cannot be applied to systems containing a thick metal (top) layer.

RISE has been applied to determine the Al-induced crystallization temperature of a-Si as a function of the ultrathin Al overlayer thickness [29].

1.2.5 Other Techniques

Besides the major (in situ) techniques discussed above, other techniques have been employed in MIC investigations as well. These can be briefly summarized as follows:

- Microscopy techniques: focused ion beam imaging, scanning electron microscopy and optical microscopy
- Spectroscopic techniques: Raman spectroscopy and AES depth profiling (mainly for studying the diffusion process in association with MIC; see, e.g., Refs. [21, 23])

- Other (indirect) in situ heating techniques: In situ heating optic reflectometry (see, e.g., Refs. [49–51]) and in situ heating electric resistometry (see, e.g., Ref. [2])

1.3 Metal-Induced Crystallization Processes in Crystalline Metal/Amorphous Semiconductor Systems: An Overview

1.3.1 Amorphous Semiconductors Showing MIC

Typical amorphous semiconductors for which MIC has been observed are a-Si, a-Ge, and a-SiGe alloys.

Amorphous semiconductors can be prepared (usually as thin films) at low temperatures [52] by chemical vapor deposition (CVD) (including, e.g., low-pressure chemical vapor deposition [LPCVD], plasma-enhanced chemical vapor deposition [PECVD], and hot-wire chemical vapor deposition [HWCVD]), sputtering (including, e.g., direct current sputtering, magnetron sputtering), and evaporation (including, e.g., thermal evaporation, electron beam evaporation). Besides these deposition methods, amorphous semiconductors can also be produced by high-energy ion irradiation of the corresponding crystalline semiconductors [53].

The microstructure of an amorphous semiconductor can be very different depending on the preparation/growth method, and thus different crystallization temperatures, crystallization energies, and crystallization behaviors can occur [52, 54]. The crystallization of amorphous semiconductors can further be influenced by the presence of incorporated hydrogen and doping elements (e.g., B, P) [54]. The predominant method for producing a-Si and a-Ge in industry is PECVD using gaseous silane(s) or germane(s) as the source material. The a-Si or a-Ge, grown by PECVD, is usually hydrogenated, which can be beneficial for electronic and optoelectronic applications [55]. Such hydrogenated amorphous semiconductors usually possess more medium-range order than those prepared by other techniques and may even already contain ordered (Si) clusters, thus providing precursors for crystallization. As a result, amorphous semiconductors prepared by PECVD are usually more prone to crystallization [54].

a-Si has a crystallization enthalpy of about 11.95 kJ/mol [53], with reported "bulk" crystallization temperatures in the range of 650°C to 800°C. a-Ge has a crystallization enthalpy of about 11.50 kJ/mol [53], with a "bulk" crystallization temperature in the range of 400°C to 650°C.

Besides a-Si and a-Ge, other amorphous materials of a covalently bonding nature, such as amorphous boron, amorphous carbon, and amorphous SiC, have also been investigated (or considered) concerning possible occurrence of MIC. It was found that metals can indeed induce the crystallization of amorphous carbon at low temperatures [56]. However, no MIC effect could be observed for amorphous boron [2] and amorphous SiC (own unpublished work).

1.3.2 Crystalline Metals Inducing MIC

A rich family of metals has been found to be able to promote the crystallization of amorphous semiconductors such that the crystallization temperatures are reduced (very) considerably as compared to the corresponding "bulk" crystallization temperatures.

- Post-transition metals:
  - Group 13 metals: Al, In
  - Group 14 metals: Sn, Pb
  - Group 15 metals: Sb, Bi
- Transition metals:
  - Group 10 metals: Ni, Pd, Pt
  - Group 11 metals: Cu, Ag, Au
  - Group 12 metal: Zn
  - Refractory metals: Mn, Rh, V, Co, Fe, Hf, Cr, Ti, Ir, Ru, Zr, Nb, Mo, W, Ta, Re

Post-transition metals (groups 13, 14, 15) constitute simple eutectic binary systems with Si or Ge. The MIC process using these metals does not involve the formation of any compound phase(s). The MIC temperatures are very low when using these metals (see Tables 1.1 and 1.2).

The transition metals Ag and Au constitute simple eutectic binary systems with Si or Ge. Thus, Ag- or Au-induced crystallization of a-Si or a-Ge is similar to that induced by post-transition metals (see above) and occurs at very low temperatures (see Tables 1.1 and 1.2). Differently, transition metals like Ni, Pd, Pt, and Cu form multiple compounds with Si or Ge. The MIC process using these metals thus involves the formation of one or more compounds and occurs at (only) moderately low temperatures (see Tables 1.1 and 1.2).

Refractory metals can also induce the crystallization of a-Si or Ge but only at quite high temperatures, only slightly (a few tens of degrees Celsius) lower than the corresponding "bulk" crystallization temperatures. Also in this case, the formation of compound phases often occurs before MIC is initiated (see Tables 1.1 and 1.2).

The (strongly metallic) group 1, 2, and 3 metals (e.g., Na, Mg, Y) have *never* been reported to induce the crystallization of amorphous semiconductors. This can be due to the high reactivity of these metal elements in air; they rarely occur as pure metals. It remains a very interesting (scientific) question whether such (near) perfect free-electron metals can induce the crystallization of amorphous semiconductors (this might be tested by, e.g., performing annealing experiments in ultrahigh vacuum to avoid oxidation).

The metal phase capable of inducing the MIC process in adjacent amorphous semiconductor material is in most cases provided in the form of thin films. An alternative way of offering such metal material is to dope, or "insert," it homogenously (meaning that the amount of metal introduced may surpass the solubility limit and the metal can even be present as clusters) into the amorphous semiconductor. This can also lead to the occurrence of MIC, although usually at higher temperatures than the MIC process occurring in *layered* metal/amorphous semiconductor structures [57, 58]. An extensive review of the MIC process induced by doped/inserted metal is given in Chapter 4.

1.3.3 Categories of Metal-Induced Crystallization

The MIC temperatures and occurrence of any silicide phase(s), as reported in the literature for MIC of a-Si with different metals, have been gathered and indicated in Table 1.1. Similar results for a-Ge have been presented in Table 1.2. The experimentally determined activation energies for the MIC process of a-Si and a-Ge, if available, have also been given in these tables.

Table 1.1 Crystallization temperature, silicide phase formation, and activation energy for MIC of a-Si in contact with different metals.

	Metal	(Metal-induced) crystallization temperature (°C)			Silicide formation before MIC	Activation energy for crystallization (eV)
		in situ XRD [42]	(in situ) TEM	DSC (10°C/min)		
	Pure a-Si	790	710 [60]	741 [61]	-	3.4 [60], 4.2 [42]
Metals (eutectic system with Si)	Au	207	130 [62]	180 [45]	Metastable Au_4Si [62], Au_3Si [45]; no silicide [42]	1.7 [42]
	Al	277	150 [36], 175 [8]	170 [8]	No [8, 36]	1.2 [8], 1.23 (own work), 1.4 [42]
	Ag	539	350 [44], 385 [10]	385 [10]	No [10, 44]	1.8 [42]
Metals (silicide forming)	Cu	429	485 [63]	n.a.	Cu_3Si [42, 63, 64]	2.2 [42]
	Pd	540	350 [65]	n.a.	Pd_2Si [65], PdSi [42]	0.6 [42]
	Ni	589	430 [66], 484 [14]	n.a.	$NiSi_2$ [14, 42, 66]	3.5 [42]
	Pt	625	n.a.	n.a.	PtSi [42]	1.9 [42]
	Mn	698	n.a.	n.a.	MnSi and $Mn_{17}Si_{47}$ [42]	n.a.
	Rh	708	n.a.	n.a.	RhSi [42]	n.a.
	V	741	n.a.	n.a.	Metal (V)-rich silicide [42]	n.a.
	Refractory metals: Co, Fe, Hf, Cr, Ti, Ir, Ru, Zr, Nb, Mo, W, Ta, Re	723–765	n.a.	n.a.	Si-rich silicide phase M_xSi_y ($x < y$) [42]	n.a.

n.a. = not available.

Table 1.2 Crystallization temperature, germanide phase formation, and activation energy for MIC of a-Ge in contact with different metals

	Metal	(Metal-induced) crystallization temperature (°C)			Germanide formation before MIC	Activation energy for crystallization (eV)
		in situ XRD [43]	(in situ) TEM	DSC (10°C/min)		
	Pure a-Ge	586	650 [67]	501[61]	–	3.1 [67], 3.8 [43]
Metals (eutectic system with Ge)	Au	173	100 [68, 69], 120 [46, 70], 130 [71]	n.a.	No [46, 68, 69]	1.8 [43] 1.87 [72], 2.03 [72]
	Al	228	125 [73]	150 [32]	No [32, 73, 74]	1.8 [43], 1.5 (own work)
	Ag	340	237 [75], 260 [9]	275 [9]	No [9, 75]	2.0 [43], 1.75 [9]
	Bi	n.a.	150 [76]	n.a.	No [76]	n.a.
	Sb	n.a.	300 [77]	n.a.	No [77, 78]	n.a.
Metals (germanide forming)	Cu	271	250 [79], 310 [80]	n.a.	Cu_3Ge [80, 81]	1.8 [43], 2.75 [80]
	Pd	395	300 [82, 83]	n.a.	Pd_2Ge and PdGe[82–84]	2.2 [43]
	Ni	392	360 [85]	n.a.	NiGe and Ni_5Ge_3[50, 86]	1.9 [43], 2.4 [50]
	Pt	472	n.a.	n.a.	$PtGe_2$ [43]	2.5 [43]
	Mn	389	n.a.	n.a.	Mn_5Ge_3 [43]	n.a.
	Refractory metals: Co, Fe, Cr, Ru, Zr, Nb, V, Rh	428–570	n.a.	n.a.	Co_5Ge_7, $FeGe_2$, Cr_5Ge_3, Ru_2Ge_3, $ZrGe_2$, Nb_5Ge_3, V_5Ge_3, $Rh_{17}Ge_{22}$ [43]	n.a.

n.a. = not available.

Depending on whether metal-semiconductor compound phases occur or don't, the possible MIC processes can be subdivided into two categories, **A** and **B** (see Tables 1.1 and 1.2 and also the discussion in Section 1.3.2).

- **Category A**

  This is MIC in simple metal semiconductor eutectic binary systems (e.g., Al, Au, Ag, and Bi with a-Si or a-Ge).

  The MIC temperatures in such systems are typically low (Tables 1.1 and 1.2). A MILE process often occurs in association with the MIC process in such systems.

- **Category B**

  This is MIC in compound(s)-forming metal semiconductor binary systems (e.g., Ni, Pd, Pt, and Cu with a-Si or a-Ge).

  The MIC temperatures in such systems are typically relatively high (see Tables 1.1 and 1.2). A so-called MILC process (see Sections 1.1 and 1.3.4) can often be realized for such systems.

1.3.4 Metal-Induced Crystallization Mechanisms

The underlying mechanisms for the two categories of MIC processes (sketched in Section 1.3.3) differ strikingly. The MIC process of type A is schematically shown in Fig. 1.1A for the case of the c-Al/a-Si layer system. In such a non-compound-forming system, the MIC process was found to initiate at the metal/amorphous semiconductor interface and/or at the grain boundaries (GBs) of the metal [29, 30, 33]. The exact crystallization site(s) and the minimum (metal-induced) crystallization temperature are sensitively controlled by the balance of the semiconductor crystallization energy and the change of interface energetics upon crystallization [29, 30, 33].

For the c-Al/a-Si system, calculations based on interface thermodynamics have shown that the first step of the MIC process involves wetting of the high-angle Al GBs by a-Si (step 1 of Fig. 1.1A) [30]. At temperatures above a critical temperature of about 140°C, crystallization of a-Si initiates *exclusively* at the high-angle Al GBs (step 2 of Fig. 1.1A) [30]. The nucleation of c-Si at the c-Al/a-Si interface is not possible at such low temperatures [30]. This MIC mechanism has been directly confirmed (visualized) by in situ heating

TEM experiments [36]. The wetting of a high-angle Al GB by a-Si at a temperature below 140°C is shown in the plasmon loss mapping images (obtained in an in situ heating valence energy–filtered TEM experiment) in Fig. 1.1B. The (cross-sectional view of) nucleation of c-Si at a high-angle Al GB at 150°C, as imaged by in situ heating HRTEM, is shown in Fig. 1.1C.

The continued crystallization and growth of Si at the original Al GBs can result in a stress gradient buildup in the bilayer system (compressive stress in the c-Al sublayer and tensile stress in the a-Si sublayer) [37]. A diffusion potential gradient thus exists for Al atoms in the Al sublayer such that they tend to move toward the adjacent, free-volume-rich a-Si sublayer. Thus, Al grains in the original c-Al sublayer are gradually dissolved. These dissolved Al atoms form Al plumes in the a-Si sublayer (see Fig. 1.1D) [37]. The Al plumes move upward within the solid a-Si layer and at the same time grow in size by collecting further Al atoms from the Al bottom layer. As a consequence of such a solid-state convection process, layer exchange can occur (see Fig. 1.1D) [37].

The MIC process of type B is schematically shown in Fig. 1.2A for the case of a c-Ni/a-Si layer system. In such a compound-forming system, the MIC process was found to be mediated by the formation and migration (into/within the amorphous semiconductor material) of one or more metal semiconductor compound phases, preferably with a small lattice mismatch with the corresponding crystalline semiconductor [13, 14, 59].

As schematically shown in Fig. 1.2A, upon heating of a c-Ni/a-Si bilayer, Ni reacts with a-Si to form $NiSi_2$ at the c-Ni/a-Si interface. c-Si then nucleates at the interface between $NiSi_2$ and a-Si, and the crystallization of c-Si proceeds by the apparent migration of $NiSi_2$ into/within the a-Si film, leaving the formation of needle-like c-Si in its wake [13, 14, 20, 59]. A direct TEM observation of such a growth mechanism is given in Fig. 1.2B [59]. Continued crystallization of a-Si then involves the repeated $NiSi_2$-induced growth of needle-like Si crystallites until the whole a-Si layer has become crystallized (see, e.g., experimental observations given in Fig. 1.2C [59]).

As a consequence of such a growth mechanism, a so-called metal-induced lateral growth process of the crystalline semiconductor often occurs in a type B MIC process, which is mediated by the lateral migration of crystallization-inducing silicide parallel to the film plane (see, e.g., Figs. 1.2D and 1.2E for the experimental observations of the MILC in the Pd/a-Si [18] and the Ni/a-Si [20] system, respectively).

References

1. Oki, F., Ogawa, Y., Fujiki, Y. (1969) Effect of deposited metals on crystallization temperature of amorphous germanium film. *Japanese Journal of Applied Physics*, **8**, 1056.
2. Bosnell, J. R., Voisey, U. C. (1970) The influence of contact materials on the conduction crystallization temperature and electrical properties of amorphous germanium, silicon and boron films. *Thin Solid Films*, **6**, 161–166.
3. Herd, S. R., Chaudhari, P., Brodsky, M. H. (1972) Metal contact induced crystallization in films of amorphous silicon and germanium. *Journal of Non-Crystalline Solids*, **7**, 309–327.
4. Ottavian, G., Sigurd, D., Marrello, V., McCaldin, J. O., Mayer, J. W. (1973) Crystal-growth of silicon and germanium in metal-films. *Science*, **180**, 948–949.
5. Ottaviani, G., Sigurd, D., Marrello, V., Mayer, J. W., McCaldin, J. O. (1974) Crystallization of Ge and Si in metal films. I. *Journal of Applied Physics*, **45**, 1730–1739.
6. Sigurd, D., Ottaviani, G., Arnal, H. J., Mayer, J. W. (1974) Crystallization of Ge and Si in metal films. II. *Journal of Applied Physics*, **45**, 1740–1745.
7. Brodsky, M. H., Turnbull, D. (1971) *Bulletin of the American Physical Society*, **16**, 304.
8. Konno, T. J., Sinclair, R. (1992) Crystallization of silicon in aluminum amorphous-silicon multilayers. *Philosophical Magazine B: Physics of Condensed Matter Statistical Mechanics Electronic Optical and Magnetic Properties*, **66**, 749–765.
9. Sinclair, R., Konno, T. J. (1994) In situ HREM: application to metal-mediated crystallization. *Ultramicroscopy*, **56**, 225–232.
10. Konno, T. J., Sinclair, R. (1995) Metal-mediated crystallization of amorphous-silicon in silicon silver layered systems. *Philosophical Magazine B: Physics of Condensed Matter Statistical Mechanics Electronic Optical and Magnetic Properties*, **71**, 163–178.
11. Konno, T. J., Sinclair, R. (1994) Metal-contact-induced crystallization of semiconductors. *Materials Science and Engineering a-Structural Materials Properties Microstructure and Processing*, **179**, 426–432.

12. Sinclair, R., Morgiel, J., Kirtikar, A. S., Wu, I. W., Chiang, A. (1993) Direct observation of crystallization in silicon by in situ high-resolution electron microscopy. *Ultramicroscopy*, **51**, 41–45.

13. Hayzelden, C., Batstone, J. L., Cammarata, R. C. (1992) In situ transmission electron microscopy studies of silicide-mediated crystallization of amorphous silicon. *Applied Physics Letters*, **60**, 225–227.

14. Hayzelden, C., Batstone, J. L. (1993) Silicide formation and silicide-mediated crystallization of nickel-implanted amorphous silicon thin films. *Journal of Applied Physics*, **73**, 8279–8289.

15. Nast, O., Puzzer, T., Koschier, L. M., Sproul, A. B., Wenham, S. R. (1998) Aluminum-induced crystallization of amorphous silicon on glass substrates above and below the eutectic temperature. *Applied Physics Letters*, **73**, 3214–3216.

16. Nast, O., Hartmann, A. J. (2000) Influence of interface and Al structure on layer exchange during aluminum-induced crystallization of amorphous silicon. *Journal of Applied Physics*, **88**, 716–724.

17. Nast, O., Wenham, S. R. (2000) Elucidation of the layer exchange mechanism in the formation of polycrystalline silicon by aluminum-induced crystallization. *Journal of Applied Physics*, **88**, 124–132.

18. Lee, S. W., Jeon, Y. C., Joo, S. K. (1995) Pd induced lateral crystallization of amorphous Si thin-films. *Applied Physics Letters*, **66**, 1671–1673.

19. Lee, S.-W., Joo, S.-K. (1996) Low temperature poly-Si thin-film transistor fabrication by metal-induced lateral crystallization. *IEEE Electron Device Letters*, **17**, 160–162.

20. Jin, Z. H., Bhat, G. A., Yeung, M., Kwok, H. S., Wong, M. (1998) Nickel induced crystallization of amorphous silicon thin films. *Journal of Applied Physics*, **84**, 194–200.

21. Wang, J. Y., Zalar, A., Zhao, Y. H., Mittemeijer, E. J. (2003) Determination of the interdiffusion coefficient for Si/Al multilayers by Auger electron spectroscopical sputter depth profiling. *Thin Solid Films*, **433**, 92–96.

22. Zhao, Y. H., Wang, J. Y., Mittemeijer, E. J. Microstructural changes in amorphous Si/crystalline Al thin bilayer films upon annealing. *Applied Physics A: Materials Science & Processing*, **79**, 681–690 (2004).

23. Wang, J. Y., Mittemeijer, E. J. (2004) A new method for the determination of the diffusion-induced concentration profile and the interdiffusion coefficient for thin film systems by Auger electron spectroscopical sputter depth profiling. *Journal of Materials Research*, **19**, 3389–3397.

24. He, D., Wang, J. Y., Mittemeijer, E. J. (2005) The initial stage of the reaction between amorphous silicon and crystalline aluminum. *Journal of Applied Physics*, **97**, 093524.
25. He, D., Wang, J. Y., Mittemeijer, E. J. (2005) Reaction between amorphous Si and crystalline Al in Al/Si and Si/Al bilayers: microstructural and thermodynamic analysis of layer exchange. *Applied Physics A: Materials Science & Processing*, **80**, 501–509.
26. Wang, J. Y., He, D., Zhao, Y. H., Mittemeijer, E. J. (2006) Wetting and crystallization at grain boundaries: origin of aluminum-induced crystallization of amorphous silicon. *Applied Physics Letters*, **88**, 061910.
27. Wang, Z. M., Wang, J. Y., Jeurgens, L. P. H., Mittemeijer, E. J. (2006) "Explosive" crystallisation of amorphous germanium in Ge/Al layer systems: comparison with Si/Al layer systems. *Scripta Materialia*, **55**, 987–990.
28. Wang, J. Y., Wang, Z. M., Mittemeijer, E. J. (2007) Mechanism of aluminum-induced layer exchange upon low-temperature annealing of amorphous Si/polycrystalline Al bilayers. *Journal of Applied Physics*, **102**, 113523.
29. Wang, Z. M., Wang, J. Y., Jeurgens, L. P. H., Mittemeijer, E. J. (2008) Tailoring the ultrathin Al-induced crystallization temperature of amorphous Si by application of interface thermodynamics. *Physical Review Letters*, **100**, 125503.
30. Wang, Z. M., Wang, J. Y., Jeurgens, L. P. H., Mittemeijer, E. J. (2008) Thermodynamics and mechanism of metal-induced crystallization in immiscible alloy systems: experiments and calculations on Al/a-Ge and Al/a-Si bilayers. *Physical Review B*, **77**, 045424.
31. Wang, Z. M., Wang, J. Y., Jeurgens, L. P. H., Mittemeijer, E. J. (2008) Investigation of metal-induced crystallization in amorphous Ge/crystalline Al bilayers by Auger microanalysis and selected-area depth profiling. *Surface and Interface Analysis*, **40**, 427–432.
32. Wang, Z. M., Wang, J. Y., Jeurgens, L. P. H., Phillipp, F., Mittemeijer, E. J. (2008) Origins of stress development during metal-induced crystallization and layer exchange: annealing amorphous Ge/crystalline Al bilayers. *Acta Materialia*, **56**, 5047–5057.
33. Wang, Z. M., Jeurgens, L. P. H., Wang, J. Y., Mittemeijer, E. J. (2009) Fundamentals of metal-induced crystallization of amorphous semiconductors. *Advanced Engineering Materials*, **11**, 131–135.

34. Wang, Z. M., Jeurgens, L. P. H., Wang, J. Y., Phillipp, F., Mittemeijer, E. J. (2009) High-resolution transmission-electron-microscopy study of ultrathin Al-induced crystallization of amorphous Si. *Journal of Materials Research*, **24**, 3294–3299.

35. Wang, Z. M., Gu, L., Jeurgens, L. P. H., Mittemeijer, E. J. (2011) Thermal stability of Al/nanocrystalline-Si bilayers investigated by in situ heating energy-filtered transmission electron microscopy. *Journal of Materials Science*, **46**, 4314–4317.

36. Wang, Z. M., Gu, L., Phillipp, F., Wang, J. Y., Jeurgens, L. P. H., Mittemeijer, E. J. (2011) Metal-catalyzed growth of semiconductor nanostructures without solubility and diffusivity constraints. *Advanced Materials*, **23**, 854–859.

37. Wang, Z. M., Gu, L., Jeurgens, L. P. H., Phillipp, F., Mittemeijer, E. J. (2012) Real-time visualization of convective transportation of solid materials at nanoscale. *Nano Letters*, **12**, 6126–6132.

38. Gordon, I., Carnel, L., Van Gestel, D., Beaucarne, G., Poortmans, J. (2008) Fabrication and characterization of highly efficient thin-film polycrystalline-silicon solar cells based on aluminium-induced crystallization. *Thin Solid Films*, **516**, 6984–6988.

39. Mittemeijer, E. J., Welzel, U. (eds.) (2012) *Modern Diffraction Methods* (Wiley-VCH, Weinheim).

40. Wohlschlögel, M., Welzel, U., Maier, G., Mittemeijer, E. J. (2006) Calibration of a heating/cooling chamber for x-ray diffraction measurements of mechanical stress and crystallographic texture. *Journal of Applied Crystallography*, **39**, 194–201.

41. Welzel, U., Ligot, J., Lamparter, P., Vermeulen, A. C., Mittemeijer, E. J. (2005) Stress analysis of polycrystalline thin films and surface regions by x-ray diffraction. *Journal of Applied Crystallography* **38**, 1–29.

42. Knaepen, W., Detavernier, C., Van Meirhaeghe, R. L., Sweet, J. J., Lavoie, C. (2008) In-situ x-ray diffraction study of metal induced crystallization of amorphous silicon. *Thin Solid Films*, **516**, 4946–4952.

43. Knaepen, W., Gaudet, S., Detavernier, C., Van Meirhaeghe, R. L., Sweet, J. J., Lavoie, C. (2009) In situ x-ray diffraction study of metal induced crystallization of amorphous germanium. *Journal of Applied Physics*, **105**, 083532.

44. Bian, B., Yie, J., Li, B., Wu, Z. (1993) Fractal formation in a-Si:H/Ag/a-Si:H films after annealing. *Journal of Applied Physics*, **73**, 7402–7406.

45. Seibt, M., Buschbaum, S., Gnauert, U., Schroter, W., Oelgeschlager, D. (1998) Nanoscale observation of a grain boundary related growth mode in thin film reactions. *Physical Review Letters*, **80**, 774–777.

46. Zhang, S., Wang, X., Chen, Z., Wu, Z., Jin-Phillipp, N. Y., Kelsch, M., Phillipp, F. (1999) In situ TEM study of fractal formation in amorphous Ge/Au bilayer films. *Physical Review B*, **60**, 5904–5908.

47. Chromik, R. R., Zavalij, L., Johnson, M. D., Cotts, E. J. (2002) Calorimetric investigation of the formation of metastable silicides in Au/a-Si thin film multilayers. *Journal of Applied Physics*, **91**, 8992–8998.

48. Hazra, S., Sakata, I., Yamanaka, M., Suzuki, E. (2004) Evolution of an amorphous silicon network from silicon paracrystallites studied by spectroscopic ellipsometry. *Physical Review B*, **69**, 235204.

49. Her, Y.-C., Wu, C.-L. (2004) Crystallization kinetics of Cu/a-Si bilayer recording film under thermal and pulsed laser annealing. *Journal of Applied Physics*, **96**, 5563–5568.

50. Her, Y.-C., Chen, J.-H., Tsai, M.-H., Tu, W.-T. (2009) Nickel-induced crystallization of amorphous Ge film for blue-ray recording under thermal annealing and pulsed laser irradiation. *Journal of Applied Physics*, **106**, 023530.

51. Her, Y.-C., Tu, W.-T., Tsai, M.-H. (2012) Phase transformation and crystallization kinetics of a-Ge/Cu bilayer for blue-ray recording under thermal annealing and pulsed laser irradiation. *Journal of Applied Physics*, **111**, 043503.

52. Fritzsche, H. (ed.) (1989) *Amorphous Silicon and Related Materials* (World Scientific, Singapore).

53. Donovan, E. P., Spaepen, F., Turnbull, D., Poate, J. M., Jacobson, D. C. (1985) Calorimetric studies of crystallization and relaxation of amorphous Si and Ge prepared by ion implantation. *Journal of Applied Physics*, **57**, 1795–1804.

54. Spinella, C., Lombardo, S., Priolo, F. (1998) Crystal grain nucleation in amorphous silicon. *Journal of Applied Physics*, **84**, 5383–5414.

55. Street, R. A. (2005) *Hydrogenated Amorphous Silicon* (Cambridge University Press, Cambridge, UK).

56. Anton, R. (2009) In situ TEM investigations of reactions of Ni, Fe and Fe–Ni alloy particles and their oxides with amorphous carbon. *Carbon*, **47**, 856–865.

57. Zanatta, A. R., Chambouleyron, I. (2005) Low-temperature Al-induced crystallization of amorphous Ge. *Journal of Applied Physics*, **97**, 094914.

58. Ferri, F. A., Zanatta, A. R., Chambouleyron, I. (2006) Metal-induced nanocrystalline structures in Ni-containing amorphous silicon thin films. *Journal of Applied Physics*, **100**, 094311.

59. Yoon, S. Y., Park, S. J., Kim, K. H., Jang, J. (2001) Metal-induced crystallization of amorphous silicon. *Thin Solid Films*, **383**, 34–38.
60. Batstone, J. L. (1993) In situ crystallization of amorphous silicon in the transmission electron microscope. *Philosophical Magazine A: Physics of Condensed Matter Structure Defects and Mechanical Properties*, **67**, 51–72.
61. Fan, J. C. C., Carl H. Anderson, J. (1981) Transition temperatures and heats of crystallization of amorphous Ge, Si, and Ge[sub 1 - x]Si[sub x] alloys determined by scanning calorimetry. *Journal of Applied Physics*, **52**, 4003–4006.
62. Hultman, L., Robertsson, A., Hentzell, H. T. G., Engstrom, I., Psaras, P. A. (1987) Crystallization of amorphous-silicon during thin-film gold reaction. *Journal of Applied Physics*, **62**, 3647–3655.
63. Russell, S. W., Li, J., Mayer, J. W. (1991) In situ observation of fractal growth during a-Si crystallization in a Cu3Si matrix. *Journal of Applied Physics*, **70**, 5153–5155.
64. Lee, S. B., Choi, D. K., Phillipp, F., Jeon, K. S., Kim, C. K. (2006) In situ high-resolution transmission electron microscopy study of interfacial reactions of Cu thin films on amorphous silicon. *Applied Physics Letters*, **88**, 083117.
65. Lee, S. W., Lee, B. I., Kim, T. K., Joo, S. K. (1999) Pd2Si-assisted crystallization of amorphous silicon thin films at low temperature. *Journal of Applied Physics*, **85**, 7180–7184.
66. Miyasaka, M., Makihira, K., Asano, T., Polychroniadis, E., Stoemenos, J. (2002) In situ observation of nickel metal-induced lateral crystallization of amorphous silicon thin films. *Applied Physics Letters*, **80**, 944–946.
67. Cao, Z. H., Liu, P., Meng, X. K., Tang, S. C., Lu, H. M. (2009) In situ transmission electron microscopy observations of the crystallization of amorphous Ge films. *Applied Physics A: Materials Science & Processing*, **94**, 393–398.
68. Chen, Z. W., Wang, X. P., Tan, S., Zhang, S. Y., Hou, J. G., Wu, Z. Q. (2001) Multifractal behavior of crystallization on Au/Ge bilayer films. *Physical Review B*, **63**, 165413.
69. Chen, Z. W., Zhang, S. Y., Tan, S., Hou, J. G., Zhang, Y. H., Sekine, H. (2001) Fractal crystallization and nonlinear V-I behavior of Au/Ge bilayer film. *Journal of Applied Physics*, **89**, 783–785.
70. Chen, Z. W., Lai, J. K. L., Shek, C. H. (2006) Microstructural changes and fractal Ge nanocrystallites in polycrystalline Au/amorphous Ge thin

bilayer films upon annealing. *Journal of Physics D: Applied Physics*, **39**, 4544–4548.

71. Bian, B., Tanaka, T., Ohkubo, T., Hirotsu, Y. (1998) Plan-view and cross-sectional TEM observations of interfacial reactions and fractal formation in a-Ge/Au films. *Philosophical Magazine A: Physics of Condensed Matter Structure Defects and Mechanical Properties*, **78**, 157–170.
72. Bian, B., Ohkubo, T., Hirotsu, Y. (1995) Crystallization and fractal formation in annealed a-Ge/Au bilayer films. *Journal of Electron Microscopy*, **44**, 182–190.
73. Katsuki, F., Hanafusa, K., Yonemura, M., Koyama, T., Doi, M. (2001) Crystallization of amorphous germanium in an Al/a-Ge bilayer film deposited on a SiO2 substrate. *Journal of Applied Physics*, **89**, 4643–4647.
74. Doi, M., Suzuki, Y., Koyama, T., Katsuki, F. (1998) Pattern evolution of crystalline Ge aggregates during annealing of an Al/Ge bilayer film deposited on a SiO2 substrate. *Philosophical Magazine Letters*, **78**, 241–245.
75. Sugawara, A., Kuwana, Y., Nittono, O. (1994) Growth dynamics of polycrystalline Ge clusters formed on surfaces during annealing of co-sputtered Ge�Ag films. *Thin Solid Films*, **251**, 10–13.
76. Missana, T., Afonso, C. N., Petford-Long, A. K., Doole, R. C. (1998) The role of interface roughness in the metal-induced crystallization of amorphous Ge in contact with Bi. *Philosophical Magazine A: Physics of Condensed Matter Structure Defects and Mechanical Properties*, **77**, 769–779.
77. Catalina, F., Ollacarizqueta, M. A., Afonso, C. N. (1995) Triggering of diffusion and crystallization of amorphous germanium in contact with antimony. *Philosophical Magazine Part B: Physics of Condensed Matter Statistical Mechanics, Electronic, Optical and Magnetic Properties*, **71**, 437–444.
78. Petford-long, A. K., Doole, R. C., Afonso, C. N. (1996) Kinetics of grain-boundary reactions at semimetal-semiconductor interfaces observed during in-situ transmission electron microscope annealing. *Philosophical Magazine A: Physics of Condensed Matter Structure Defects and Mechanical Properties*, **74**, 907–918.
79. Sadoh, T., Kurosawa, M., Hagihara, T., Toko, K., Miyao, M. (2011) Low-temperature (similar to 250 degrees C) Cu-induced lateral crystallization of amorphous Ge on insulator. *Electrochemical and Solid State Letters*, **14**, H274–H276.

80. Her, Y. C., Tu, W. T., Tsai, M. H. (2012) Phase transformation and crystallization kinetics of a-Ge/Cu bilayer for blue-ray recording under thermal annealing and pulsed laser irradiation. *Journal of Applied Physics*, **111**, 043503.

81. Doyle, J. P., Svensson, B. G., Johansson, S. (1995) Morphological instability of bilayers of copper germanide films and amorphous germanium. *Applied Physics Letters*, **67**, 2804–2806.

82. Chen, Z. W., Zhang, S. Y., Tan, S., Hou, J. G., Zhang, Y. H. (1998) Effect of fractal crystallization on the depositing sequence of a Pd/Ge thin film system. *Journal of Vacuum Science & Technology A: Vacuum Surfaces and Films*, **16**, 2292–2294.

83. Chen, Z. W., Li, Q. B., Wang, J., Pan, D. Y., Jiao, Z., Wu, M. H., Shek, C. H., Wu, C. M. L., Lai, J. K. L. (2011) Probing into interesting effects of fractal Ge nanoclusters induced by Pd nanoparticles. *Inorganic Chemistry*, **50**, 6756–6761.

84. Wu, X. H., Feng, Y. Z., Li, B. Q., Wu, Z. Q., Zhang, S. Y. (1994) Annealing behavior of Pd/a-Ge bilayer films. *Journal of Applied Physics*, **75**, 2415–2417.

85. Park, J.-H., Kapur, P., Saraswat, K. C., Peng, H. (2007) A very low temperature single crystal germanium growth process on insulating substrate using Ni-induced lateral crystallization for three-dimensional integrated circuits. *Applied Physics Letters*, **91**, 143107.

86. Patterson, J. K., Park, B. J., Ritley, K., Xiao, H. Z., Allen, L. H., Rockett, A. (1994) Kinetics of Ni/a-Ge bilayer reactions. *Thin Solid Films*, **253**, 456–461.

Chapter 2

Thermodynamics and Atomic Mechanisms of Metal-Induced Crystallization of Amorphous Semiconductors at Low Temperatures

Zumin Wang,[a] **Lars P. H. Jeurgens,**[a,*] **and Eric J. Mittemeijer**[a,b]
[a]*Max Planck Institute for Intelligent Systems, Heisenbergstraße 3, D-70569 Stuttgart, Germany*
[b]*Institute for Materials Science, University of Stuttgart, D-70569 Stuttgart, Germany*
z.wang@is.mpg.de

The thermodynamics and atomic-scale mechanisms of metal-induced crystallization (MIC) of amorphous semiconductors are described in detail. It is shown that the MIC effect for a wide range of metal/amorphous semiconductor systems is in general an *interface-controlled phenomenon*, which includes three major aspects: (i) metal-induced weakening of covalent bonds at the interface, (ii) fast atomic transportation along the interface, and (iii) *interface*

**Present address: Swiss Federal Laboratories for Materials Science and Technology,* Überlandstrasse *129, 8600 Dübendorf, Switzerland*

Metal-Induced Crystallization: Fundamentals and Applications
Edited by Zumin Wang, Lars P. H. Jeurgens, and Eric J. Mittemeijer

ISBN 978-981-4463-40-9 (Hardcover), 978-981-4463-41-6 (eBook)
www.panstanford.com

thermodynamics, which critically controls whether low-temperature crystal nucleation and crystal growth can occur or not. By quantitative calculation of the interface thermodynamics, recognizing aspects (i) and (ii), the (very) different MIC temperatures and behaviors observed for various metal/amorphous semiconductor systems can now be understood and predicted on a unified basis. The theoretical predictions have been confirmed in particular by in situ heating high-resolution transmission electron microscopy (TEM) and valence energy–filtered TEM experiments, which also revealed the atomic-scale mechanisms of the MIC process. The fundamental understanding reached may lead to pronounced technological progress in applications of the MIC process, in particular regarding the low-temperature manufacturing of high-efficiency solar cells and other electronic components on cheap and flexible substrates such as glasses and plastics.

2.1 Introduction

Amorphous semiconductors such as silicon and germanium can crystallize at a temperature much lower than their "bulk" crystallization temperatures when they are put in direct contact with a metal, such as Al [1–17], Au [18–22], Ag [23–25], Ni [26–33], Cu [34–36], and Pd [37, 38]. This phenomenon, which was firstly observed more than 40 years ago for amorphous germanium [39], is now commonly referred to as metal-induced crystallization (MIC) [40–43]. In the past decade, the MIC process in various metal/amorphous semiconductor systems has been extensively investigated, which has largely been driven by its many (potential) applications, for example, in the low-temperature production of high-performance crystalline semiconductor-based solar cells, flat-panel displays, and high-density data storage devices (see Chapters 5, 6, and 7 of the book).

As a result of numerous investigations from different research groups all over the world, the MIC characteristics in various metal/amorphous semiconductor systems have now been disclosed in great detail, as reported in the literature (see, e.g., Refs. [1–42]). It has been found that in MIC the (reduced) crystallization temperatures as well as the crystallization behaviors of amorphous semiconductors are strongly dependent on the type of the contacting metal (e.g., Refs. [40–43]).

For compound (e.g., silicide)-forming metals such as Ni, Cu, and Pd, the MIC of amorphous semiconductors usually occurs at a relatively high temperature of about 500°C [26–38]. The MIC process in such systems is often associated with initial formation of one or more compound phases at the metal/semiconductor interface and subsequent motion of the compound phase(s) reaction front into the amorphous semiconductor phase, leaving behind a crystalline semiconductor phase in the wake of the compound phase reaction front [26–38].

For non-compound-forming metals such as Al, Ag, and Au, the MIC of amorphous semiconductors can occur at much lower temperatures (as low as 150°C for amorphous Si [a-Si] in contact with Al [5, 11] or Au [21]). The MIC process in such systems is often associated with transportation of semiconductor material into the metal phase, and a crystalline semiconductor phase is formed at the location of the original metal phase [3, 9]. In such cases, even very small variations in the detailed microstructure of the metal phase (e.g., single-crystalline or polycrystalline, fine-grained or coarse-grained, presence of defects and/or contaminants, film thickness) can result in very different crystallization temperatures and crystallization behaviors [2, 9].

Since most previous studies on MIC are largely application orientated, the fundamental, scientific aspects of the MIC process have often been discussed only briefly and qualitatively and sometimes have even been fully discarded in the literature. The microscopic, atomistic mechanisms of the MIC process have remained largely unclear.

In the last years, the present authors have devoted considerable effort to attain a fundamental understanding of the thermodynamics and atomic mechanisms of the MIC process in various metal/amorphous semiconductor systems. On the one hand theoretical modeling of the (interface) thermodynamics was performed by employing the macroscopic atom approach [10, 11, 44]. On the other hand dedicated, in particular in situ heating high-resolution transmission electron microscopy (HRTEM) and valence energy–filtered transmission electron microscopy (VEFTEM), experiments were performed to reveal the operating atomistic mechanisms [14, 15]. It has been found that MIC is in general an *interface-controlled phenomenon* [10, 11, 14, 40]. The crucial roles of interfaces (including, especially, grain boundaries [GBs] in the metal phase) in MIC involve (1) bond weakening of the semiconductor material at the interface with the metal, leading to enhanced mobility of semiconductor atoms at the interface; (2) the interfaces (including GBs in the metal) serving as fast, short-circuit diffusion paths for material transportation; and (3) very importantly, the interface energetics (thermodynamics), which is decisive for whether crystal nucleation and crystal growth will occur. As will be described in detail in this chapter, the MIC process as observed in a wide range of metal/amorphous semiconductor systems can be described on the above basis. The theoretical predictions have been confirmed by in situ heating X-ray diffraction (XRD), HRTEM, and VEFTEM experiments, where the HRTEM and VEFTEM methods also revealed the atomistic mechanisms of the MIC process.

The thus achieved fundamental understanding of the MIC process makes possible highly sophisticated application(s) of MIC in various technologies. For example, direct application of the model for interface thermodynamics enables us to tailor the crystallization temperature of a-Si systematically from 700°C down to 150°C (and any temperature in between) by controlling the Al overlayer thickness [10]. Such fundamental understanding has led, for example, to the development of a novel process for the growth of crystalline Si (c-Si) nanowires at exceedingly low temperatures [14].

2.2 Metal-Induced Covalent Bond Weakening

Nature strives to crystallize "bulk" amorphous semiconductors to lower the system's (Gibbs) energy in view of the intrinsically higher (Gibbs) energy of the amorphous state as compared to the crystalline state. However, energy barriers (activation energies) must be overcome to rearrange the semiconductor atoms into (tiny) crystalline nuclei to initiate the crystallization process. The formation of initial crystalline nuclei in bulk amorphous semiconductors is hindered by the strong covalent bonds of the semiconductor atoms (2.30 eV for Si–Si, 1.95 eV for Ge–Ge [45]). Consequently, the activation energy for the initiation of the crystallization process in bulk amorphous semiconductors can only be surmounted (i.e., thermally activated) at high temperatures (e.g., as high as 700°C for a-Si and 500°C for amorphous Ge [a-Ge]) [41].

In the 1970s, it was discovered by Hiraki et al. [46, 47] that the strong covalent bonds in semiconductors such as Si, Ge, GaAs, InP, and InSb can be weakened at interfaces with metals as Au, Ag, and Pd. In their original work, it was shown that about 100 nm thick SiO_2 can grow above a gold-covered Si wafer in air at a low temperature of only about 100°C [46] (see Fig. 2.1 [48]). This observation strongly implies the occurrence of metal-induced bond weakening of semiconductor atoms (see further later). Besides the low-temperature oxidation of semiconductors, the metal-induced bond weakening often leads to interfacial reactions and/or intermixing of the metal and the semiconductor at low temperatures [46–48]. One of the microscopic explanations for this metal-induced bond-weakening phenomenon concerns the so-called Coulomb screening effect [48–50]: When a covalently bonded semiconductor (e.g., Si) is in contact with a metal (e.g., Au) layer, the wave function of the free-electron gas of the metal extends into the adjacent semiconductor, thereby screening the covalent bonds present therein [48–50]. These screened bonds are thereby weakened (and/or broken). This (interfacial) bond-weakening effect is a very local electronic interaction phenomenon [48]: the thickness of the "weakened" interfacial layer has been estimated to be only about ~2 monolayers (ML) [11, 48].

Due to the interfacial bond-weakening effect, the 2 ML thick layer of interfacial semiconductor atoms with enhanced mobility (further designated as "free" semiconductor atoms in the following) is generally believed to provide the agent for initiation of crystallization of amorphous semiconductors at low temperatures: (i) crystallization can occur directly at the interface with the metal, and/or (ii) diffusion of the free semiconductor atoms can take place along the GBs in the metal (layer) and the diffused semiconductor atoms can crystallize there. Whether process (i) or process (ii), or both, is energetically favored depends on a sensitive energy balance weighting the change of the "bulk" Gibbs energies against the change of the corresponding surface and interface energies upon initiation of MIC. As will be shown in the sections later, the different MIC temperatures/behaviors in various (immiscible) metal/amorphous semiconductor systems can thus be understood and predicted.

Figure 2.1 SiO_2 formation at ~100°C in air on a Au film (about 90 nm) on top of a Si substrate. (a) Just after Au deposition, the golden color of Au is observed. (b) SiO_2 growth (layer thickness ≈ 100 nm) changes the color to brown. Note that the silver-colored bare Si surface is unchanged. This observation strongly suggests the occurrence of Au-induced covalent bond weakening of Si. Reprinted with permission from Ref. [48], Copyright 1983, Elsevier.

2.3 Interface Thermodynamics of Metal-Induced Crystallization

2.3.1 Grain Boundary Wetting

The metal layer, as deposited to catalyze the crystallization of amorphous semiconductors, is usually polycrystalline and contains a high density of GBs. These GBs can be wetted by the free interfacial semiconductor (e.g., Si or Ge) atoms in the contacting amorphous semiconductor layer (and thereby eventually mediate the MIC process). The possibility for the occurrence of this GB-wetting process requires, thermodynamically, that the total energy of the system, as defined by the amorphous semiconductor or polycrystalline metal bilayer, be reduced by replacing the original GBs in the metal layer with two amorphous semiconductor/metal interphase boundaries.

For a quantitative evaluation, values of the surface and interface energies in various metal (M)/semiconductor (S) systems must be assessed. The surface energies and interface energies in solid–solid (e.g., crystalline–crystalline, crystalline–amorphous, amorphous–amorphous) systems as a function of temperature (T) can be calculated quantitatively employing methods recently developed on the basis of the macroscopic atom approach, as reviewed in Ref. [44]. For example, such calculated surface energies and interface energies as a function of T for the Al/Si and Al/Ge systems are shown in Fig. 2.2 (see Ref. [11]).

On this basis, the energetics for possible GB-wetting processes in various crystalline metal/amorphous semiconductor (further designated as <M>|{S}; <>: crystalline, {}: amorphous) systems can be evaluated quantitatively. A comparison of the interface energies of metal high-angle GBs ($\gamma_{<\mathrm{M}>}^{\mathrm{GB}}$, here M: Al, Ag, Au) and two times the <M>/{S} (here S: Si, Ge) interphase boundaries ($2 \times \gamma_{\langle \mathrm{M}\rangle|\{\mathrm{S}\}}$) is provided in Fig. 2.3. It follows that the wetting of the high-angle metal GBs by amorphous semiconductors is indeed energetically favored for most metal/semiconductor systems (e.g., Al/a-Si, Al/a-Ge, Au/a-Si, and Ag/a-Si) because this reduces the Gibbs energy of the system (see the corresponding schematic illustration in Fig. 2.3a). The driving force ($\Delta\gamma_{\{\mathrm{S}\}\rightarrow\langle \mathrm{M}\rangle\mathrm{GB}}^{\mathrm{wetting}}$) for such GB wetting as function temperature (T) is given by

$$\Delta\gamma_{\{\mathrm{S}\}\rightarrow\langle \mathrm{M}\rangle\mathrm{GB}}^{\mathrm{wetting}}(T) = \gamma_{<\mathrm{M}>}^{\mathrm{GB}}(T) - 2\times\gamma_{\langle \mathrm{M}\rangle|\{\mathrm{S}\}}(T) \tag{2.1}$$

As shown in Figs. 2.3b and 2.3c, the (positive) driving forces are in the range of 0.1–0.6 J m^{-2} (at 150°C) for most metal/amorphous semiconductor systems. This GB-wetting process can play an important role in the initiation of the MIC process (see Section 2.3.2).

The kinetics of the GB wetting is controlled by the diffusion of semiconductor atoms along the GBs in the metal. GB diffusion is known to be very fast at even low temperatures [51]. For example, on the basis of experimental diffusivity data for Si diffusion along Al GBs [52], for annealing for 15 minutes at a temperature as low as 120°C (150°C), the diffusion lengths possibly covered by Si atoms have

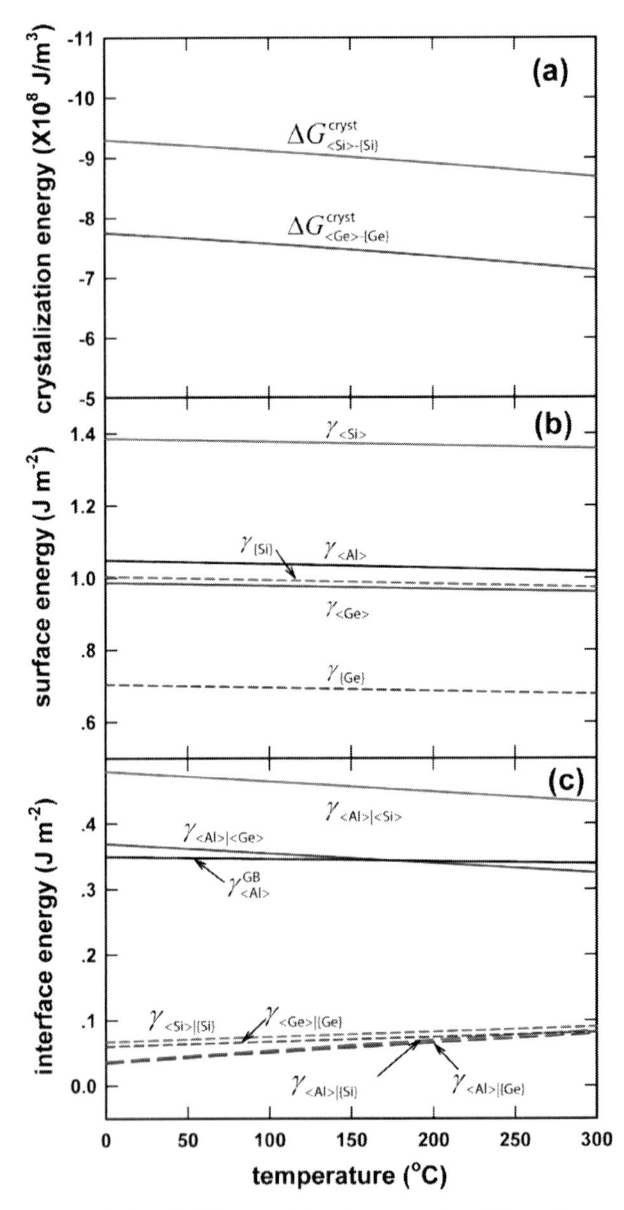

Figure 2.2 Calculated (a) crystallization energies, (b) surface energies, and (c) interface energies related to Al/Si and Al/Ge layer systems. Reprinted with permission from Ref. [11], Copyright 2008, American Physical Society. The methods for these calculations have been reviewed in Ref. [44].

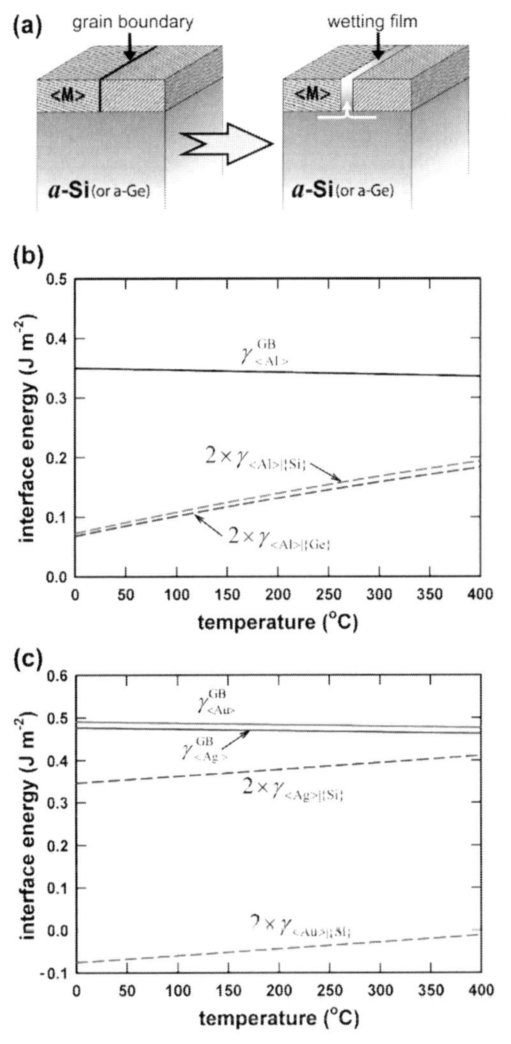

Figure 2.3 (a) Schematic illustration of the wetting of a metal (<M>) GB by amorphous semiconductor material. Reprinted with permission from Ref. [40], Copyright 2009, John Wiley and Sons.(b) Energetics for wetting of high-angle <Al> GBs by a-Ge and a-Si. A positive driving force $\gamma_{\langle \mathrm{Al} \rangle}^{\mathrm{GB}} - 2\gamma_{\langle \mathrm{Al} \rangle | \{\mathrm{Si\ or\ Ge}\}}$ is evident for wetting of the <Al> GBs by {Si} or {Ge}. (c) Energetics for wetting of high-angle <Ag> and <Au> GBs by {Si}. It follows {Si} can also wet both Ag GBs and Au GBs. Reprinted with permission from Ref. [11], Copyright 2008, American Physical Society.

been calculated to be 21 nm (136 nm) along Al GBs (as compared to the corresponding diffusion length possibly covered by volume diffusion of less than 1 nm) [14]. The large thermodynamic driving force (see previous paragraph) and the fast GB diffusion enable the occurrence of metal GB wetting by amorphous semiconductors at relatively low temperatures.

Very recently, the wetting of metal GBs by an amorphous semiconductor has been directly observed in the crystalline Al (c-Al)/a-Si bilayer system by in situ heating VEFTEM experiments [14] (see Fig. 2.4). The in situ heating VEFTEM method exploits the distinct differences of the plasmon loss characteristics (i.e., plasmon loss peak energy and peak full width at half-maximum [FWHM]) of the metal (Al) and the semiconductor (Si) [53]. An as-grown 150 nm a-Si/100 nm c-Al bilayer was heated in situ in an analytic transmission electron microscope at temperatures from 100°C to 150°C in steps of 10°C. As shown in Fig. 2.4a, at temperatures up to 110°C the bilayer structure remains intact. During the in situ annealing step at 120°C, the transportation of Si atoms into the Al GB initiates, as evidenced by increases in both the plasmon loss peak energy and the FWHM at the location of the initial Al GB near the a-Si/c-Al interface (Fig. 2.4b). During subsequent annealing steps at 130°C (Fig. 2.4c) and 140°C (Fig. 2.4d), the Si atoms diffuse progressively into the Al GB and subsequently have wetted the Al GB completely at 140°C (Fig. 2.4d). Subsequent ex situ HRTEM investigations of the wetted high-angle Al GBs (by Si) showed no c-Si phase, confirming that the wetting Si phase at high-angle Al GBs is amorphous. Simultaneous in situ VEFTEM observations showed that the wetting process occurs exclusively at high-angle Al GBs at temperatures below 150°C, whereas low-angle Al GBs, which possess a relatively low Gibbs energy, are inert for GB wetting at temperatures below 150°C (see Fig. 2.5) [14]. The in situ heating VEFTEM method can be applied to investigate the GB-wetting processes occurring in many metal/amorphous semiconductor systems, recognizing the inherently pronounced differences of the plasmon loss characteristics of a metal and a semiconductor [53].

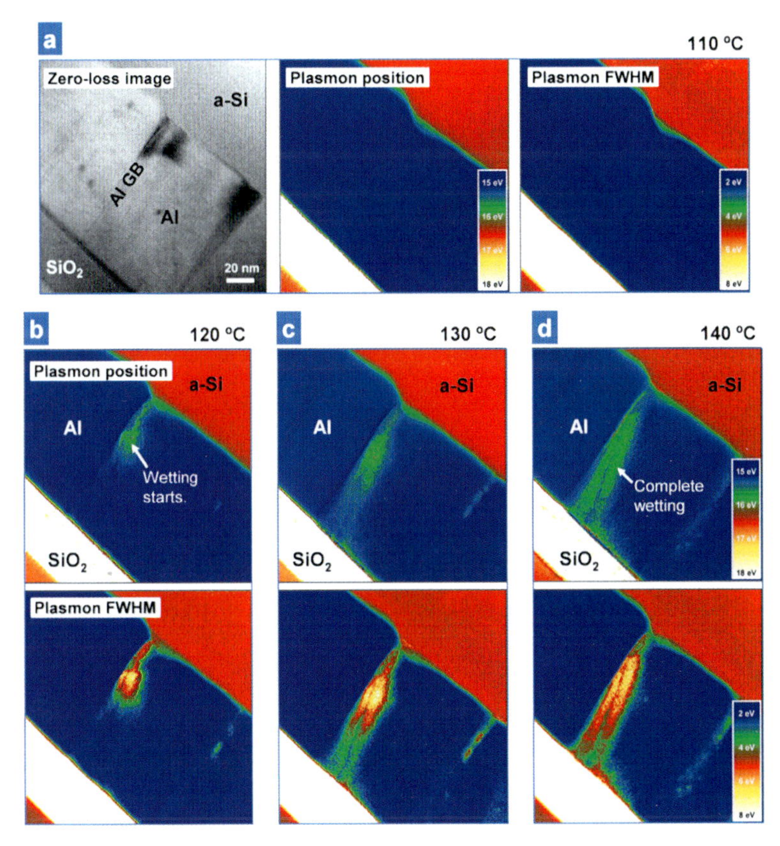

Figure 2.4 In situ heating VEFTEM observation of wetting of an Al GB by a-Si during annealing of a 100 nm c-Al/150 nm a-Si bilayer. (a) Zero-loss bright-field TEM image of the cross section of the bilayer specimen upon in situ annealing at 110°C for 15 minutes. The corresponding mappings of the plasmon loss peak energy and FWHM, as acquired by VEFTEM measurements, are shown as well. (b–d) Evolution of the mappings of the plasmon loss peak energy and FWHW at the same location of the specimen upon in situ annealing at temperatures increasing in steps of 10°C (15 minutes at each temperature). The increases in both the plasmon loss peak energy and the FWHW at the location of the Al GB near the c-Al/a-Si interface observed at 120°C (b) prove the initial wetting of the Al GB by Si. The complete wetting of the Al GB by Si has been realized at 140°C (d). Reprinted with permission from Ref. [14], Copyright 2011, John Wiley and Sons.

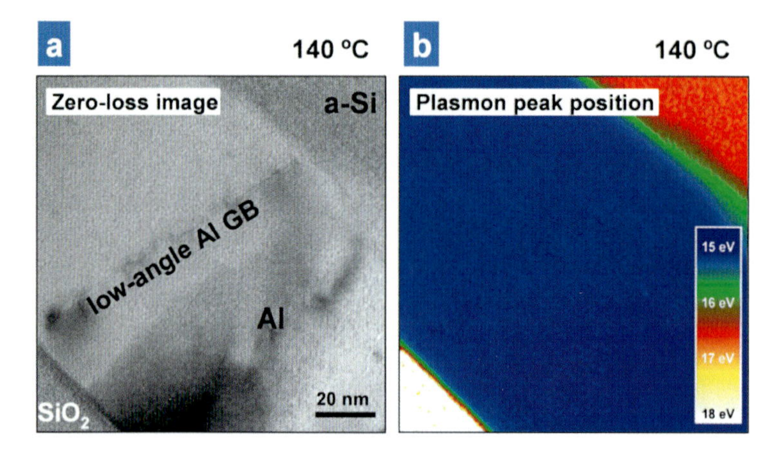

Figure 2.5 (a) Zero-loss bright-field TEM image at a location containing a low-angle Al GB in the Al sublayer in the c-Al/a-Si bilayer specimen annealed at 140°C (i.e., under the same conditions as in Fig. 2.4d). (b) The corresponding mapping of the plasmon loss peak energy. The low-angle Al GB is not wetted in contrast with what is observed for high-angle Al GBs (see Fig. 2.4). Reprinted with permission from Ref. [14], Copyright 2011, John Wiley and Sons.

2.3.2 Nucleation of Crystalline Semiconductors

The nucleation of c-Si or crystalline Ge (c-Ge) from the free (bond-weakened) interfacial atoms could occur at low temperatures heterogeneously at the interface with the metal and/or at the wetted metal GBs (see Fig. 2.6a and discussions before). The interface energies of the initial crystalline/amorphous interfaces ($\gamma_{\langle M \rangle | \{S\}}$) are usually lower than those of the created crystalline/crystalline interface energies ($\gamma_{\langle M \rangle | \langle S \rangle}$); see Fig. 2.2b (Refs. [11, 44]). This implies that a *thin enough* amorphous semiconductor layer can be thermodynamically stable at a surface and/or interface as long as the higher bulk energy of the amorphous phase, as compared to that of the corresponding bulk crystalline phase, is overcompensated by a lower sum of the surface and/or interfaces energies. Only beyond a certain critical thickness of the amorphous semiconductor film (at the surface/interface/GB), the release of bulk energy (i.e., $-\Delta G^{\mathrm{cryst}}_{\langle S \rangle - \{S\}}$: the so-called bulk crystallization energy) will be higher than the associated energy penalty for creating less energetically favorable crystalline/crystalline interfaces.

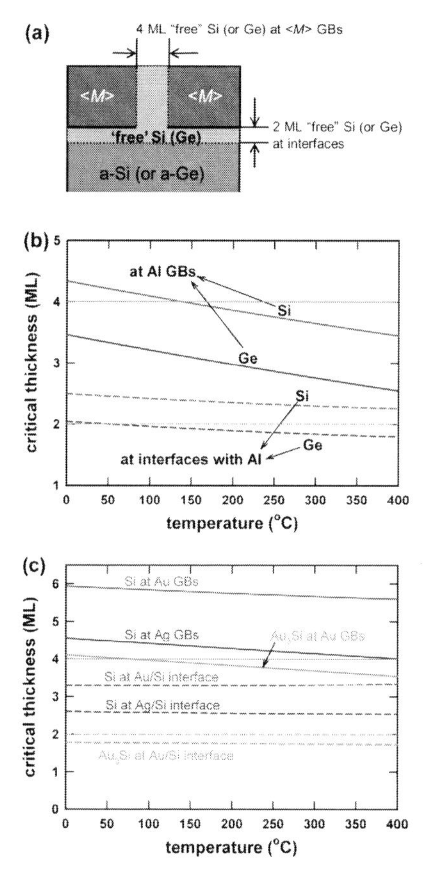

Figure 2.6 (a) Schematic illustration of the two types of sites for low-temperature nucleation of crystallization of amorphous semiconductors: (1) metal/semiconductor interface (including 2 ML of "free" atoms) and (2) wetted metal GBs (including 4 ML of "free" atoms). (b) Energetics of the nucleation of c-Ge (or Si) at Al GBs and at the c-Al/a-Ge (or c-Al/a-Si) interface. It follows that c-Ge can nucleate both at the Al GBs and the c-Al/a-Ge interface (above 50°C), whereas c-Si can nucleate only at the Al GBs (at above 140°C). Reprinted with permission from Ref. [11], Copyright 2008, American Physical Society. (c) Energetics of the nucleation of c-Si at Ag GBs and Au GBs and at the c-Ag/a-Si and c-Au/a-Si interfaces. It follows that c-Si can nucleate at the Ag GBs at (above) ~400°C. For the c-Au/a-Si system, it follows that c-Si cannot nucleate directly at Au GBs and at the c-Au/a-Si interface. MIC in c-Au/a-Si is, however, mediated by the formation of metastable c-Au_3Si nucleated at Au GBs. Reprinted with permission from Ref. [11], Copyright 2008, American Physical Society.

The critical thickness, $h_{\langle\mathrm{M}\rangle|\{\mathrm{S}\}}^{\mathrm{critical}}(T)$, for nucleation of a crystalline semiconductor (<S>) at the crystalline metal/amorphous semiconductor interface (<M>|{S}) at temperature *T*, can be calculated by dividing the increase of the interface energy (in J m^{-2}) upon crystallization (in this case, the original <M>|{S} interface is split up into an <M>|<S> interface and an <S>|{S} interface) by the corresponding decrease in bulk energy (i.e., the bulk crystallization energy, $-\Delta G_{\langle\mathrm{S}\rangle-\{\mathrm{S}\}}^{\mathrm{cryst}}$, in J m^{-3}) of the interfacial semiconductor layer:

$$h_{\langle\mathrm{M}\rangle|\{\mathrm{S}\}}^{\mathrm{critical}}(T)=\frac{\gamma_{\langle\mathrm{M}\rangle|\langle\mathrm{S}\rangle}(T)+\gamma_{\langle\mathrm{S}\rangle|\{\mathrm{S}\}}(T)-\gamma_{\langle\mathrm{M}\rangle|\{\mathrm{S}\}}(T)}{-\Delta G_{\langle\mathrm{S}\rangle-\{\mathrm{S}\}}^{\mathrm{cryst}}(T)} \tag{2.2}$$

As a further constraint, the critical thickness should be smaller than 2 ML in order that crystallization can initiate at an <M>|{S} interface at relatively low temperatures, because the layer of free semiconductor atoms at the interface has a maximal thickness of only 2 ML (see Section 2.2).

Alternatively, the initiation of crystallization of an amorphous semiconductor {S} could also occur at a metal <M> GB if the <M> GB is wetted by {S} (see Section 2.3.1). Considering a wetting {S} layer present at the metal <M> GBs, the critical thickness, $h_{\{\mathrm{S}\}\ \mathrm{in}\ \langle\mathrm{M}\rangle\mathrm{GB}}^{\mathrm{critical}}(T)$, for nucleation of crystallization at temperature *T*, can be calculated by dividing the increase of the interface energy (in J m^{-2}) upon crystallization (in this case, two original <M>|{S} interfaces are replaced by two <M>|<S> interfaces) by the decrease in the "bulk" energy of the wetting semiconductor film ($-\Delta G_{\langle\mathrm{S}\rangle-\{\mathrm{S}\}}^{\mathrm{cryst}}$, in J m^{-3}):

$$h_{\{\mathrm{S}\}\ \mathrm{in}\ \langle\mathrm{M}\rangle\mathrm{GB}}^{\mathrm{critical}}(T)=\frac{2\times\left[\gamma_{\langle\mathrm{M}\rangle|\langle\mathrm{S}\rangle}(T)-\gamma_{\langle\mathrm{M}\rangle|\{\mathrm{S}\}}(T)\right]}{-\Delta G_{\langle\mathrm{S}\rangle-\{\mathrm{S}\}}^{\mathrm{cryst}}(T)} \tag{2.3}$$

The wetting amorphous semiconductor {S} layer at the metal GBs is sandwiched between two metal grains, and consequently, the maximum thickness of the free semiconductor layer at the metal GBs is about 2 × 2 ML = 4 ML. Hence, as a further constraint, $h_{\{\mathrm{S}\}\ \mathrm{in}\ \langle\mathrm{M}\rangle\mathrm{GB}}^{\mathrm{critical}}(T)$ must be smaller than (or maximally equal to) 4 ML in order that crystallization of amorphous semiconductor {S} can initiate at the metal GBs.

Thus, using the calculated values for the crystallization energies and interface energies (see Fig. 2.2 [11]), the critical thicknesses for initiation of crystallization of {Ge} and {Si} at the wetted metal (<Al>, <Ag>, <Au>) GBs and at the interfaces with the metal have been calculated as a function of temperature T by applying Eqs. 2.2 and 2.3. The results are shown in Fig. 2.6b for c-Al/a-Si and c-Al/a-Ge systems and in Fig. 2.6c for crystalline Au (c-Au)/a-Si and crystalline Ag (c-Ag)/a-Si systems [11], where the critical thickness data is presented in units of ML, where 1 ML Si ≈ 2.2 Å and 1 ML Ge ≈ 2.5 Å [54].

For a-Ge, the critical thickness for nucleation of c-Ge is below 2 ML at the <Al>|{Ge} interface (above 50°C). Therefore, the 2 ML free Ge layers can crystallize at the <Al>|{Ge} interfaces at low temperatures (above 50°C). This prediction is in accordance with experimental observations reported in Refs. [12, 55, 56]. At the Al GBs, the critical thickness for nucleation of crystallization of a-Ge varies from about 3.5 ML at 0°C to about 2.6 ML at 400°C (Fig. 2.6b), which is below the thickness (4 ML) of the free Ge at Al GBs. Hence, the low-temperature initiation of the crystallization of a-Ge can *also* occur at the Al GBs. This prediction is fully consistent with corresponding experimental observations [57].

For a-Si, the critical thickness for nucleation of crystallization at the <Al>|{Si} interface is *above* 2 ML even at 400°C. Therefore, the 2 ML of free Si at the interface stay amorphous and initiation of crystallization *cannot* occur at this location. At the Al GBs, the critical thickness of {Si} for initiation of crystallization is below 4 ML at temperatures above 140°C. Hence, the only site for c-Si to nucleate at low temperatures is the Al GB, and the crystallization temperature should be higher than 140°C for the c-Al/a-Si layer system. The predicted crystallization temperature of about 140°C is in accordance with experimental values reported in Refs. [1, 5, 41]

A recent in situ heating HRTEM experiment confirmed the *exclusive* nucleation of c-Si at the Al high-angle GB [14]. As, for example, shown in Fig. 2.7, a c-Si nucleus has formed at a high-angle Al GB upon in situ heating at 150°C [14]. At the same time, during the onset of nucleation and initial growth of the c-Si nucleus at the Al GB, the structure of the <Al>|{Si} interface, as well as the interface/Al-GB triple junction, stayed unchanged (see Fig. 2.7 [14]). Real-time in situ HRTEM experiments performed using other as-grown a-Si/c-Al specimens confirmed that the nucleation and growth of c-Si occur exclusively at high-angle Al GBs, at temperatures equal to or above

150°C. The exclusive nucleation (and growth) of c-Si at the Al GBs is further observed by an in situ plan-view VEFTEM experiment [14]. As shown in Fig. 2.8, c-Si nanowalls (or nanowires), with a width of only about 15 nm, develop at the original Al GBs in the Al sublayer upon in situ annealing of a 50 nm c-Al/30 nm a-Si up to 170°C. Therefore, by employing this Al-induced crystallization of the a-Si process, Si nanowalls can now be prepared at exceedingly low temperatures, without the conventional constraints of high solubility and bulk diffusivity (both of which demand high processing temperatures) [14].

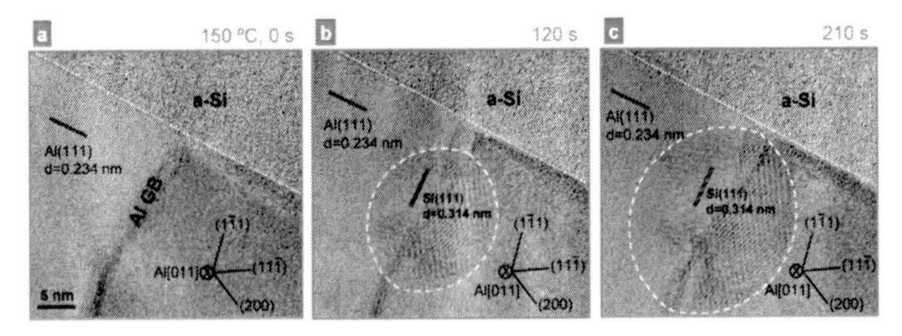

Figure 2.7 In situ heating HRTEM observation (cross-sectional views) of the nucleation of c-Si at a high-angle Al GB at 150°C in a 100 nm c-Al/150 nm a-Si bilayer. (a) Before nucleation of c-Si (0 seconds). (b–c) Initial nucleation and growth of a c-Si nucleus at the Al GB (120 seconds and 210 seconds at 150°C, respectively). Reprinted with permission from Ref. [14], Copyright 2011, John Wiley and Sons.

According to the thermodynamic model calculations, the most striking difference between the c-Al/a-Ge and c-Al/a-Si layer systems is that c-Ge can (also) nucleate at the <Al>|{Ge} interface, whereas c-Si *cannot* nucleate at the <Al>|{Si} interface. Al GBs are preferred sites for nucleation of crystallization in both layer systems. This difference in nucleation behaviors leads to very different MIC kinetics in the c-Al/a-Ge and c-Al/a-Si layer systems. As evidenced by the in situ XRD measurement results shown in Fig. 2.9, the a-Ge in the c-Al/a-Ge layer system exhibits an "explosive"-like crystallization behavior as the temperature reaches 150°C [7], which is ascribed to the occurrence of abundant nucleation in this system [7, 11]. On the other hand, the a-Si in the c-Al/a-Si shows a gradual crystallization with increasing temperature [9, 11], owing to the much more limited number of nucleation sites (wetted Al GBs).

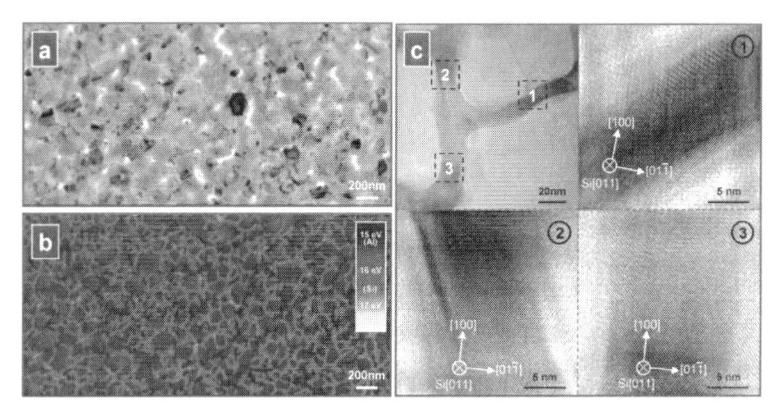

Figure 2.8 Plan-view TEM images of a 50 nm c-Al/30 nm a-Si bilayer annealed in situ up to 170°C in an analytic TEM microscope, showing the formation of c-Si "nanowalls" at the high-angle GBs in the columnar, nanocrystalline Al sublayer. (a) Zero-loss bright-field TEM image. (b) Mapping of the plasmon loss peak energy corresponding to (a); red is c-Si (see also [c]), and green corresponds to the (overlapping) c-Al/a-Si sublayers. (c) HRTEM of the c-Si nanowalls formed at the original Al GBs in the Al sublayer. Reprinted with permission from Ref. [14], Copyright 2011, John Wiley and Sons.

The thermodynamic model calculations, as applied above to the c-Al/a-Si and c-Al/a-Ge systems, can be performed in general for predicting the MIC behaviors in crystalline metal/amorphous semiconductor bilayer systems. For example, such results for the c-Ag/a-Si and c-Au/a-Si systems are shown in Fig. 2.6c [11]. As follows from Fig. 2.6c, the critical thickness for a-Si to crystallize at Ag GBs (which are wetted by a-Si; see Section 2.3.1 and Fig. 2.3c) is smaller than 4 ML for a temperature larger than about 400°C, whereas at the <Ag>|{Si} interface, the critical thickness for crystallization of a-Si is about 2.5 ML (>2 ML). Hence, with reference to the discussion above, it can be predicted that the (Ag-induced) crystallization of a-Si in the c-Ag/a-Si system will initiate *exclusively* at Ag GBs, at a temperature larger than about 400°C. These predictions agree very well with the experimental results: the crystallization of a-Si starts at GBs in the Ag layer at 390°C–410°C [23, 41].

For the c-Au/a-Si system, model calculations reveal that the critical thickness for nucleation of crystallization of a-Si is ~5.5 ML to 6 ML (for $T < 400$°C) at Au GBs (which are wetted by a-Si; see

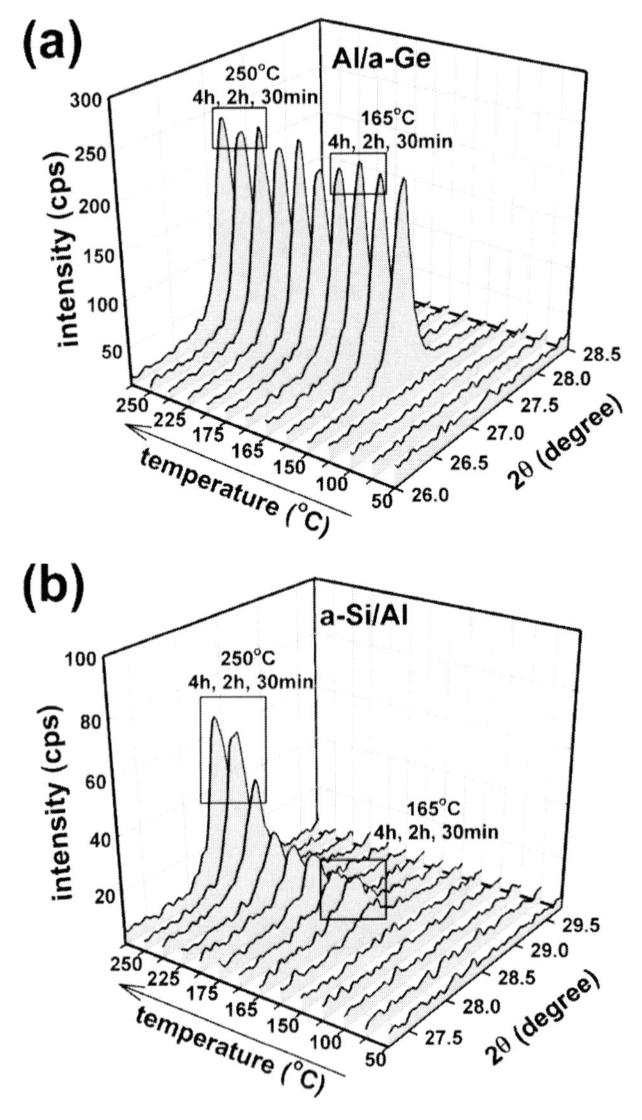

Figure 2.9 (a) The evolution of the Ge(111) diffraction peak upon in situ heating of an Al/a-Ge specimen in a diffractometer. An "explosive"-like, full crystallization of the a-Ge layer is observed at 150°C. (b) The evolution of the Si(111) diffraction peak upon similar in situ heating of an a-Si/Al specimen. A gradual crystallization of a-Si is observed with increasing temperature and annealing time. Reprinted with permission from Ref. [11], Copyright 2008, American Physical Society.

Section 2.3.1 and Fig. 2.3c) and ~3.3 ML at the Au/a-Si interface (see Fig. 2.6c). Hence, it is predicted that c-Si cannot nucleate directly (i.e., without the aid of an intermediate phase; see later), neither at the Au GBs, nor at the <Au>|{Si} interface. Instead, experiments have shown that the MIC process in the c-Au/a-Si system is mediated by the formation of a (metastable) interfacial <Au_3Si> phase at a very low temperature of ~100°C [21] (see Fig. 2.10). To understand this observation, thermodynamic calculations on the possible nucleation of <Au_3Si> at the <Au> GBs and at the <Au>|{Si} interface have also been carried out [11] within the framework provided by Ref. [44]. It follows that the critical thickness of a-Si at <Au> GBs to form <Au_3Si> is close to 4 ML at a low temperature of 80°C (and is smaller at higher temperatures), which indeed explains the experimentally observed formation of <Au_3Si> at Au GBs in the temperature range of 100°C–200°C [21]. At the Au/a-Si interface, the critical thickness is ~1.8 ML (<2 ML), indicating that <Au_3Si> can also nucleate at the Au/a-Si interface. Recognizing the very large positive driving force for Si atoms diffusing into (i.e., wetting) the Au GBs (see Fig. 2.3c) and also the high mobility of Au atoms at Au GBs, it can be conceived that the formation of <Au_3Si> will be more pronounced at Au GBs than at the <Au>|{Si} interface, which is consistent with experimental observations [21].

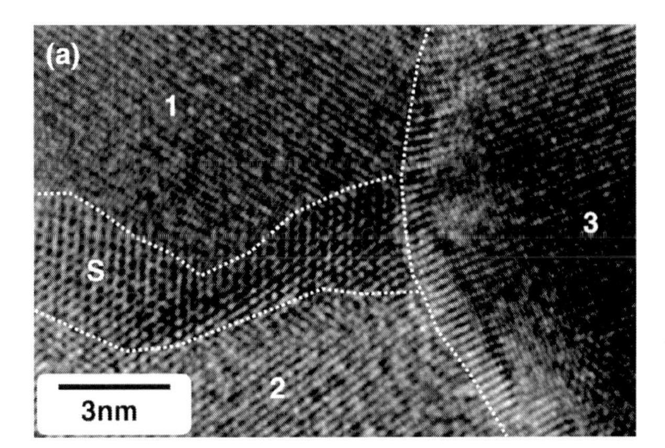

Figure 2.10 Plan-view HRTEM observation of metastable Au_3Si nucleation at a GB of the c-Au:Au_3Si crystallite (labeled "S") and three adjacent Au grains (labeled "1", "2", "3"). Reprinted with permission from Ref. [21], Copyright 1998, American Physical Society.

Very recently, the thermodynamic model calculations have also been applied to metal (Al)-induced crystallization of amorphous semiconductor *alloy* (a-SiGe) [17]. As discussed in Section 2.3.1, high-angle Al GBs can be wetted by both a-Si and a-Ge (see Fig. 2.3b). It is not surprising that the high-angle Al GBs can also be wetted by a-SiGe alloy, which is indeed supported by the corresponding thermodynamic model calculation (see Fig. 2.11a) [17]. Accordingly, the critical thicknesses for initiation of crystallization of a-SiGe at the wetted <Al> GBs and at the <Al>|{SiGe} interface have been calculated quantitatively as a function of temperature T by applying Eqs. 2.2 and 2.3. The results are shown in Fig. 2.11c. Interestingly, the calculation for the c-Al/a-SiGe system predicts a transition from the *exclusive* Al GB–mediated nucleation of crystallization to (also) the interface-mediated nucleation of crystallization, at a critical temperature of about 320°C (Fig. 2.11c) [17]. The predicted MIC behaviors are schematically shown in Fig. 2.11d. The predictions are compatible with the experimental observations: c-SiGe phases of two different compositions develop at annealing temperatures above 325°C (i.e., a little above the predicted transition temperature), whereas at a temperature below 300°C a c-SiGe phase of a single composition develops (see the XRD results in Fig. 2.11e [17]). Obviously, at temperatures below the transition temperature, c-SiGe nucleates exclusively at Al GBs. Since the diffusivity of Ge is higher than Si at Al GBs, a Ge-rich c-SiGe phase thus forms at the Al GBs. At temperatures higher than the transition temperature, a c-SiGe phase also develops at the <Al>|{SiGe} interface, in addition to the formation of a Ge-rich c-SiGe phase at the Al GBs. Since the (local) nucleation of a c-SiGe phase at the <Al>|{SiGe} interface does not require long-range diffusion of Si and Ge atoms, the composition of this c-SiGe resembles the bulk composition of the a-SiGe layer (or is relatively Si rich due to depletion of Ge diffused to Al GBs). Thus, at a temperature above the transition temperature of 320°C, the formation of dual-phase c-SiGe occurs.

2.3.3 Continued Crystallization and Layer Exchange

As discussed in the preceding sections, for many metal/semiconductor (M/S) systems such as c-Al/a-Si, c-Ag/a-Si, and c-Al/a-Ge, nucleation of a crystalline <S> was theoretically predicted and

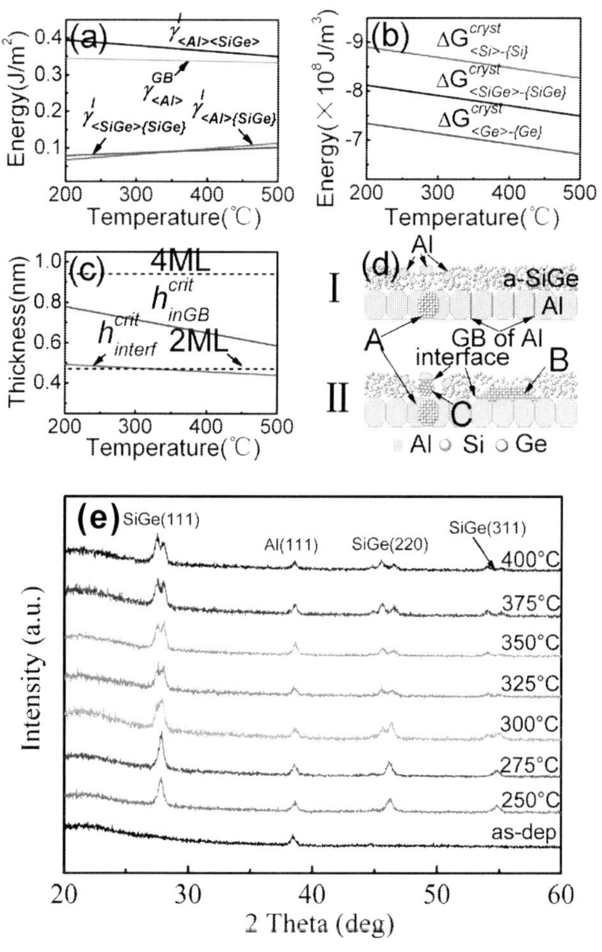

Figure 2.11 Calculated (a) interface energies, (b) bulk crystallization energies, and (c) critical thicknesses for nucleation of crystallization of a-SiGe at the <Al>|{SiGe} interface and at <Al> GBs. (d) Schematic illustration of crystallization processes at a temperature below (panel I) or above (panel II) the critical temperature (~320°C) for nucleation of <SiGe> at the <Al>|{SiGe} interface. (e) XRD patterns of the c-Al/a-SiGe samples as deposited and annealed at different temperatures from 250°C to 400°C. After annealing at 250°C, the diffraction peaks, namely, 111, 220, and 311, can be observed, indicating crystallization of a-SiGe. Double XRD peaks start to appear when the annealing temperature is raised to 300°C, indicating the formation of a *dual-phase* c-SiGe structure. Reprinted with permission from Ref. [17], Copyright 2012, AIP Publishing LLC.

experimentally confirmed to occur at the wetted <M> GBs at low temperatures. After the formation of an <S> nucleus at an <M> GB, the GB in the metal layer has been replaced by two <M>|<S> interphase boundaries. To proceed with the crystallization process of {S} at this location, the atoms in the amorphous semiconductor {S} layer must continue to diffuse, now into the <M>|<S> interphase boundaries (continued "wetting") and crystallize there. The driving force for this secondary wetting of the <M>|<S> interphase boundaries by {S} is given by (see Eq. 2.1)

$$\Delta\gamma^{\text{wetting}}_{\{S\}\rightarrow\langle M\rangle|\langle S\rangle}(T)=\gamma_{\langle M\rangle|\langle S\rangle}(T)-\left[\gamma_{\langle M\rangle|\{S\}}(T)+\gamma_{\langle S\rangle|\{S\}}(T)\right] \tag{2.4}$$

As an example, for the c-Al/a-Si system in which the nucleation of c-Si occurs *exclusively* at Al GBs (see Section 2.3.2), this driving force is calculated to be positive (~0.3 J m^{-2} at 150°C; see Fig. 2.12a) [11], indicating that a-Si is capable of continuing to wet the <Al>|<Si> boundaries. As also shown in Fig. 2.12a, such a secondary wetting process can occur, too, in the c-Al/a-Ge system.

Once wetting {S} layers are formed at the <M>|<S> interphase boundaries, two cases are possible: (i) the wetting {S} layer joins with the adjacent <S> grains to crystallize, as a result of which the <S> grains grow laterally, perpendicular to the <M>|<S> boundaries and (ii) new grains of <S> nucleate. The critical thickness in case (i), that is, for continued lateral <S> grain growth perpendicular to the <M>|<S> boundaries (in this case, the original <M>|{S} and <S>|{S} interfaces are replaced by an <M>|<S> interface), is given by

$$h^{\text{critical}}_{\langle S\rangle\text{ grain growth}}(T)=\frac{\gamma_{\langle M\rangle|\langle S\rangle}(T)-\left[\gamma_{\langle M\rangle|\{S\}}(T)+\gamma_{\langle S\rangle|\{S\}}(T)\right]}{-\Delta G^{\text{cryst}}_{\langle S\rangle\text{-}\{S\}}(T)} \tag{2.5}$$

The critical thickness in case (ii), that is, for the formation of new <S> grains at the <M>|<S> boundaries (in this case, the original <M>|{S} and <S>|{S} interfaces are replaced by an <M>|<S> interface and an <S>|<S> interface [i.e., an <S> GB]), is given by

$$h^{\text{critical}}_{\langle S\rangle\text{ new nucleation}}(T)=\frac{\gamma_{\langle M\rangle|\langle S\rangle}(T)+\gamma_{\langle S\rangle|\langle S\rangle}(T)-\left[\gamma_{\langle M\rangle|\{S\}}(T)+\gamma_{\langle S\rangle|\{S\}}(T)\right]}{-\Delta G^{\text{cryst}}_{\langle S\rangle\text{-}\{S\}}(T)} \tag{2.6}$$

Again taking the c-Al/a-Si bilayer as an example system, the critical thicknesses for the two cases have been calculated as a

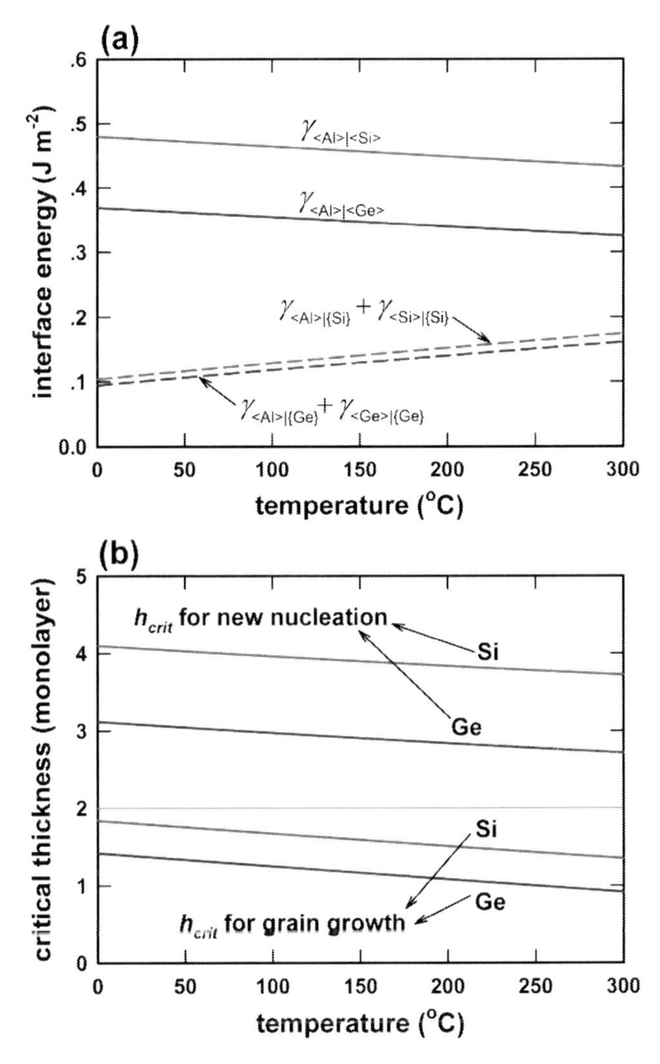

Figure 2.12 (a) Energetics of the continued wetting after completing the initial nucleation of c-Si or c-Ge at Al GBs. A positive driving force occurs for continued wetting of the formed <Al>|<Ge> and <Al>|<Si> boundaries by {Ge} and {Si}, respectively. (b) Energetics for continued lateral grain growth of c-Ge and c-Si in the original Al layer (perpendicular to the original Al GBs). Continued grain growth, rather than repeated nucleation of crystallization, is favored for both Ge and Si. The formation of new crystalline nuclei is impossible. Reprinted with permission from Ref. [11], Copyright 2008, American Physical Society.

function of temperature T; the results are shown in Fig. 2.12b. The critical thickness for the formation of new c-Si nuclei (Eq. 2.6) is as large as ~4 ML. Recognizing that the thickness of the free Si atoms adjacent to Al is only about 2 ML, it follows that the formation of new <Si> nuclei at the <Al>|<Si> boundaries is *impossible*. Instead, the critical thickness for continued c-Si grain growth (Eq. 2.5) is only ~1.5 ML (at $T > 150°C$). Hence, continued lateral growth of the <Si> grains initially formed at the original Al GBs is possible. Additionally recognizing the large positive driving force $\Delta\gamma^{\text{wetting}}_{\{S\}\to\langle M\rangle|\langle S\rangle}$ for the continued wetting of <Al>|<Si> boundaries by {Si} (Fig. 2.12a), this result strongly indicates that the lateral growth of the initially nucleated <Si> grains at the original Al GBs is realized by continuously incorporating Si atoms diffusing from the a-Si layer into the <Al>|<Si> boundaries [11]. This is in accordance with the in situ and ex situ VEFTEM and HRTEM analyses (i.e., only growth of the c-Si grains initially formed at Al GBs is observed; see Fig. 2.13). A schematic illustration of the sequential stages of the low-temperature Al-induced crystallization of a-Si process as mediated by an Al GB is given in Fig. 2.14 [40].

The continuous inward diffusion, crystallization, and grain growth of Si within the original Al GBs can result in (i) compressive stress buildup in the Al grains and (ii) tensile stress buildup (and enrichment/aggregation of free volume) in the a-Si sublayer. As a result of both effects, Al tends to migrate toward the free-volume-rich, tensilely stressed Si layer (in situ XRD stress measurements have indicated [12, 15], indeed, the relaxation of the compressive stress in the Al phase during MIC). Thus, the Al layer is gradually "replaced" by the c-Si grains growing inside it, and the Al migrates to the original location of the a-Si layer, as revealed by the development of convective Al "plumes" comprising many Al nanocrystals (see Fig. 2.13a,b). As a consequence of this process, a continuous c-Si layer is formed at the position of the original Al layer: layer exchange has occurred (see Fig. 2.13c). Almost all experimental results on the c-Al/a-Si layer system indicate that MIC in c-Al/a-Si is associated with a layer exchange of Al and Si [2–4, 6, 9, 13]. Due to the thermodynamic impossibility of new c-Si nucleation after the initial nucleation of c-Si at the Al GBs (see before), it is predicted that the final lateral c-Si grain sizes are similar to the original Al lateral grain sizes, which well agrees with experimental observations for the Al/Si layer systems [4, 6, 9].

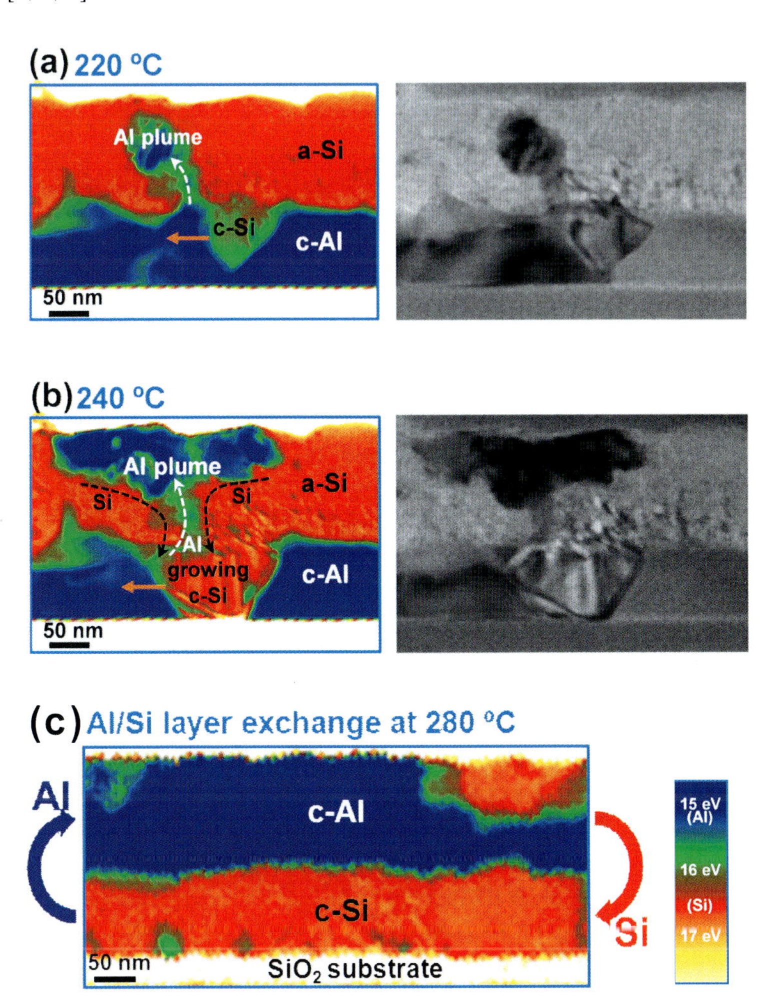

Figure 2.13 In situ VEFTEM observation of the lateral growth of a c-Si grain nucleated at the c-Al bottom layer during heating of a 150 nm a-Si/100 nm c-Al bilayer (a) at 220°C and (b) at 240°C. The left-hand side shows the plasmon loss energy mappings, and the right-hand side shows the corresponding bright-field TEM images of the same location. (c) Upon heating at 280°C, Al and Si sublayers have practically exchanged their locations: a layer exchange has occurred.

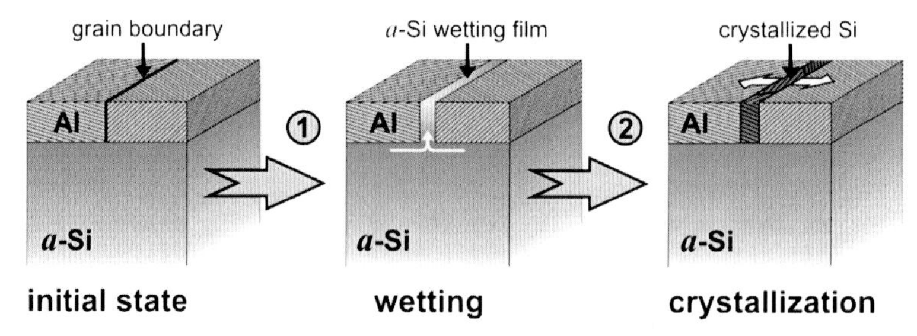

Figure 2.14 Schematic illustration of the process of Al-induced low-temperature crystallization of a-Si. Initially, GBs in the Al overlayer get "wetted" by a-Si. Beyond a critical thickness of the wetting a-Si film, crystallization can initiate at these wetted Al GBs and the formed c-Si grain subsequently grows laterally in the Al overlayer. Reprinted with permission from Ref. [40], Copyright 2009, John Wiley and Sons.

In contrast with the c-Al/a-Si layer system, c-Ge can nucleate both at the Al GBs and at the interface with the Al layer (Section 2.3.2). The two possible cases of continued crystallization of a-Ge at the original Al GBs can be considered similarly as for the c-Al/a-Si system (see before). The results of the corresponding calculations are also shown in Figs. 2.12a and 2.12b, respectively. It follows that a-Ge is capable of wetting the <Al>|<Ge> boundaries ($\Delta\gamma^{\text{wetting}}_{\{\text{Ge}\}\to\langle\text{Al}\rangle|\langle\text{Ge}\rangle} > 0$) and subsequently crystallizes there by joining the pre-existing c-Ge nuclei ($h^{\text{critical}}_{\langle\text{Ge}\rangle\text{ grain growth}} \approx 1$ ML), thereby establishing lateral growth of the initially nucleated c-Ge grains. Similar to the c-Al/a-Si layer system, the formation of new <Ge> nuclei at the <Al>|<Ge> boundaries is impossible ($h^{\text{critical}}_{\langle\text{Ge}\rangle\text{ new nucleation}} \approx 3$ ML). Since <Ge> is also capable of nucleating at the c-Al/a-Ge interface, c-Ge grain growth can also occur at the same time within the a-Ge layer itself, starting from the c-Ge grains nucleated at the interface. The interface-mediated and GB-mediated crystallizations of a-Ge occur simultaneously. As a result, the a-Ge phase is very fast fully crystallized, long before a complete layer exchange, as induced by crystallization at the original Al GBs, has been realized, which agrees with the experimental observations [7, 11, 12, 57]. Furthermore, in a recent study [16, 58], by creating an interfacial Ge oxide layer between the c-Al and a-Ge sublayers, the interface-mediated nucleation of c-Ge in c-Al/a-Ge bilayers can be suppressed due to the modified interface energetics by the presence of the interfacial oxide layer. Then the Al GBs exclusively mediate the low-temperature crystallization of a-Ge, and layer exchange of

Al and Ge sublayers upon Al-induced crystallization of a-Ge was observed indeed (similar to that observed in the c-Al/a-Si bilayers) [58].

Similar to the c-Al/a-Si bilayer system (see before), lateral growth of c-Si in the metal layers during MIC has recently also been observed in the c-Ag/a-Si [24] and c-Au/a-Si [22] bilayer systems. In both bilayer systems an eventual layer exchange of the metal and silicon sublayers was observed as well, in association with the continued crystallization of a-Si, thereby fully supporting the above model interpretation.

2.4 Ultrathin Metal–Induced Crystallization: Tailoring the Crystallization Temperature of Amorphous Semiconductors

The theoretical predictions for and the real-time in situ heating HRTEM and VEFTEM experiments performed with respect to the MIC process discussed before have convincingly demonstrated that the different MIC temperatures and behaviors for the different metal/amorphous semiconductor systems can be rationalized on the basis of interface thermodynamics, while accounting for the bond-weakening effect (see Sections 2.2 and 2.3). The success of the model calculations implies that the MIC temperatures and MIC behaviors can in principle be tailored through controlled adjustments of the contributions of the interface energies. This idea can, for example, be dramatically illustrated considering *ultrathin* MIC of amorphous semiconductors: through controlled adjustment of the metal overlayer thickness, and thus the governing role of the overall interface thermodynamics, the onset (crystallization) temperature of MIC processes can be precisely tailored [10].

Consider the nucleation of <S> at <M> GBs wetted by an amorphous semiconductor {S}. If the metal <M> layer thickness is very small and comparable with the thickness of the {S} wetting layer, the energetics for the nucleation of <S> at <M> GBs differ from that given by Eq. 2.3. As shown in Fig. 2.15a for the example of ultrathin-Al-induced crystallization of a-Si, for comparably small values of the (columnar) Al overlayer thickness h_{Al}, it follows that upon initiation of crystallization of the wetting a-Si film of uniform thickness l_{Si} at the Al GBs, not only the interface energy change $2h_{Al}\cdot(\gamma_{\langle Al\rangle|\langle Si\rangle}-\gamma_{\langle Al\rangle|\{Si\}})$, associated with the replacement of the two original <Al>|{Si} interfaces by two <Al>|<Si> interfaces, but also the surface and interface energy changes $l_{Si}\cdot(\gamma_{<Si>}-\gamma_{\{Si\}})$ and $l_{Si}\cdot(\gamma_{<Si>}-\gamma_{<Si>|\{Si\}})$, associated with the formation of the crystalline <Si> surface and the <Si>|{Si} interface, respectively, have to be considered. Consequently, the corresponding critical thickness for initiation of crystallization of a-Si at the Al GBs as function of both T and h_{Al} is now given by [10]

$$l^{\text{critical}}_{\{\text{Si}\}\,\text{in}\,\langle\text{Al}\rangle\,\text{GB}}(T,h_{\text{Al}})=\frac{2\times\left[\gamma_{\langle\text{Al}\rangle|\langle\text{Si}\rangle}(T)-\gamma_{\langle\text{Al}\rangle|\{\text{Si}\}}(T)\right]}{-\Delta G^{\text{cryst}}_{\langle\text{Si}\rangle-\{\text{Si}\}}(T)-\dfrac{\left[\gamma_{\langle\text{Si}\rangle}(T)-\gamma_{\{\text{Si}\}}(T)\right]+\gamma_{\langle\text{Si}\rangle|\{\text{Si}\}}(T)}{h_{\text{Al}}}} \quad (2.7)$$

The critical thickness $l^{\text{critical}}_{\{Si\}\,\text{in}\,<Al>\,GB}$ has been calculated as a function of both h_{Al} and T according to Eq. 2.7, using the values for the surface and interface energies given in Figs. 2.2b and 2.2c; the results are presented as a contour plot in Fig. 2.15b. As follows from the discussion in Section 2.3.2, for the c-Al/a-Si layer system the only possible sites for initiation of crystallization of a-Si are at the Al GBs, and $l^{\text{critical}}_{\{Si\}\,\text{in}\,<Al>\,GB}$ must be smaller than (or maximally equal to) 4 ML so that nucleation of crystallization can occur at the Al GBs. Hence, the thermodynamic limit for the onset crystallization temperature of a-Si as a function of the Al overlayer thickness is given by the solid line $l^{\text{critical}}_{\{Si\}\,\text{in}\,<Al>\,GB}$ = 4 ML in Fig. 2.15b. It follows that the crystallization temperature is in the range of 150°C–200°C for h_{Al} > 20 nm. For h_{Al} < 20 nm, the crystallization temperature increases rapidly with decreasing h_{Al} up to the "bulk" crystallization temperature when h_{Al} is smaller than 2 nm.

These theoretical predictions were experimentally confirmed by monitoring the crystallization of a-Si as a function of the c-Al overlayer thickness under ultrahigh vacuum conditions by real-time in situ spectroscopic ellipsometry (RISE; see, e.g., Fig. 2.16) [10]. The RISE technique is highly sensitive to changes in optical properties as, for example, caused by crystallization of a-Si [59]. The optical absorption of a-Si shows a broad peak around a photon energy of 3.5 eV. This broad absorption peak separates into two sharper prominent peaks at about 3.4 eV and 4.2 eV for bulk Si upon crystallization [59]. The measured onset crystallization temperatures of a-Si as a function of the Al overlayer thickness, based on RISE measurements of a series of c-Al/a-Si specimens with different c-Al overlayer thicknesses, have also been indicated in Fig. 2.15b [10]. The crystallization temperature of a-Si is indeed highly sensitive to the Al overlayer thickness below 20 nm: the corresponding crystallization temperature of a-Si decreases from about 700°C to 180°C for an increase of the thickness h_{Al} of the covering Al film from $h_{Al} < 1$ nm to $h_{Al} = 20$ nm (see Fig. 2.15b).

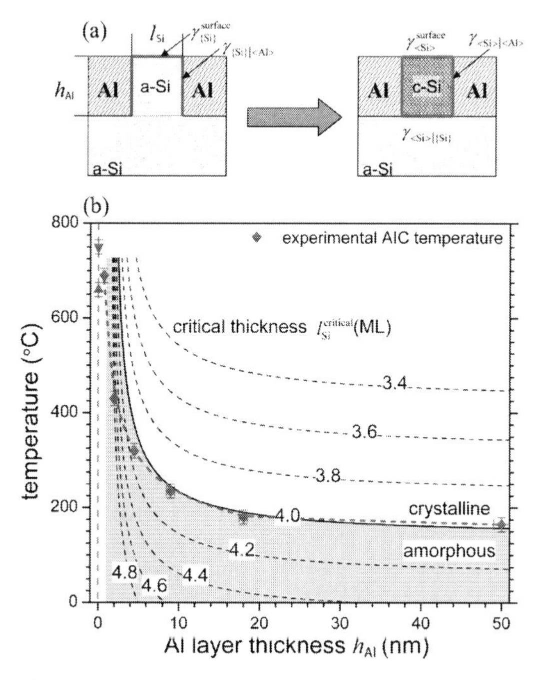

Figure 2.15 (a) Schematic representation of the thermodynamic model for the initiation of crystallization of a-Si at Al GBs in the ultrathin, columnar Al overlayer of thickness h_{Al}, recognizing that h_{Al} is comparable with the thickness of the Al GB-wetting layer of a-Si, l_{Si}. (b) Calculated critical thickness for initiation of crystallization of a-Si at Al GBs ($l_{Si}^{critical}$) as a function of both Al layer thickness h_{Al} and temperature. The line $l_{Si}^{critical} = 4$ ML gives a theoretical prediction of the crystallization temperature of a-Si as a function of the Al overlayer thickness. The corresponding experimental results for the dependence of the crystallization temperature of a-Si on h_{Al} have also been indicated in the figure. Reprinted with permission from Ref. [10], Copyright 2008, American Physical Society.

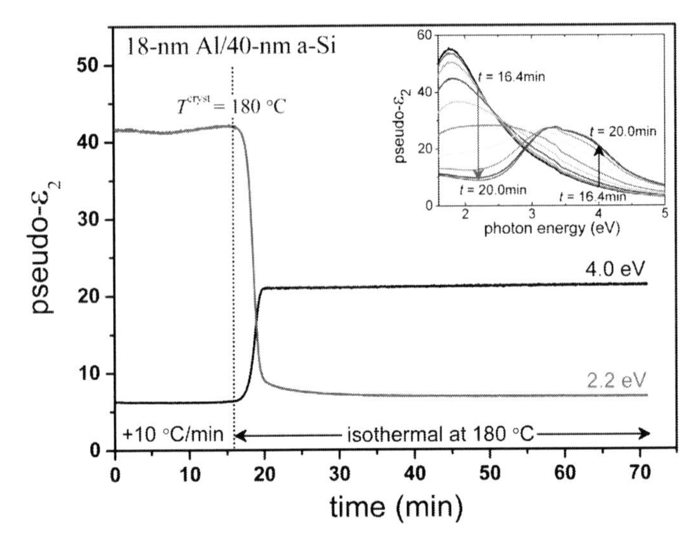

Figure 2.16 Real-time evolutions of the imaginary part ε_2 of the pseudodielectric function of an 18 nm Al/40 nm a-Si specimen at photon energies of 2.2 eV and 4.0 eV, as determined by real-time spectroscopic ellipsometry during in situ annealing. The corresponding detailed pseudo ε_2 spectra (i.e., as a function of photon energy) between 1.6 eV and 5.0 eV are shown in the inset. The observed fast increase in pseudo ε_2 at photon energies around 4.0 eV at 180°C clearly indicates the formation of c-Si. The simultaneous decrease in pseudo ε_2 at photon energies around 2.2 eV is not only related to the crystallization of a-Si but is also partly ascribed to the inward "movement" of Al upon continued crystallization of the a-Si phase (i.e., a layer exchange process of Al and Si; see Section 2.3.3). It thus follows that the onset temperature for initiation of Al-induced crystallization is 180°C. Reprinted with permission from Ref. [10], Copyright 2008, American Physical Society.

The above-discussed approach can be applied to precisely tailor the crystallization temperature in many metal/semiconductor systems where MIC is exclusively initiated at the metal GBs (this holds, e.g., for c-Ag/a-Si, c-Al/a-Si, and c-Ag/a-Ge).

2.5 Conclusions and Outlook

In this chapter, the cardinal role of interface thermodynamics in the MIC processes occurring in metal/amorphous semiconductor layer systems has been exposed. The atomic-scale mechanisms have been revealed on the basis of dedicated in situ heating HRTEM and VEFTEM experiments.

MIC occurring in various metal/amorphous semiconductor systems is in general an *interface-controlled phenomenon*.

Firstly, the interfacial covalent bond–weakening effect, in the semiconductor adjacent to the metal, generates a (limited) amount of high mobility ("free") interfacial semiconductor atoms.

Secondly, these "free" semiconductor atoms may migrate, in particular, along short-circuit fast diffusion paths such as metal/semiconductor interfaces and metal GBs at low temperatures.

Thirdly, the occurrence of (intermediate) wetting, the nucleation of crystallization, and the continued semiconductor crystal growth are *all* governed by the *interface thermodynamics* under the constraint of a *limited amount (layer thickness) of available "free" semiconductor atoms*.

This fundamental understanding, acquired in recent years, now allows correct prediction of the (very) different MIC temperatures and behaviors in a wide range of metal/amorphous semiconductor systems. On this basis, one may now be able to control and tailor the growth of crystalline semiconductor materials with an optimized structure, composition, or morphology under desirable growth conditions such as low growth temperatures.

We further foresee important applications of the models presented here in the case of MIC of alloy or compound semiconductors in which an additional thermodynamic parameter (composition) and an additional kinetic parameter (different diffusivities of [alloying] elements) have to be incorporated, thus facilitating more flexible control and engineering of low-temperature semiconductor crystal growth.

Acknowledgments

We are indebted to Prof. J. Y. Wang for his contributions during his stay in the department of Prof. Mittemeijer at the Max Planck Institute for Intelligent Systems (formerly Max Planck Institute for Metals Research). We are grateful to Prof. F. Sommer for valuable discussions and to Dr. F. Phillipp and Prof. L. Gu (all at Max Planck Institute for Intelligent Systems) for their collaboration in the in situ TEM research performed in this project in the past years.

References

1. Konno, T. J., Sinclair, R. (1992) Crystallization of silicon in aluminum amorphous-silicon multilayers. *Philosophical Magazine B: Physics of Condensed Matter Statistical Mechanics Electronic Optical and Magnetic Properties*, **66**, 749–765.
2. Nast, O., Hartmann, A. J. (2000) Influence of interface and Al structure on layer exchange during aluminum-induced crystallization of amorphous silicon. *Journal of Applied Physics*, **88**, 716–724.
3. Nast, O., Wenham, S. R. (2000) Elucidation of the layer exchange mechanism in the formation of polycrystalline silicon by aluminum-induced crystallization. *Journal of Applied Physics*, **88**, 124–132.
4. Zhao, Y. H., Wang, J. Y., Mittemeijer, E. J. (2004) Microstructural changes in amorphous Si/crystalline Al thin bilayer films upon annealing. *Applied Physics A: Materials Science & Processing*, **79**, 681–690.
5. He, D., Wang, J. Y., Mittemeijer, E. J. (2005) The initial stage of the reaction between amorphous silicon and crystalline aluminum. *Journal of Applied Physics*, **97**, 093524.
6. He, D., Wang, J. Y., Mittemeijer, E. J. (2005) Reaction between amorphous Si and crystalline Al in Al/Si and Si/Al bilayers: microstructural and thermodynamic analysis of layer exchange. *Applied Physics A: Materials Science & Processing*, **80**, 501–509.
7. Wang, Z. M., Wang, J. Y., Jeurgens, L. P. H., Mittemeijer, E. J. (2006) "Explosive" crystallisation of amorphous germanium in Ge/Al layer systems; comparison with Si/Al layer systems. *Scripta Materialia*, **55**, 987–990.
8. Wang, J. Y., He, D., Zhao, Y. H., Mittemeijer, E. J. (2006) Wetting and crystallization at grain boundaries: origin of aluminum-induced crystallization of amorphous silicon. *Applied Physics Letters*, **88**, 061910.
9. Wang, J. Y., Wang, Z. M., Mittemeijer, E. J. (2007) Mechanism of aluminum-induced layer exchange upon low-temperature annealing of amorphous Si/polycrystalline Al bilayers. *Journal of Applied Physics*, **102**, 113523.
10. Wang, Z. M., Wang, J. Y., Jeurgens, L. P. H., Mittemeijer, E. J. (2008) Tailoring the ultrathin Al-induced crystallization temperature of amorphous Si by application of interface thermodynamics. *Physical Review Letters*, **100**, 125503.
11. Wang, Z. M., Wang, J. Y., Jeurgens, L. P. H., Mittemeijer, E. J. (2008) Thermodynamics and mechanism of metal-induced crystallization in

immiscible alloy systems: experiments and calculations on Al/a-Ge and Al/a-Si bilayers. *Physical Review B*, **77**, 045424.

12. Wang, Z. M., Wang, J. Y., Jeurgens, L. P. H., Phillipp, F., Mittemeijer, E. J. (2008) Origins of stress development during metal-induced crystallization and layer exchange: annealing amorphous Ge/crystalline Al bilayers. *Acta Materialia*, **56**, 5047–5057.
13. Wang, Z. M., Jeurgens, L. P. H., Wang, J. Y., Phillipp, F., Mittemeijer, E. J. (2009) High-resolution transmission-electron-microscopy study of ultrathin Al-induced crystallization of amorphous Si. *Journal of Materials Research*, **24**, 3294–3299.
14. Wang, Z. M., Gu, L., Phillipp, F., Wang, J. Y., Jeurgens, L. P. H., Mittemeijer, E. J. (2011) Metal-catalyzed growth of semiconductor nanostructures without solubility and diffusivity constraints. *Advanced Materials* **23**, 854–859.
15. Wang, Z. M., Gu, L., Jeurgens, L. P. H., Phillipp, F., Mittemeijer, E. J. (2012) Real-time visualization of convective transportation of solid materials at nanoscale. *Nano Letters*, **12**, 6126–6132.
16. Hu, S., McIntyre, P. C. (2012) Nucleation and growth kinetics during metal-induced layer exchange crystallization of Ge thin films at low temperatures. *Journal of Applied Physics*, **111**, 044908.
17. Zhang, T.-W., Ma, F., Zhang, W.-L., Ma, D.-Y., Xu, K.-W., Chu, P. K. (2012) Diffusion-controlled formation mechanism of dual-phase structure during Al induced crystallization of SiGe. *Applied Physics Letters*, **100**, 071908.
18. Hultman, L., Robertsson, A., Hentzell, H. T. G., Engstrom, I., Psaras, P. A. (1987) Crystallization of amorphous-silicon during thin-film gold reaction. *Journal of Applied Physics*, **62**, 3647–3655.
19. Hou, J. G., Wu, Z. Q. (1990) Experimental demonstration of the role of local latent-heat in Ge pattern-formation. *Physical Review B*, **42**, 3271–3274.
20. Tan, Z. Q., Heald, S. M., Rapposch, M., Bouldin, C. E., Woicik, J. C. (1992) Gold-induced germanium crystallization. *Physical Review B*, **46**, 9505–9519.
21. Seibt, M., Buschbaum, S., Gnauert, U., Schroter, W., Oelgeschlager, D. (1998) Nanoscale observation of a grain boundary related growth mode in thin film reactions. *Physical Review Letters*, **80**, 774–777.
22. Park, J.-H., Kurosawa, M., Kawabata, N., Miyao, M., Sadoh, T. (2011) Au-induced low-temperature (similar to 250 degrees C) crystallization of Si on insulator through layer-exchange process. *Electrochemical and Solid State Letters*, **14**, H232–H234.

23. Konno, T. J., Sinclair, R. (1995) Metal-mediated crystallization of amorphous-silicon in silicon silver layered systems. *Philosophical Magazine B: Physics of Condensed Matter Statistical Mechanics Electronic Optical and Magnetic Properties*, **71**, 163–178.

24. Scholz, M., Gjukic, M., Stutzmann, M. (2009) Silver-induced layer exchange for the low-temperature preparation of intrinsic polycrystalline silicon films. *Applied Physics Letters*, **94**, 012108.

25. Zhang, C., Bates, C. W., Jr. (2009) Metal-mediated crystallization in Si-Ag systems. *Thin Solid Films*, **517**, 5783–5785.

26. Hayzelden, C., Batstone, J. L. (1993) Silicide formation and silicide-mediated crystallization of nickel-implanted amorphous silicon thin films. *Journal of Applied Physics*, **73**, 8279–8289.

27. Jang, J., Oh, J. Y., Kim, S. K., Choi, Y. J., Yoon, S. Y., Kim, C. O. (1998) Electric-field-enhanced crystallization of amorphous silicon. *Nature*, **395**, 481–483.

28. Jin, Z. H., Bhat, G. A., Yeung, M., Kwok, H. S., Wong, M. (1998) Nickel induced crystallization of amorphous silicon thin films. *Journal of Applied Physics*, **84**, 194–200.

29. Yoon, S. Y., Oh, J. Y., Kim, C. O., Jang, J. (1998) Low temperature solid phase crystallization of amorphous silicon at 380 degrees C. *Journal of Applied Physics*, **84**, 6463–6465.

30. Zhonghe, J., Gururaj, A. B., Milton, Y., Hoi, S. K., Man, W. (1998) Nickel induced crystallization of amorphous silicon thin films. *Journal of Applied Physics*, **84**, 194–200.

31. Guliants, E. A., Anderson, W. A. (2001) Study of dynamics and mechanism of metal-induced silicon growth. *Journal of Applied Physics*, **89**, 4648–4656.

32. Guliants, E. A., Anderson, W. A., Guo, L. P., Guliants, V. V. (2001) Transmission electron microscopy study of Ni silicides formed during metal-induced silicon growth. *Thin Solid Films*, **385**, 74–80.

33. Miyasaka, M., Makihira, K., Asano, T., Polychroniadis, E., Stoemenos, J. (2002) In situ observation of nickel metal-induced lateral crystallization of amorphous silicon thin films. *Applied Physics Letters*, **80**, 944–946.

34. Lee, S. B., Choi, D. K., Lee, D. N. (2005) Transmission electron microscopy observations of Cu-induced directional crystallization of amorphous silicon. *Journal of Applied Physics*, **98**, 114911.

35. Lee, S. B., Choi, D. K., Phillipp, F., Jeon, K. S., Kim, C. K. (2006) In situ high-resolution transmission electron microscopy study of interfacial

reactions of Cu thin films on amorphous silicon. *Applied Physics Letters*, **88**, 083117.

36. Her, Y. C., Tu, W. T., Tsai, M. H. (2012) Phase transformation and crystallization kinetics of a-Ge/Cu bilayer for blue-ray recording under thermal annealing and pulsed laser irradiation. *Journal of Applied Physics*, **111**, 043503.
37. Lee, S. W., Jeon, Y. C., Joo, S. K. (1995) Pd induced lateral crystallization of amorphous Si thin-films. *Applied Physics Letters*, **66**, 1671–1673.
38. Lee, S. W., Lee, B. I., Kim, T. K., Joo, S. K. (1999) Pd2Si-assisted crystallization of amorphous silicon thin films at low temperature. *Journal of Applied Physics*, **85**, 7180–7184.
39. Oki, F., Ogawa, Y., Fujiki, Y. (1969) Effect of deposited metals on crystallization temperature of amorphous germanium film. *Japanese Journal of Applied Physics*, **8**, 1056.
40. Wang, Z. M., Jeurgens, L. P. H., Wang, J. Y., Mittemeijer, E. J. (2009) Fundamentals of metal-induced crystallization of amorphous semiconductors. *Advanced Engineering Materials*, **11**, 131–135.
41. Konno, T. J., Sinclair, R. (1994) Metal-contact-induced crystallization of semiconductors. *Materials Science and Engineering: A*, **179**, 426–432.
42. Knaepen, W., Detavernier, C., Van Meirhaeghe, R. L., Sweet, J. J., Lavoie, C. (2008) In-situ x-ray diffraction study of metal induced crystallization of amorphous silicon. *Thin Solid Films*, **516**, 4946–4952.
43. Knaepen, W., Gaudet, S., Detavernier, C., Van Meirhaeghe, R. L., Sweet, J. J., Lavoie, C. (2009) In situ x-ray diffraction study of metal induced crystallization of amorphous germanium. *Journal of Applied Physics*, **105**, 083532.
44. Jeurgens, L. P. H., Wang, Z., Mittemeijer, E. J. (2009) Thermodynamics of reactions and phase transformations at interfaces and surfaces. *International Journal of Materials Research*, **100**, 1281–1307.
45. Cottrell, T. L. (1958) *The Strengths of Chemical Bonds* (Butterworths Publications, London).
46. Hiraki, A., Nicolet, M. A., Mayer, J. W. (1971) Low-temperature migration of silicon in thin layers of gold and platinum. *Applied Physics Letters*, **18**, 178–181.
47. Hiraki, A., Shuto, K., Kim, S., Kammura, W., Iwami, M. (1977) Room-temperature interfacial reaction in Au-semiconductor systems. *Applied Physics Letters*, **31**, 611–612.
48. Hiraki, A. (1983) Low temperature reactions at Si/metal interfaces: what is going on at the interfaces? *Surface Science Reports*, **3**, 357–412.

49. Hiraki, A. (1980) A model on the mechanism of room-temperature interfacial intermixing reaction in various metal-semiconductor couples: what triggers the reaction. *Journal of the Electrochemical Society*, **127**, 2662–2665.

50. Okuno, K., Ito, T., Iwami, M., Hiraki, A. (1980) Presence of critical Au-film thickness for room-temperature interfacial reaction between Au(film) and Si(crystal substrate). *Solid State Communications*, **34**, 493–497.

51. Mehrer, H. (2007) *Diffusion in Solids: Fundamentals, Methods, Materials, Diffusion-Controlled Processes* (Springer, Heidelberg).

52. Wang, J. Y., Mittemeijer, E. J. (2004) A new method for the determination of the diffusion-induced concentration profile and the interdiffusion coefficient for thin film systems by Auger electron spectroscopical sputter depth profiling. *Journal of Materials Research*, **19**, 3389–3397.

53. Egerton, R. F. (1996) *Electron Energy-Loss Spectroscopy in the Electron Microscope* (Plenum Press, New York).

54. Slater, J. C. (1964) Atomic radii in crystals. *Journal of Chemical Physics*, **41**, 3199–3204.

55. Li, B. Q., Zheng, B., Zhang, S. Y., Wu, Z. Q. (1993) Dependence of fractal formation on the thickness ratio in Al/a-Ge bilayers. *Physical Review B*, **47**, 3638–3641.

56. Katsuki, F., Hanafusa, K., Yonemura, M., Koyama, T., Doi, M. (2001) Crystallization of amorphous germanium in an Al/a-Ge bilayer film deposited on a SiO2 substrate. *Journal of Applied Physics*, **89**, 4643–4647.

57. Wang, Z. M., Wang, J. Y., Jeurgens, L. P. H., Mittemeijer, E. J. (2008) Investigation of metal-induced crystallization in amorphous Ge/crystalline Al bilayers by Auger microanalysis and selected-area depth profiling. *Surface and Interface Analysis*, **40**, 427–432.

58. Hu, S., Marshall, A. F., McIntyre, P. C. (2010) Interface-controlled layer exchange in metal-induced crystallization of germanium thin films. *Applied Physics Letters*, **97**, 082104.

59. Hazra, S., Sakata, I., Yamanaka, M., Suzuki, E. (2004) Evolution of an amorphous silicon network from silicon paracrystallites studied by spectroscopic ellipsometry. *Physical Review B*, **69**, 235204.

Chapter 3

Diffusion, Crystallization, and Layer Exchange upon Low-Temperature Annealing of Amorphous Si/Polycrystalline Al–Layered Structures

Jiang Yong Wang,[a] Zumin Wang,[b] Lars P. H. Jeurgens,[b,*] and Eric J. Mittemeijer[b,c]

[a]*Department of Physics, Shantou University, 243 Daxue Road, Shantou, 515063 Guangdong, China*

[b]*Max Planck Institute for Intelligent Systems, Heisenbergstraße 3, D-70569 Stuttgart, Germany*

[c]*Institute for Materials Science, University of Stuttgart, D-70569 Stuttgart, Germany*

wangjy@stu.edu.cn

The occurrence of (1) diffusion, (2) metal-induced crystallization, and (3) layer exchange, upon annealing of amorphous Si/polycrystalline Al (a-Si/c-Al) layered structures at temperatures below 250°C, has been investigated in depth in recent years employing techniques as Auger electron spectroscopy depth profiling, differential scanning

Present address: Swiss Federal Laboratories for Materials Science and Technology, Überlandstrasse 129, 8600 Dübendorf, Switzerland

Metal-Induced Crystallization: Fundamentals and Applications
Edited by Zumin Wang, Lars P. H. Jeurgens, and Eric J. Mittemeijer

ISBN 978-981-4463-40-9 (Hardcover), 978-981-4463-41-6 (eBook)
www.panstanford.com

calorimetry, X-ray diffraction, and transmission electron microscopy. The insights obtained by this research have been reviewed here. Upon annealing the diffusion process is initiated by Si diffusing into the Al sublayer along Al grain boundaries (GBs), while Al diffusing into the a-Si sublayer can be neglected. The aluminum-induced crystallization of a-Si takes place only at Al GBs in the Al phase and not at the a-Si/c-Al interface. Eventually, Al↔Si layer exchange occurs as a consequence of the lateral growth of crystalline Si grains formed at the GBs in the Al sublayer. A fundamental understanding of these processes has been derived from a combination of interface thermodynamics and kinetics.

3.1 Introduction

Amorphous Si/polycrystalline Al (a-Si/c-Al) layered structures, as important contacts in semiconductor devices, have been investigated intensively with respect to their response to annealing. The annealing-induced phenomena, including diffusion, aluminum-induced crystallization (AIC) of a-Si, and layer exchange of Si and Al sublayers, have attracted much attention in particular, not only because of technological interest, but especially also because of their importance for acquiring fundamental insights into dynamic processes in the solid state.

Diffusion in nanosized systems, as thin films and layered structures, can be governed by defects, crystal imperfections, and stresses. So, diffusion processes in thin films can most appropriately reveal such microstructure–diffusion interactions. However, this requires the availability of experimental methods to trace diffusion for diffusion lengths of the order of 1 nm.

The study of failure mechanisms and instabilities of a-Si/c-Al contacts in semiconductor devices led to the discovery of AIC of a-Si [1]: Whereas pure a-Si crystallizes at a temperature around 700°C, a-Si in intimate contact with a crystalline aluminum phase can crystallize at temperatures much lower than the eutectic temperature of 577 °C of the SiAl system (see Section 3.3).

The layer exchange of Si and Al sublayers upon annealing an a-Si/c-Al layered structure, that is, Al-induced layer exchange (ALILE), was first observed in 1976 [2, 3]. It has drawn increasing attention since the end of the 1990s because, as a result of ALILE, a continuous crystalline Si (c-Si) layer can be formed on a foreign substrate (e.g., glass) at low temperatures [4], and this continuous c-Si layer can subsequently be used as a seed layer for further (epitaxial) thickening [5–7]. ALILE is regarded as a very promising method for producing low-cost solar cells and flexible thin-film transistors (TFTs) [8–10].

This chapter is organized as follows. New methods for reliable determination of diffusion coefficients from measured Auger electron spectroscopy (AES) depth-profiling data, and application of these new methods to a-Si/c-Al bilayered or multilayered structures, are presented in Section 3.2. Experimental observations and the obtained fundamental understanding, on that basis, of the AIC and ALILE processes are presented in Section 3.3 and Section 3.4, respectively. General, conclusive remarks are provided in Section 3.5.

3.2 Diffusion in a-Si/c-Al Layered Structures upon Low-Temperature Annealing

To determine the diffusion coefficient in a layered structure (such as an a-Si/c-Al layered structure), AES depth profiling is often used because of its high depth resolution (as compared with X-ray photoelectron spectroscopy [XPS] depth profiling) and its well-developed quantification methods (e.g., as compared with secondary ion mass spectrometry [SIMS]) [11, 12]. Various methods have been developed in the past decades for determining the diffusion coefficients from measured AES depth-profiling data—for example, the Hwang–Balluffi method [13], the plateau rise method [14], the center gradient method [15], the interface width method [16], and the Fourier series method [17]. However, in all these methods the concentration profile–broadening effects during experimental AES depth profiling and the intrinsic surface/interface roughness of the layered structure are not taken into account. These profile-broadening effects cannot be neglected as compared to the diffusion-induced profile broadening, in particular if the diffusion length is only a few nanometers, as pertains to the cases studied in this chapter. Thereby, the methods mentioned before are unsatisfactory.

In this section, two new methods for extracting accurate diffusion coefficients from measured AES depth profile data are presented, which take into account the instrumental concentration profile broadening and intrinsic surface/interface roughness of the layered structure. Using these newly developed methods, the diffusion occurring upon annealing in a-Si/c-Al bilayered and multilayered structures has been analyzed quantitatively.

3.2.1 Methods for Determination of Diffusion Coefficients by AES Depth Profiling of Layered Structures

The not-diffusion-induced depth profile–broadening effects exhibited in AES depth-profiling data are caused mainly by the sputtering-induced atomic mixing and interface roughness, the intrinsic surface/interface roughness of the layered structure, the escape depth of Auger electrons, and preferential sputtering. These broadening effects can be taken together in a smearing function, g, such that the measured depth profile can be conceived as the convolution of the true concentration–depth profile, C, with the smearing function, g, that is

$$I(z)/I_0 = \int_{-\infty}^{+\infty} C(z') \times g(z-z')dz' \tag{3.1}$$

where $I(z)/I_0$ is the measured and normalized Auger electron intensity at the sputter depth z. The smearing function $g(z - z')$ can be described by the so-called mixing-roughness-information (MRI) depth model with four well-defined parameters [18]: the atomic mixing parameter w, the surface/interface roughness σ, the escape depth of the Auger electrons λ, and the preferential sputtering ratio r. From a purely mathematical point of view, two approaches can be distinguished to fit a calculated AES depth profile to the measured one: either varying $g(z - z')$ (*method 1* next) or varying $C(z')$ (*method 2* later).

3.2.1.1 Method 1

For an initial stage of diffusion the concentration–depth profile may be written as the convolution of the initial concentration profile with a Gaussian function ($\sim \exp[-z^2/4Dt]$, with D as the diffusion coefficient). Then, recognizing that in the MRI model for the smearing function the interface roughness is described by a Gaussian function ($\sim \exp[-z^2/2\sigma^2]$, with σ as the roughness parameter), and taking $C(z')$ as the initial (i.e., before interdiffusion) concentration–depth profile, the occurring diffusion can be described by a change of $g(z - z')$, as expressed by an increase of the interface roughness parameter. It thus follows

$$2Dt = \Delta\sigma^2 = \sigma_T^2 - \sigma_0^2 \qquad (3.2)$$

where t is the annealing time, and σ_T and σ_0 are the values of the roughness parameter after annealing at temperature T and before annealing, respectively.

3.2.1.2 Method 2

If the smearing function $g(z - z')$ is known, $C(z')$ can be calculated from Eq. 3.1 by an iterative convolution procedure, that is, assuming a concentration–depth profile $C(z')$, by adopting an appropriate diffusion model and comparing the calculated profile with the measured profile until an optimum fit has been achieved.

It has been demonstrated that very small diffusion coefficients characterized by a diffusion length of around 1 nm can be determined according to the above-mentioned methods [11, 19].

3.2.2 Application to a-Si/c-Al Bilayered and Multilayered Structures

Specimens of 4× a-Si (15 nm)/c-Al (15 nm) multilayered and a-Si (150 nm)/c-Al (50 nm) bilayered structures were prepared by magnetron sputter deposition in an ultrahigh vacuum (UHV) chamber with a base pressure of 10^{-7} Pa in a single run on Si substrates. The specimens were then annealed isothermally at different temperatures for different times under a protective Ar gas atmosphere and thereafter cooled down to room temperature. The concentration–depth profiles for the as-deposited and annealed specimens were measured using a JEOL JAMP-7830F scanning Auger microscope. Details of the sputter deposition, the AES depth profiling, and the characterization of the layered structures can be found in Refs. [19, 20].

3.2.2.1 Multilayered structure 4× a-Si (15 nm)/c-Al (15 nm)

Method 1 has been applied for determination of the interdiffusion coefficient operating in a *multilayered* structure [20]. The measured Al concentration–depth-profiling data (open circles) and the fitted, smeared concentration–depth profiles (solid lines) as calculated by least-squares fitting to the measured data according to *method 1* are shown in Figs. 3.1a and 3.1b, respectively, for the as-deposited specimen and the annealed (at 150°C for 20 minutes) specimen, respectively. The dashed line indicated in Fig. 3.1a represents the as-deposited layered structure and is adopted as the true concentration–depth profile for the as-deposited specimen and acts as the initial concentration–depth profile for the annealed layered structure.

For the as-deposited specimen, a close examination of Fig. 3.1a reveals that the maximum (minimum) value of the Al concentration measured for each sublayer decreases (increases) with sputter depth. This is due to the ion bombardment–induced roughness for the polycrystalline material in a stationary (nonrotating) mode, which can be remedied by applying a specimen-rotating mode during ion bombardment [21]. The measured AES depth-profiling data for the as-deposited specimen could be fitted well, assuming that the roughness parameter (σ_0) *increases* linearly with the sputter depth z (see the result indicated in Fig. 3.1a) and taking the other three MRI parameters (w, λ, and r) as depth independent [22], as indicated in Fig. 3.1a.

Upon annealing, the measured maximum Al concentration in the near-surface Al sublayer has decreased more pronouncedly than that in the deeper Al sublayers, as shown in Fig. 3.1b. This indicates a depth dependence of the diffusion coefficient. Adopting *method 1*, that is, assuming that the occurrence of diffusion can be described as (additional) "roughening" (of the interfaces) and adopting the initial concentration–depth profile (the dashed line in Fig. 3.1a) as $C(z')$, the measured depth profiles were fitted, adopting a "roughness" parameter (σ_T) that *decreases* linearly with the sputter depth. The other three MRI parameters (w, λ, and r) were taken the same as for the as-deposited specimen (Fig. 3.1a). The thus determined linear depth dependence of the roughness parameter σ_T for the diffusion-annealed specimen has been indicated in Fig. 3.1b. Now, the *local*

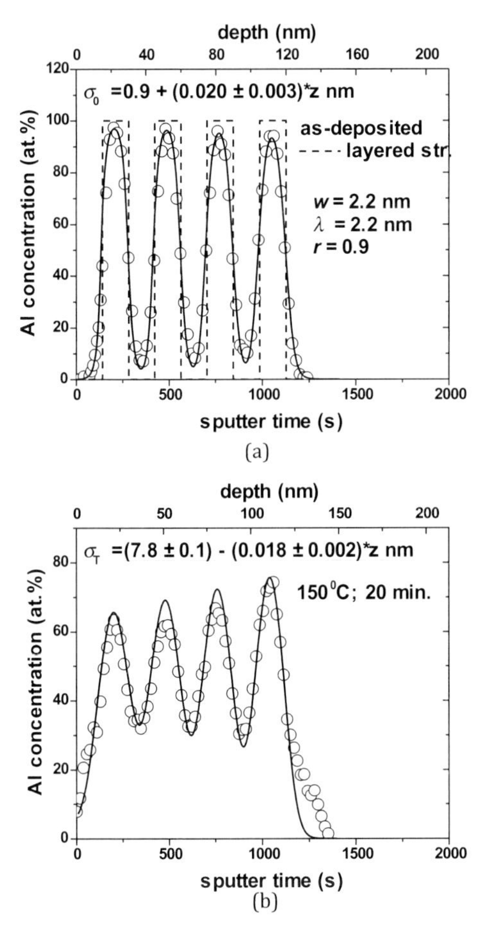

Figure 3.1 Measured Al concentration–depth profiles (open circles) and the corresponding optimum fits (solid lines), as calculated according by *method 1* described in Section 3.2.1 for the 4× a-Si (15 nm)/c-Al (15 nm) multilayered specimens (a) as deposited (the dashed line represents the as-deposited Al layered structure, and the measured depth profile was fitted assuming that the roughness parameter σ_0 *increases* linearly with the sputter depth z) and (b) annealed at 150°C for 20 minutes (the measured depth profile was fitted assuming that the roughness parameter σ_T *decreases* linearly with the sputter depth z). Values for the parameters of the smearing function, that is, the atomic mixing length (w), the information depth (λ), and the preferential sputter ratio (r), have been indicated in Fig. 3.1a. Reprinted with permission from Ref. [20], Copyright 2004, Elsevier.

(i.e., at fixed depth) diffusion coefficient for the specimen annealed at 150°C for 20 minutes can be determined from the fitted depth dependences of σ_0 and σ_T by application of Eq. 3.2. In the same way, the depth-dependent diffusion coefficients for anneals at 120°C for 240 minutes and at 135°C for 60 minutes can be determined from the measured AES depth-profiling data (not shown here). Next, (the pre-exponential factor D_0 and) the activation energy Q for the diffusion upon annealing this multilayered structure can be obtained from an Arrhenius plot applied to the diffusion coefficient data at a fixed depth [20]. The values of (D_0 and) Q at the first and last Si/Al interfaces in the multilayered structure have been listed in Table 3.1.

Table 3.1 Pre-exponential factor D_0 and activation energy Q for interdiffusion as determined for (different) *a-Si/c-Al layered structures* and as obtained from the Arrhenius plots applied to the diffusion coefficient data as determined by the methods (*method 1* and *method 2*) described in Section 3.2.1. The values for D_0 have been included here to allow the reader to calculate diffusion coefficients at specific temperatures.

Layered structure	D_0 (m^2/s)	Q (eV)	T range (°C)	Method	Ref.
4× a-Si (15 nm)/c-Al (15 nm)	4.5×10^{-6} 2.8×10^{-7}	1.21 (1st interface) 1.14 (last interface)	120–150	*Method 1*	[20]
4× a-Si (85 nm)/c-Al (45 nm)	1.0×10^{-1} 2.4×10^{-5}	1.57 (1st interface) 1.31 (last interface)	150–180	*Method 1*	[22]
a-Si (150 nm)/c-Al (50 nm)	7.6×10^{2}	1.81	120–150	*Method 2*	[19]

A similar analysis of the depth dependence of the interdiffusion has been made for a 4× a-Si (85 nm)/c-Al (45 nm) multilayered structure, which has a much larger average (lateral) grain size of the c-Al phase—51 nm as compared to 19 nm in the 4× a-Si (15 nm)/c-Al (15 nm) multilayered specimen [22, 23]. The fittings of the measured AES depth-profiling data yield a similar activation energy value (see Table 3.1). The observed depth dependence of the interdiffusion coefficient has been ascribed to the depth dependence of the microstructure of the sublayers developing during the a-Si/c-Al multilayer deposition [20, 22].

3.2.2.2 Bilayered structure of a-Si (150 nm)/c-Al (50 nm)

Method 2 has been applied for determination of the interdiffusion coefficient operating in a *bilayered* structure. To apply this method, the smearing function g, for the measurement conditions applied, can be determined by the optimum fit of the measured AES depth-profiling data for the as-deposited specimen, for which the true concentration–depth profile, $C(z')$, is known (see Fig. 3.2a). For the bilayer considered, the trial concentration–depth profile C can be taken as [19]

$$C(z,t) = \frac{C_0}{2}\left[erf\left(\frac{h+z}{2\sqrt{Dt}} \right) + erf\left(\frac{h-z}{2\sqrt{Dt}} \right) \right] \tag{3.3}$$

where *erf* denotes the error function; C_0 and h are the initial concentration and thickness of the considered Al sublayer, respectively; and D and t are the corresponding interdiffusion coefficient and annealing time, respectively. This type of concentration–depth profile (Eq. 3.3) can be well used as an appropriate diffusion model if only the diffusion of Si into Al grain boundaries (GBs) is significant at an initial diffusion stage (see discussion in Section 3.2.3).

The measured Al concentration–depth profiles for the specimen annealed at 150°C for 20 minutes are shown by the open data points in Fig. 3.2b. Using the MRI parameters as obtained for the as-deposited specimen (Fig. 3.2a), that is, the thereby established smearing function $g(z - z')$, and taking only the diffusion coefficient D in Eq. 3.3 as a fit parameter, the measured Al concentration–depth profile for the annealed specimen could be well fitted (least-squares analysis) by the solid line shown in Fig. 3.2b. The corresponding *true* diffusion concentration–depth profile is shown as a dashed line in the figure. In the same way, the diffusion coefficients pertaining to the bilayers annealed at 120°C for 240 minutes and at 135°C for 30 minutes could be determined from the measured AES depth-profiling data (not shown here). The values of (the pre-exponential factor D_0 and) the activation energy Q for the diffusion in this bilayered structure can then be obtained from an Arrhenius plot. The results have been included in Table 3.1.

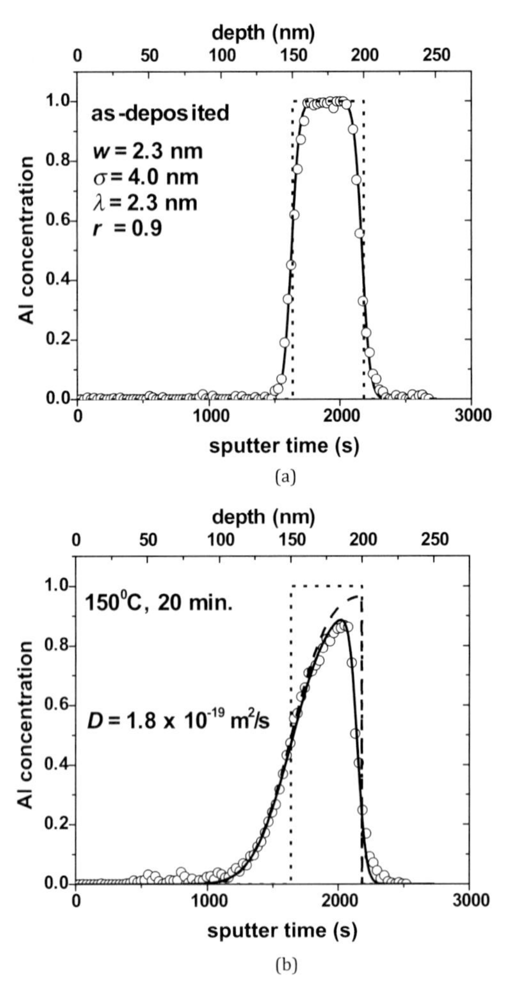

Figure 3.2 Measured Al concentration–depth profiles (open circles) and the corresponding optimum fits (solid lines), as calculated according to *method 2* described in Section 3.2.1 for the a-Si (150 nm) /c-Al (50 nm) bilayered specimens with high Al GB density (a) as deposited and (b) annealed at 150°C for 20 minutes. The dotted line indicates the as-deposited Al layered structure. The dashed line in Fig. 3.2b represents the accordingly determined diffusion-induced concentration–depth profile. Values for the parameters of the smearing function, that is, the atomic mixing length (w), the surface roughness parameter (σ), the information depth (λ), and the preferential sputter ratio (r), have been indicated in Fig. 3.2a. Reprinted with permission from Ref. [19], Copyright 2004, Cambridge University Press.

3.2.3 Operating Diffusion Mechanism

Values for the activation energy Q as reported in the literature for the diffusion of either Al into single-crystalline (or poly)-crystalline Si or Si into c-Al (no data is available for the diffusion of Si into single-crystalline Al) have been gathered in Table 3.2.

Table 3.2 Diffusion parameters (D_0 and Q) reported in the literature for the diffusion Si in Al and for the diffusion of Al in a (bulk) Si matrix. The values for D_0 have been included here to allow the reader to calculate diffusion coefficients at specific temperatures.

Diffusion of Si into an Al matrix				
D_0 (m²/s)	Q (eV)	T range (°C)	Remarks	Ref.
3.5×10^{-5}	1.28	345–631	c-Al film	[97]
2.0×10^{-4}	1.41	480–620	c-Al film	[98]
6.0×10^{-5}	1.36	450–600	Wrought Al	[99] and references therein
Diffusion of Al into a Si matrix				
D_0 (m²/s)	Q (eV)	T range (°C)	Remarks	Ref.
1.30×10^{3}	2.64	350–425	Polycrystal Si film	[13]
8.0×10^{-4}	3.47	1085–1375	Single-crystal Si	[100]
2.9×10^{-4}	3.23	800–1250	Single-crystal Si	[101]
1.4×10^{-4}	3.41	1119–1390	Epitaxial Si layer	[102]
1.8×10^{-4}	3.20	1025–1175	Single-crystal Si	[103]
7.4×10^{-4}	3.42	1000–1290	Single-crystal Si	[104]
1.1×10^{-3}	3.46	1000–1200	Single-crystal Si	[105]

Comparing the values of the activation energy determined for the layered structures in the present project at relatively low temperatures (120°C–220°C) presented in Table 3.1 with those presented in Table 3.2 determined at relatively high temperatures (in the range of 345°C to 631°C), it follows that the activation energy values determined here for diffusion in the a-Si/c-Al layered structures are similar to those for diffusion of Si into c-Al and are significantly smaller than those for diffusion of Al into either single-crystalline Si or polycrystalline Si. According to the AlSi phase diagram, the maximum solubility of Si in Al is only about 0.75 at.% at 500°C and drops to about 0.025 at.% at 250°C [24]. Hence, it is concluded that volume diffusion of Si into Al grains at temperatures below 250°C can certainly be neglected and that GB diffusion of Si into Al sublayer governs the transport of Si into Al.

To validate the above conclusion, an additional a-Si (150 nm)/c-Al (50 nm) bilayered specimen (see Section 3.2.2.2) was prepared as follows. An Al layer (50 nm) was first deposited onto an Al_2O_3(0001) (sapphire) substrate at 250°C and subsequently annealed at 250°C for two hours in the same deposition chamber in order to reduce the Al GB density: the lateral grain size of this bilayered specimen is more than seven times larger than that of the corresponding bilayered specimen considered in Section 3.2.2.2 [25]. Finally, after slowly cooling to room temperature and sputter cleaning of the surface of the annealed Al sublayer, an a-Si sublayer (150 nm) was deposited on top of the Al sublayer.

The measured Al concentration–depth profile for this additional bilayer specimen annealed at 250°C for five hours is shown in Fig. 3.3. As compared to the as-deposited state, obviously no change in the Al (Si) concentration–depth profile was detected at all, which is strikingly different from the results obtained for the corresponding specimen with high Al GB density (Fig. 3.2b). Hence, indeed (i) volume diffusion of Si into Al grains (and of Al into the a-Si layer) can be neglected and (ii) the diffusion process is initiated by Si diffusing into the Al sublayer along Al GBs.

3.3 Aluminum-Induced Crystallization of Amorphous Si upon Low-Temperature-Annealing of a-Si/c-Al Layered Structures

The AIC of a-Si has been studied extensively since the 1970s [1,26–61]. The reported crystallization temperatures vary depending on the particular experimental conditions, such as deposition method, deposition temperature and pressure, microstructure of the Al (Si) phase, annealing time, heating rate, and the technique used to detect crystallization. In any case, the reported crystallization temperatures of a-Si, when in contact with Al, are substantially lower than that of pure a-Si (see the data gathered in Table 3.3).

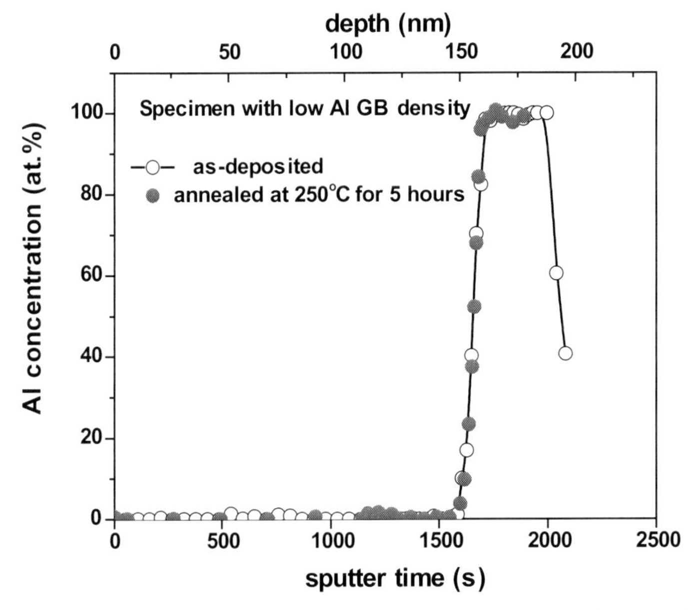

Figure 3.3 Measured Al concentration–depth profiles for the a-Si (150 nm)/c-Al (50 nm) bilayered specimens with low Al GB density as deposited (open circles) and annealed at 250°C for five hours (closed circles). Reprinted with permission from Ref. [25], Copyright 2005, AIP Publishing LLC.

Two mechanisms were suggested for understanding the nucleation of c-Si upon low-temperature annealing of a-Si/c-Al layered structures: (i) nucleation of c-Si would be initiated by the formation of a metastable Al silicide phase at the a-Si/c-Al interface as an intermediate state [33, 37, 40, 62], and (ii) nucleation and growth of c-Si would occur at favorable sites at the a-Si/c-Al interface and/or within the Al sublayer (e.g., at the Al GBs) [25, 38, 63]. Direct, convincing evidence, provided by calorimetric and/or microscopic measurements, showing the occurrence of an Al silicide phase at the a-Si/c-Al interface during AIC of a-Si, has not been obtained [34] (see also Section 3.3.1.1), in accordance with the absence of an Al silicide phase in the AlSi phase diagram. It will be demonstrated in this section, both experimentally and theoretically that, upon low-temperature annealing of a-Si/c-Al layered structures, (i) nucleation of Si crystallization is initiated by Si wetting of Al GBs due to the favorable a-Si/c-Al interface energy as compared to the Al GB energy and (ii) nucleation of Si crystallization takes place *only* at Al GBs within the Al sublayer when the wetting-induced a-Si layer at the Al GB is beyond a critical thickness.

Table 3.3 Crystallization temperatures of pure, bulk a-Si and a-Si in a-Si/c-Al layered structures

Material	Preparation method	Crystallization temperature (°C)	Detection technique	Ref.
	Evaporation	660	Ellipsometry	[106]
a-Si	Sputtering	740	DSC	[107]
	Implantation	687	DSC	[108]
	Sputtering	165	XRD	[67]
	Sputtering/	150	in situ HRTEM	[70]
	evaporation	180	Electron	[1]
	Evaporation	180	diffraction	[37]
	PECVD	150	contact resistivity	[48]
Si/Al	PECVD	200	TEM	[40]
	Glow discharge	150	TEM, SEM	[23]
	Sputtering	157	XRD	[62]
	Evaporation	167	TEM, AES	[33]
	Evaporation	180–220	TEM	[34]
	Sputtering		TEM, DSC, XRD	

Abbreviations: PECVD, plasma-enhanced chemical vapor deposition; SEM, scanning electron microscopy.

3.3.1 Experimental Observations of AIC

3.3.1.1 Multilayered structure 10× a-Si (50 nm)/c-Al (50 nm)

Multilayered films of 10× a-Si (50 nm)/c-Al (50 nm) were prepared by electron beam evaporation in a high-vacuum chamber (10^{-4} Pa) on Si(111) wafers covered with NaCl (prior evaporated onto the wafers). After floating off in distilled water and drying in air, the multilayered films were annealed isochronally with different heating rates in a Perkin–Elmer DSC-7 differential scanning calorimetry (DSC) instrument. One of the measured DSC profiles, recorded with a heating rate of 5°C/minute, is shown in the inset of Fig. 3.4 (the signal of the baseline, which is obtained by a second DSC run, was subtracted). An exothermic reaction is observed at temperatures at and around 195°C. The DSC measurements were intentionally interrupted at two specific temperatures, that is, 150°C and 195°C; the two corresponding specimens were subjected to phase analysis by X-ray diffraction (XRD) employing a Guinier camera (using Cu Kα radiation) equipped with an image plate. The resulting XRD patterns are shown in Fig. 3.4. It can be concluded that before the reaction takes place (see the XRD pattern denoted with "150°C") the only polycrystalline phase present is the Al phase and that the exothermic reaction (see the XRD pattern denoted with "195°C") involves the emergence of a polycrystalline Si phase. Hence, the exothermic reaction (see inset of Fig. 3.4) corresponds to the process of AIC of a-Si.

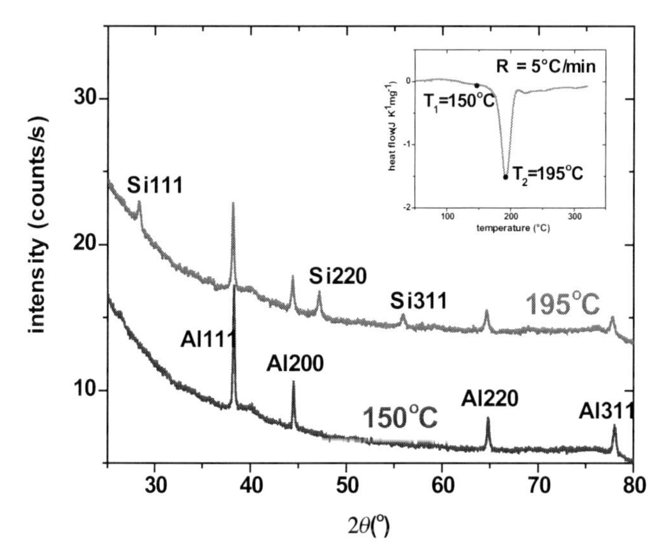

Figure 3.4 XRD patterns recorded using Cu Kα radiation for the 10× a-Si (50 nm)/c-Al (50 nm) multilayered specimen upon isochronal heating (heating rate 5°C/minute) up to 150°C and 195°C, respectively, in a DSC instrument. The inset shows the isochronal baseline corrected DSC profile at a heating rate of 5°C/minute.

By measuring DSC profiles with different heating rates, the activation energy for this AIC process in the a-Si/c-Al *multilayer* can be determined from a Kissinger plot [64, 65] (see Fig. 3.5). The determined value of the activation energy of 1.2 eV for AIC of a-Si is similar to the activation energy pertaining to Si diffusing into Al GBs upon annealing a-Si/c-Al *multilayered* structures (see Section 3.2.2 and Table 3.1), which strongly suggests that the kinetics of the AIC process is governed by diffusion of Si into the GBs of the Al sublayer.

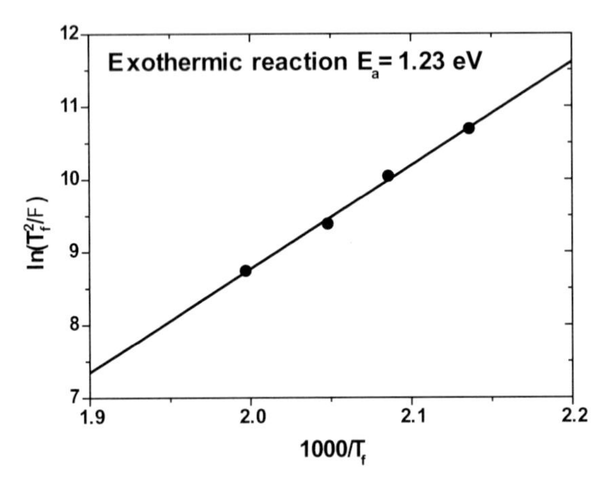

Figure 3.5 Kissinger plot as determined from the peak-maximum temperatures in the baseline-corrected DSC profiles of the 10× a-Si (50 nm)/c-Al (50 nm) multilayered specimen measured under different heating rates, showing that the activation energy of the observed exothermic reaction (or the AIC process) is 1.23 eV. (Φ is the heating rate; T_f is the observed peak temperature at the heating rate Φ; see Ref. [64]).

3.3.1.2 Bilayered structure a-Si (150 nm)/c-Al (50 nm)

The Al concentration–depth profiles and XRD patterns obtained from an a-Si (150 nm)/c-Al (50 nm) bilayered specimen with high Al GB density annealed at 250°C for various times are shown in Figs. 3.6a and 3.6b, respectively [66]. The XRD patterns were measured using a Philips MRD diffractometer and applying Cu Ka radiation (for details, see Ref. [67]). After the specimen had been subjected to XRD analysis after a first anneal, the same specimen was encapsulated again and annealed for an additional period of time. This procedure was repeated up to a total, cumulative annealing time of five hours at 250°C; after each additional period of annealing the specimen was subjected to exactly the same diffraction conditions, that is, the position, shape, and size of the diffracting volume of the specimen were kept the same.

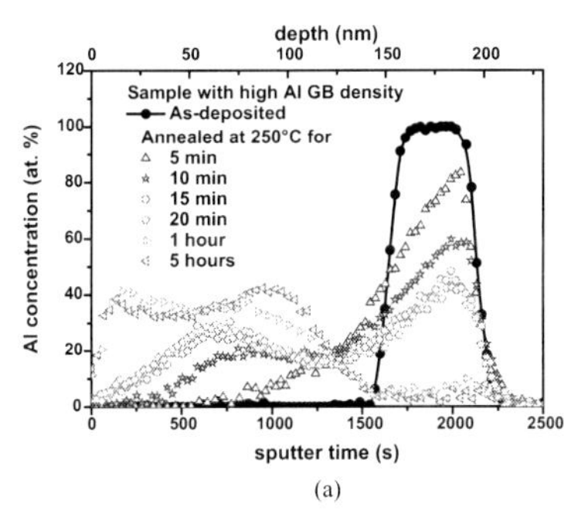

(a)

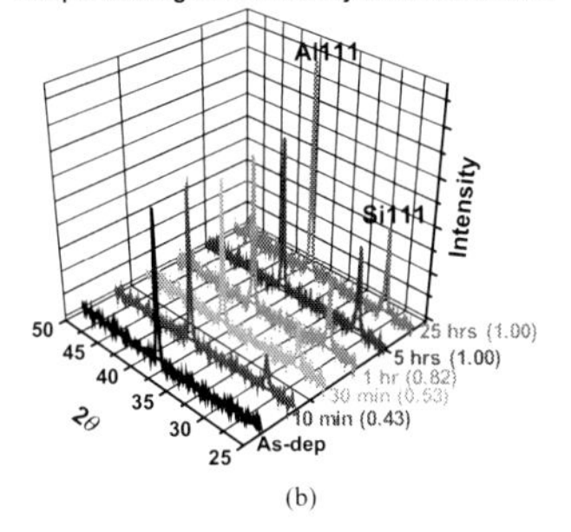

(b)

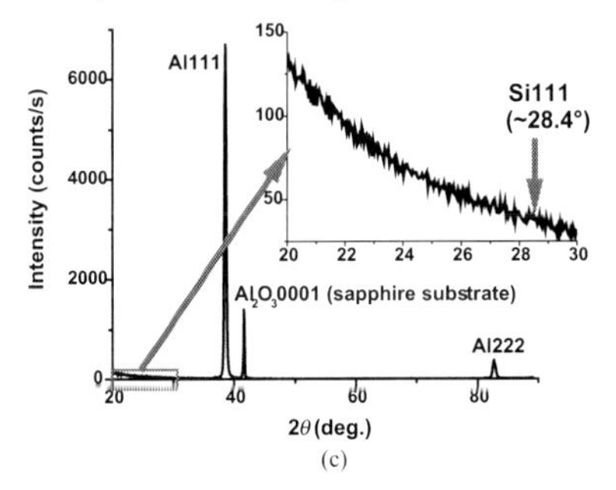

(c)

Figure 3.6 (a) Measured Al concentration–depth profiles and (b) XRD patterns (recorded using Cu Kα radiation) for the a-Si (150 nm)/c-Al (50 nm) bilayered specimens with high Al GB density in the Al phase as deposited and annealed at 250°C for various times (the numbers given within the parentheses in Fig. 3.6b represent normalized values of the integrated intensity of the Si 111 peak). (c) XRD pattern (recorded using Cu Kα radiation) for an a-Si (150 nm)/c-Al (50 nm) bilayered specimen, prepared on an Al_2O_3(0001) sapphire substrate, with very low Al GB density in the Al phase annealed at 250°C for five hours. Reprinted with permission from Ref. [66], Copyright 2009, American Scientific Publishers.

It follows from Figs. 3.6a and 3.6b that both the amount of Si transported to the location of the original Al sublayer and the integrated intensity of the Si 111 diffraction peak at $2\theta = 28.4^{o}$ increase with increasing annealing time (the numbers given within the parentheses in Fig. 3.6b represent normalized values of the integrated intensity of the Si 111 peak). In accordance with the conclusion of Section 3.3.1.1, the results suggest that the kinetics of the AIC process is governed by (the amount of) Si diffusing into the Al sublayer. It also follows from Fig. 3.6b that the integrated intensity of the Si 111 peak after 5 hours of annealing is larger than that after 1 hour and is the same as that after 25 hours. The crystallization of Si proceeds in parallel with the replacement of the Al sublayer by an c-Si layer. This layer exchange is already largely established after one hour of annealing at 250°C (see Fig. 3.6a). A detailed discussion of the layer exchange is presented in Section 3.4.

Upon annealing of the additional a-Si (150 nm)/c-Al (50 nm) bilayered specimen with low Al GB density (see Section 3.2.3) at 250°C for five hours, no diffraction peaks indicative of c-Si could be detected (see Fig. 3.6c) and no replacement of the Al sublayer by Si took place (see the corresponding Al concentration–depth profile shown in Fig. 3.3).

Hence, the AES and XRD results obtained for the multilayered (Section 3.3.1.1) and bilayered (present section) a-Si/c-Al specimens provide unambiguous evidence that AIC of a-Si takes place only if a significant amount of (high-angle) GBs in the adjacent Al is present. This implies that nucleation of Si crystallization initiates at the Al GBs and not at the a-Si/c-Al interface.

The activation energy for AIC of a-Si upon annealing of a-Si/c-Al bilayer has been determined as 1.8 eV [68, 69]. This value is exactly the same as the activation energy obtained for Si diffusing into Al GBs upon annealing the a-Si (150 nm)/c-Al (50 nm) bilayered structure investigated in this study (see Table 3.2). Again, this suggests that the AIC process is governed by the GB diffusion of Si into the Al sublayer.

3.3.1.3 Direct visualization of Al-induced crystallization of a-Si

The formation of c-Si at Al GBs at low temperatures in a-Si/c-Al bilayered specimens has been revealed in a direct way by employing in situ heating transmission electron microscopy (TEM) experiments, applying both cross-sectional views and plan views.

An in situ heating high-resolution transmission electron microscopy (HRTEM) experiment [70] was performed in a JEOL ARM1250 high-voltage microscope, which is operated at 1.25 MeV with a point-to-point resolution of 0.12 nm and which is equipped with a drift compensator. The in situ heating was carried out by using a Gatan 652-Ta double-tilt heating stage. An HRTEM micrograph, taken from a cross-sectional thin foil of a 150 nm a-Si/100 nm c-Al bilayer specimen after in situ annealing at 150°C, is shown in Fig. 3.7a. A small Si crystallite, which has nucleated at a high-angle Al GB, is clearly observable in the HRTEM image. No formation (nucleation) of c-Si could be observed at the interface of the c-Al and a-Si sublayers. The wedge-shaped c-Si growth front (tip) at the Al GB suggests that the c-Si crystallite nucleates and grows by transport of Si from the a-Si sublayer into the Al sublayer at the location of a GB in the Al sublayer.

An in situ heating valence energy–filtered transmission electron microscopy (VEFTEM) experiment [70] was performed in the Zeiss Sub-eV-Sub-Ångstrom Microscope (SESAM; Carl Zeiss, Oberkochen, Germany) operated at 200 keV and equipped with an electrostatic monochromator and the MANDOLINE filter. A 50 nm c-Al/30 nm a-Si bilayer specimen was heated in situ using a Gatan 652 double-tilt heating stage in the SESAM. Due to the prominent differences in the plasmon absorption characteristics of Si and Al, the spatial distributions of Si and Al in the a-Si/c-Al bilayer can be revealed clearly by mapping the position of the plasmon loss peaks, with a spatial resolution better than 0.5 nm [70]. As shown in Figs. 3.7b (plan-view, bright-field TEM) and 3.7c (plasmon loss energy mapping), after in situ heating of the bilayer specimen at 170°C for 15 minutes, nanowall structures of Si are clearly observed at (exclusively) the original Al GBs. From the diffraction contrast in the corresponding bright-field TEM image (Fig. 3.7b), it follows that the Si nanowalls are crystalline, which has been confirmed by the HRTEM investigation of the same specimen [70]. Hence, the in situ heating TEM experiments have unambiguously confirmed (both cross-sectional and plan views) that AIC initiates and continues at the Al GBs and *not* at the a-Si/c-Al interface.

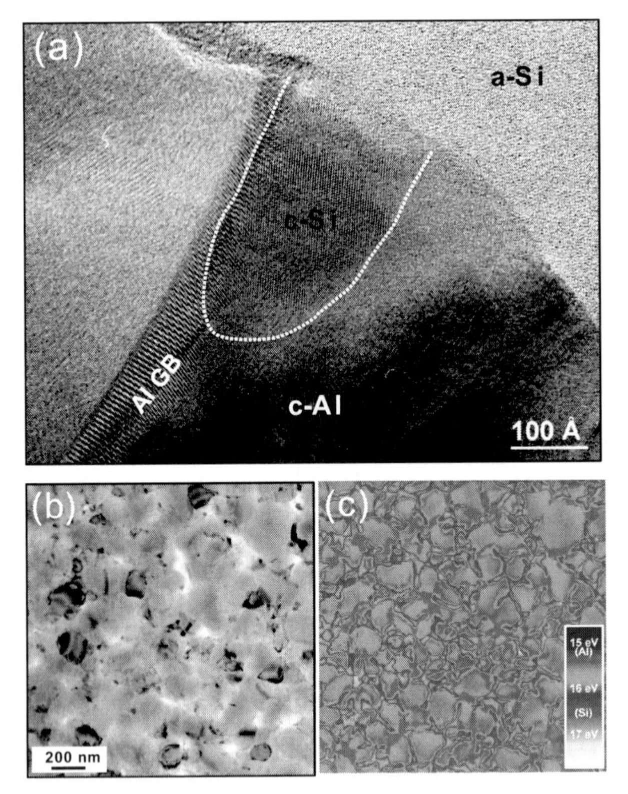

Figure 3.7 (a) In situ heating HRTEM observation (cross-sectional view) of the nucleation of c-Si at a high-angle Al GB at 150°C (for about three minutes) in a 150 nm a-Si/100 nm c-Al bilayered specimen. (b) Plan-view bright-field TEM image of a 50 nm c-Al/30 nm a-Si bilayer heated in situ up to 170°C (heating rate ≈ 10°C/minute) in the SESAM. (c) Mapping of the plasmon loss peak energy corresponding to (b), showing the formation of (crystalline) Si nanowalls (red) at the original Al GBs in the columnar, nanocrystalline Al sublayer.

3.3.2 Thermodynamics of AIC of a-Si

The thermodynamics of AIC of a-Si has been provided in a rigorous way by interface thermodynamic calculations performed in Ref. [71], using a framework for interface thermodynamics as reviewed in Ref. [72]. The (general) interface thermodynamics and atomic mechanisms of metal-induced crystallization in various metal/amorphous semiconductor systems (including c-Al/a-Si) have been reviewed in Chapter 2. In the following section, the results pertaining to the c-Al/a-Si layer system will be summarized briefly.

3.3.2.1 Inward diffusion of Si along Al GBs into the Al sublayer (Al GB wetting)

When a-Si is in contact with c-Al, the wave function of the free-electron gas in Al extends into the adjacent a-Si, thereby screening the covalent bonds present therein, leading to a weakening of the covalent bonds of the interface adjacent a-Si atoms [73]. In the following section, such generated weakly bonded interfacial Si atoms will be called "free" Si atoms. This Coulomb screening effect is a very local electronic interaction, and accordingly, the layer thickness of the free Si atoms is limited to about 2 monolayers (ML) adjacent to the Al layer [71, 74]. These relatively mobile (free) Si atoms can diffuse into the Al sublayer along GBs in the Al phase at low temperatures as induced by the corresponding driving force discussed later.

When the free Si atoms diffuse into an Al GB in the Al sublayer, the Al GB is replaced by two <Al>/{Si} interfaces (<> and {} denote the crystalline phase and the amorphous phase, respectively). The driving force for this process is given by the energy difference of the Al GB energy ($\gamma_{\mathrm{Al}}^{\mathrm{GB}}$) and two times the <Al>/{Si} interface energy ($2\times\gamma_{\langle \mathrm{Al}\rangle/\{\mathrm{Si}\}}^{\mathrm{interface}}$). This driving force has been plotted as a function of temperature in Fig. 3.8 (in its inset the corresponding Al GB energy and two times the <Al>/{Si} interface energy are shown) [71]. It follows that a positive driving force exists, in the temperature range shown (0°C–300°C), for the inward diffusion of Si atoms along Al GBs into the Al sublayer (i.e., a driving force exists for the Al GBs to "be wetted" by a-Si). Such wetting of high-angle Al GBs by a-Si has recently been directly observed by in situ heating VEFTEM [70].

For the a-Si (150 nm)/c-Al (50 nm) bilayer with very low Al GB density (see Sections 3.2.3 and 3.3.1.2) the very few occurring Al GBs are of low energy: high-angle Al GBs in the Al sublayer of this specimen have practically vanished because of the preannealing of this Al sublayer at 250°C for two hours. Against this background it can be understood that wetting of (low-angle) Al GBs by inward diffusion of Si along an Al GB does not occur in this bilayered specimen upon annealing at temperatures below 250°C (see Fig. 3.3). (In the following section, reference to a bilayered a-Si (150 nm)/c-Al (50 nm) specimen implicitly involves the specimen of high Al GB density.)

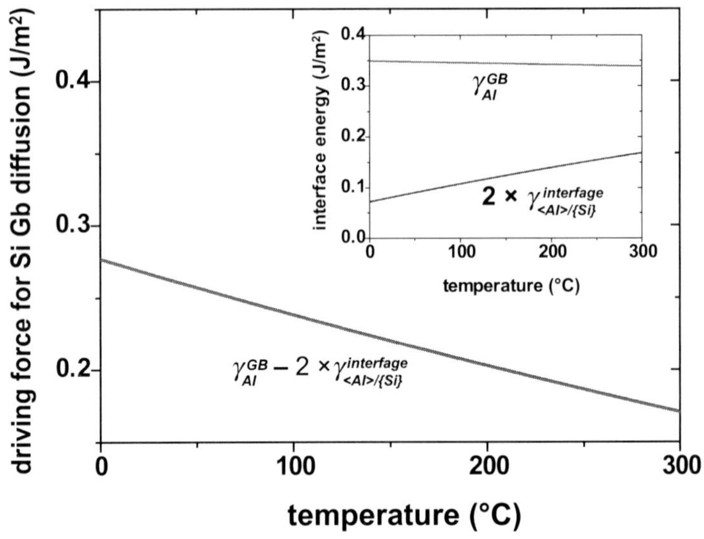

Figure 3.8 Driving force (as a function of temperature) for "wetting" of Al GBs in the Al sublayer by a-Si. The inset shows the Al GB energy and two times the <Al>/{Si} interface energy. Reprinted with permission from Ref. [66], Copyright 2009, American Scientific Publishers.

3.3.2.2 Nucleation of crystalline Si at Al grain boundaries

This wetting Si layer at the Al GB can maintain its amorphous state until reaching a critical thickness, beyond which nucleation of c-Si starts. Crystallization can occur when the increase in the energy of the interface between the crystalline Al grain and the wetting Si layer upon crystallization of a-Si is compensated by the crystallization energy ($-\Delta G_{<\text{Si}>\text{-}\{\text{Si}\}}^{\text{crystallization}}$) of this wetting Si layer. Thus, the critical thickness for nucleation of c-Si at the (original) Al GB can be calculated as

$$h_{\text{Al GB}}^{\text{critical}} = \frac{2\times(\gamma_{<\text{Al}>/<\text{Si}>}^{\text{interface}} - \gamma_{<\text{Al}>/\{\text{Si}\}}^{\text{interface}})}{-\Delta G_{<\text{Si}>\text{-}\{\text{Si}\}}^{\text{crystallization}}} \tag{3.4}$$

Similarly, the critical layer thickness for nucleation of Si crystallization at the a-Si/c-Al interface follows from

$$h_{<\text{Al}>/\{\text{Si}\}}^{\text{critical}} = \frac{\gamma_{<\text{Al}>/<\text{Si}>}^{\text{interface}} + \gamma_{<\text{Si}>/\{\text{Si}\}}^{\text{interface}} - \gamma_{<\text{Al}>/\{\text{Si}\}}^{\text{interface}}}{-\Delta G_{<\text{Si}>\text{-}\{\text{Si}\}}^{\text{crystallization}}} \tag{3.5}$$

The critical thickness values for crystallization at the Al GBs and at the a-Si/c-Al interface have been plotted as a function of temperature in Fig. 3.9 [71]. Because the screening effect of Al on a-Si is limited to 2 ML of Si atoms (see before), the total thickness of "free" Si at the Si/Al interface is only 2 ML of Si atoms, which is shown as a dotted line in Fig. 3.9. Because the wetting Si layer at the Al GBs is sandwiched between two Al grains, the possible total thickness of such a sandwiched wetting Si layer, composed of "free" Si, is 2 × 2 = 4 ML of Si atoms, which is shown as a dash-dotted line in Fig. 3.9. It follows from Fig. 3.9 that the critical thickness for nucleation of c-Si at the a-Si/c-Al interface is above 2 ML, even at 300°C. Therefore, the 2 ML of "free" Si at the interface stay amorphous and nucleation of c-Si cannot occur at this location. At the Al GBs the critical thickness for nucleation of Si crystallization is less than the total thickness of "free" Si (4 ML) when the temperature is above 140°C. Hence, nucleation of c-Si can occur at the Al GBs within the Al sublayer at temperatures above 140°C [71] in accordance with experimental observations (see Section 3.3.1). The lowest-observed crystallization temperature of a-Si for an a-Si/c-Al layered structure indeed equals about 150°C [70] (see also Table 3.3).

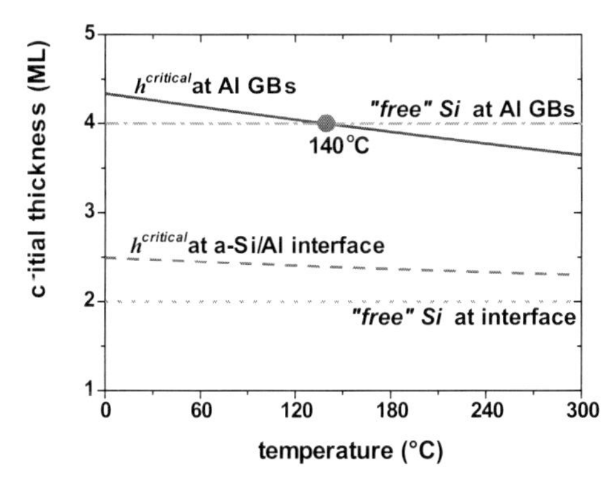

Figure 3.9 Critical thickness (in ML) of Si (1 ML = 0.22 nm; see Eq. 3.4) for nucleation of c-Si at the Al GBs (solid line) and at the a-Si/c-Al interface (dashed line). The thickness of the "free" Si layers is ~2 ML at the original interface with Al (dotted line) and ~4 ML at the Al GBs (dash-dotted line) (see text). It follows that c-Si can nucleate only at the Al GBs at temperatures above 140°C and not at the a-Si/c-Al interface in the temperature range investigated. Reprinted with permission from Ref. [66], Copyright 2009, American Scientific Publishers.

3.4 Aluminum-Induced Layer Exchange upon Low-Temperature Annealing of a-Si/c-Al Layered Structures

The ALILE process involves that the original layer sequence of a-Si/c-Al layered structure be exchanged upon annealing. For a bilayered structure, this process implies that the original c-Al sublayer is replaced by a continuous c-Si sublayer. In most studies reported in the literature, ALILE was observed at a temperature near the eutectic temperature of 577°C for the Si-Al system [4, 63, 68, 75–81], and it was claimed that an interfacial oxide layer between the a-Si and c-Al sublayers would be essential for the ALILE process [69, 80]. The currently prevalent interpretation of the kinetics of the ALILE process is as follows [63, 76, 77]: (i) Si atoms diffuse into and dissolve within the Al sublayer; (ii) nucleation of c-Si takes place when a certain concentration of Si in the Al grains is exceeded (i.e., Si supersaturation); and (iii) c-Si growth occurs in lateral directions, confined between the interfacial oxide layer and the substrate, until the Al sublayer is eventually replaced by the laterally growing c-Si grains.

The above interpretation of the ALILE process cannot hold, at least not for the case of low-temperature annealing. In particular, it has been disclosed that the ALILE process can also occur upon annealing a-Si/c-Al bilayers at a temperature as low as 165°C and in the absence of an interfacial oxide layer [82]. At such a low temperature, dissolution of Si and volume diffusion of Si into and through Al grains are negligible (see Section 3.2.3). In the following section, an interpretation of the kinetics of the low-temperature ALILE process will be given, which complies with the available experimental data.

3.4.1 Experimental Observations of ALILE

3.4.1.1 Multilayered structure 4× a-Si (15 nm)/c-Al (15 nm)

For this multilayered structure, after annealing at 165°C for 20 minutes, an apparent AlSi sublayer exchange has occurred (e.g., see Fig. 3.10; an Al-rich sublayer has developed adjacent to the surface; note that the result obtained after annealing at 150°C for 20 minutes was considered in Section 3.2.2.1 to analyze the interdiffusion [see Fig. 3.1b]). The "diffusion length" (defined as $\sqrt{Dt}$, D: diffusion coefficient; t: time) can be taken as a tool for determining the annealing conditions, that is, annealing temperature and time, for the occurrence of layer exchange. Using the diffusion coefficient parameters D_0 and Q determined in Section 3.2.2.1 (see also Table 3.1), the diffusion length, across the first Si/Al interface nearest to the surface for this multilayered structure after the anneal at 165°C for 20 minutes, is calculated as 8.8 nm. This diffusion length is larger than one-fourth of the thickness of one Si sublayer + one Al sublayer ([15 nm + 15 nm]/4 = 7.5 nm), which also equals one half of each sublayer thickness: clearly this amount of intermixing by diffusion can be compatible with layer exchange. Such layer exchange was also observed after annealing at 225°C for 20 minutes (corresponding to a longer diffusion length of about 129 nm) of a multilayered structure with larger sublayer thicknesses of 85 nm and 45 nm for a-Si and c-Al, respectively. Note that in this case the c-Al sublayer possesses a larger average lateral grain size, which reduces the rate of intermixing [83].

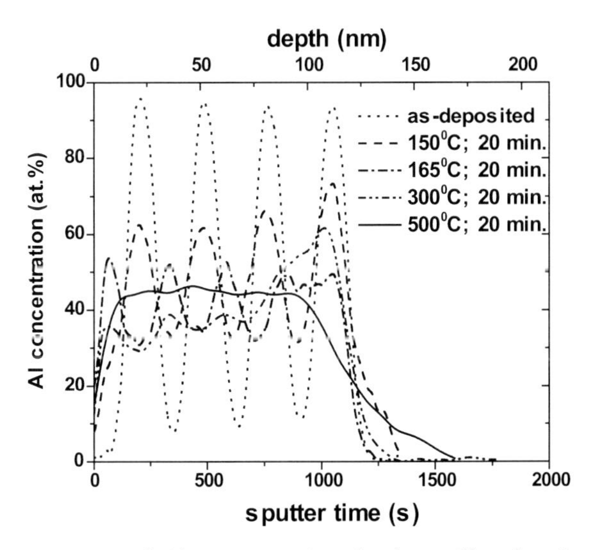

Figure 3.10 Measured Al concentration–depth profiles for the 4× a-Si (15 nm)/c-Al (15 nm) multilayered specimens as deposited and annealed at 150°C, 165°C, 300°C, and 500°C for 20 minutes. Reprinted with permission from Ref. [20], Copyright 2004, Elsevier.

Annealing of the 4× a-Si (15 nm)/c-Al (15 nm) multilayered structure for the same time at a higher temperature causes the layered structure to transform into a more or less uniform "SiAl mixture" (see the results for annealing at 300°C for 20 minutes and at 500°C for 20 minutes in Fig. 3.10). In the uniform SiAl mixture region the Si and Al AES peaks, as measured in the high-resolution (of energy) mode, show no significant change (chemical shift) as compared to the pure Si and Al Auger peaks, thereby indicating that no (metastable) aluminum-silicide phase formation occurs during the annealing process. This observation contrasts with the AES findings reported in Ref. [32] but is consistent with the in situ HRTEM observations (see Fig. 3.7) and DSC measurements (see Fig. 3.4) of a-Si/c-Al multilayers reported in Refs. [34, 38]—no occurrence of any metastable aluminum silicide phase.

3.4.1.2 Bilayered structure a-Si (150 nm)/c-Al (50 nm)

For this bilayered structure with high Al GB density in the Al sublayer, Al concentration–depth profiles, as measured after annealing for one hour at various temperatures in the range from 165°C to 250°C, have been gathered in Fig. 3.11a. The percentage of Al atoms transported to the location of the original Si sublayer can be determined from the cocnetration–depth profiles as a function of the annealing temperature. It follows that the amount of transported Al atoms increases, for one-hour annealing, from 19% at 165°C to 27% at 210°C, to 76% at 240°C, and to 92% at 250°C (see the inset in Fig. 3.11a). A practically complete exchange of the original Al and Si sublayers has taken place after one hour at 250°C. Using the diffusion coefficient parameters (D_0 and Q) determined in Section 3.2.2.2 (see also Table 3.1), the "diffusion length," $\sqrt{Dt}$, across the Si/Al interface, for this bilayered structure annealed at 250°C for one hour, is calculated as 3.2 μm. Taking this assessment of diffusion length as a criterion for the occurrence of Al→Si layer exchange upon low-temperature annealing (see the discussion below in Section 3.4.2.1), the annealing times needed for establishment of layer exchange at 165°C and 195°C are estimated as 105 days and 5 days, respectively, predictions that well agree with the experimental findings (see Fig. 3.11b). After the Al→Si layer exchange has been realized, a practically continuous c-Si layer is present at the position of the original c-Al sublayer (see the concentration–depth profiles after annealing in Fig. 3.11b). The ALILE process has also been

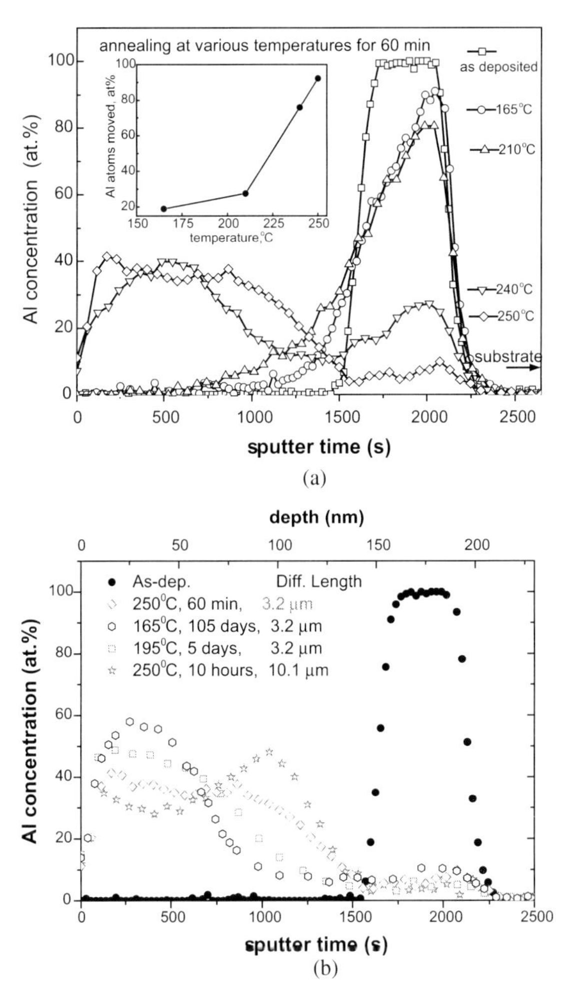

Figure 3.11 (a) Al concentration depth profiles of the a-Si (150 nm)/c-Al (50 nm) bilayer specimen as deposited and annealed for 1 hour at various temperatures in the range from 165°C to 250°C. The inset shows the percentage of Al atoms that have been transported to the location of the original Si sublayer upon annealing (from Ref. [93]). (b) Al concentration–depth profiles of the a-Si (150 nm)/c-Al (50 nm) bilayer specimen annealed at various temperatures for different times, upon which practically complete Al→Si layer exchange has occurred. The corresponding "diffusion lengths," across the Si/Al interface for this bilayered structure upon annealing, have also been given in the figure. Reprinted with permission from Ref. [91], Copyright 2007, AIP Publishing LLC.

observed upon annealing a c-Al (50 nm)/a-Si (150 nm) bilayered structure, with the Al sublayer *at the top*, at 250°C for five hours [84].

3.4.1.3 Direct visualization of Al→Si layer exchange

Very recently, the layer exchange of Al and Si sublayers upon heating of an a-Si/c-Al bilayer has been directly visualized in real time at the nanoscale by in situ heating VEFTEM [85].

The in situ heating VEFTEM experiment was performed in the Zeiss SESAM. A *cross-sectional* thin foil of a 150 nm a-Si/100 nm c-Al bilayer specimen was investigated. Due to the prominent differences in the plasmon absorption characteristics of Si and Al, the spatial distributions of Si and Al in the a-Si/c-Al bilayer (as deposited and upon annealing) can be revealed accurately by mapping the energy of the plasmon loss peak, with a spatial resolution better than 0.5 nm (see Figs. 3.12a–d; see Section 3.3.1.3 and Fig. 3.7c). The as-prepared a-Si/c-Al bilayer specimen was heated in situ using a Gatan 652 double-tilt heating stage in the SESAM at temperatures, consecutively, from 100°C to 300°C in steps of 20°C (30 minutes at each step).

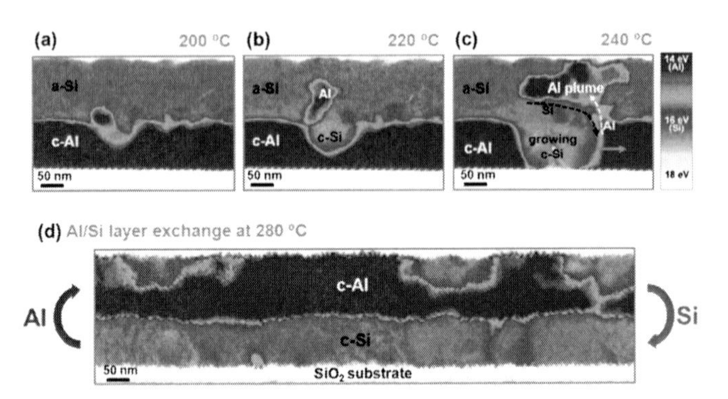

Figure 3.12 (a–c) VEFTEM observation (plasmon loss energy mapping from the cross-sectional view) of a 150 nm a-Si/100 nm c-Al bilayer during in situ annealing of the specimen at the temperatures indicated, showing the growth of c-Si nuclei in the Al bottom layer and, at the same time, the development of mushroom-shaped Al "plumes" of cloud-like morphology in the a-Si top layer. (d) Plasmon loss energy mapping of a specimen annealed at 280°C: Si and Al sublayers have practically exchanged their locations, forming a complete c-Si bottom layer at the initial location of the c-Al bottom layer. Reprinted with permission from Ref. [85], Copyright 2012, American Chemical Society.

Upon heating of the a-Si/c-Al bilayer, diffusion, wetting, and initial crystal growth of Si at the Al GBs in the Al sublayer were observed to occur already at temperatures below 160°C (see Ref. [70] and Section 3.3). As shown in Fig. 3.12a–c, the c-Si grain formed at the Al GB continues to grow upon increasing the annealing temperature (160°C→240°C). The continued growth of c-Si in Al leads to distinct changes of the morphology of the original a-Si/c-Al interface. A "plume" of the Al phase is observed to develop in the original a-Si upper layer, immediately above the growing c-Si grain in the Al bottom layer (see the in situ recorded VEFTEM images in Fig. 3.12a–c). Detailed HRTEM analyses have shown that an individual Al plume is constituted of a number of very small Al nanocrystals (as small as 10 nm in size) [85]. Upon continued heating, the Al plume continues to develop and grows within the a-Si upper layer and eventually reaches the top surface (see Fig. 3.12c). The developing, mushroom-shaped Al plume in the a-Si sublayer stays in connection with the c-Al bottom layer through a narrow Al-rich channel, which probably acts as a pathway for the continuous supply of Al material to the Al plume (see Fig. 3.12b,c).

The continuous growth of c-Si grains in the original Al sublayer and the concurrent development of Al plumes in the original a-Si sublayer eventually lead to a more or less complete exchange of the original layer sequence at an annealing temperature of 280°C (Fig. 3.12d): A continuous c-Si layer, with a thickness equal to that of the original Al bottom layer (100 nm), has thereby formed at the location of the original Al bottom layer. The Al has been driven away completely to the top layer, forming a near-continuous Al overlayer comprising some "trapped" Si phase regions (Fig. 3.12d).

3.4.1.4 Microstructural changes of the Al and Si phases upon ALILE

To derive a complete description of the driving force for the ALILE process (presented in Section 3.4.2), the microstructural changes that are associated with energy changes of the system upon ALILE have to be known. To this end, the residual macrostress and crystallite size of the Al phase and the Si phase have been evaluated by XRD, in particular for annealing the a-Si (150 nm)/c-Al (50 nm) bilayered structure at 165°C for 105 days and at 250°C for 1 hour. Note that the diffusion length is the same for these two annealing conditions.

As the c-Al (or c-Si) phase exhibits a {111} fiber texture and possesses a rotationally symmetric biaxial, planar state of stress, the residual macrostress of the c-Al (or c-Si) phase can be calculated from the slope of the plot of $d_{\{111\}}$ versus $\sin^2 y$, where $d_{\{111\}}$ is the {111} lattice spacing and y is the tilt angle [86]. Assuming that the structurally broadened profile originates from small crystallite size (often modeled with a Lorentzian broadening function) and/or microstrain (often modeled with a Gaussian broadening function), the crystallite size can be obtained by analyzing diffraction line broadening by applying the single-line Voigt method (for details, see Refs. [87–90]). Thereby, the macrostress and crystallite size of the c-Al and (once developed) c-Si phases upon annealing of the a-Si (150 nm)/c-Al (50 nm) bilayer at 165°C for 105 days and at 250°C for 1 hour can be determined accordingly [91]. The results have been gathered in Table 3.4. It follows that, upon annealing:

(i) the *macrostress* in the c-Al phase changes from compressive (−120 MPa) to tensile (187 MPa for annealing at 165°C for 105 days and 181 MPa for annealing at 250°C for 1 hour); also the formed c-Si phase possesses tensile stress.

(ii) the *c-Al crystallite (grain) size* for the as-deposited condition is equally large in perpendicular and lateral directions (about 50 nm) and of a value comparable to the Al layer thickness; after annealing at 165°C for 105 days and at 250°C for 1 hour, the Al grain size has remained about the same in the perpendicular direction but has increased laterally to about 142 nm and about 70 nm, respectively. Evidently, annealing at 165°C for 105 days (low-temperature annealing) leads to a larger Al grain size in lateral directions than annealing at 250°C for 1 hour (high-temperature annealing), which has also been verified by focused ion beam microscopy observations [91].

(iii) the c-Si crystallite (grain) size in both perpendicular and lateral directions is smaller than those of Al in the as-deposited and annealed conditions. Evidently, annealing at 165°C for 105 days leads to a larger Si grain size in both lateral and perpendicular directions than annealing at 250°C for 1 hour.

It should be recognized that the above-given stress values for the Al phase pertain to measurements performed after cooling the specimens to room temperature and thereby cannot be interpreted without more ado as stresses in the Al phase at the annealing temperatures of 165°C and 250°C. At the beginning of the annealing process, the stress in the Al sublayer is composed of the intrinsic film stress and the thermal stress due to heating up to the annealing temperature. The thermal stress is caused by the difference in thermal expansions of the substrate and the Al sublayer. Assuming that the thermal misfit is elastically and fully accommodated by the thin Al sublayer, the (in-plane) thermal stress of the Al sublayer is calculated to be −332 MPa and −545 MPa at 165°C and 250°C, respectively [67]. Thus the total stress in the Al sublayer at the beginning of annealing at 165°C and 250°C was about −452 MPa and −665 MPa, respectively, if fully elastic accommodation of the thermal misfit would occur. The observation of a tensile macrostress in Al at room temperature after annealing can then be interpreted such that during annealing at 165°C and 250°C all compressive macrostress in the Al is relaxed (for recent experimental proof of such relaxation, by in situ XRD stress measurements, see Ref. [85]). Cooling after annealing would lead to buildup of tensile thermal stress (analogous to the above reasoning). Because the cooling (as compared to the heating up) is very slow (2°C/minute) stress relaxation during cooling can be appreciable, and this explains that only part of the maximal thermal stress due to cooling is observed after cooling. For the c-Si phase, the stress values can be taken to be identical to those present after the annealing at 165°C and at 250°C as, obviously, no thermal misfit develops between the c-Si and the c-Si wafer upon cooling after annealing. The development of a (remaining) tensile stress in the c-Si phase could be ascribed to the volume decrease experienced by the Si phase upon its crystallization.

The final Si grain size in the practically continuous c-Si layer strongly depends on the microstructure of the as-deposited Al sublayer (see the discussion in Section 3.4.2). In the direction perpendicular to the surface, the Si grain size should not be larger than the thickness of the original Al sublayer regardless of the annealing temperature. In the lateral directions, the Si grain size depends on the initial lateral Al grain size and also on the annealing temperature. A larger lateral Al grain size in the original Al sublayer will lead to a larger Si grain size in the continuous c-Si layer, which agrees with experimental observations [63]. In principle, if all Al GBs provide sites for the nucleation of Si crystallization, the Si grain size in the continuous c-Si layer will be similar to the lateral Al grain size in the original Al sublayer. The smaller *overall* values for the c-Si size, as compared to the original grain size of the c-Al sublayer, as determined by XRD (Table 3.4) can be ascribed to the creation of many small Si crystallites at the location of the original a-Si sublayer upon the outward transport of Al (see Sections 3.4.1.3 and 3.4.3). As will be discussed in Section 3.4.3, low-temperature annealing (165°C) results in less nucleation sites for Si crystallization (a lesser amount of Al GBs that can be wetted by Si) as compared to high-temperature annealing (250°C). Thus, low-temperature annealing will lead to a larger Si grain size in the eventual continuous c-Si layer, which agrees with the experimental results gathered in Table 3.4 and experimental observations reported in Ref. [63].

Table 3.4 Residual macrostress and average crystallite size of the Al and Si phases for the a-Si (150 nm)/c-Al (50 nm) bilayered specimens with high Al GB density in the Al phase as deposited and annealed at 165°C for 105 days and at 250°C for 1 hour [91]

		Crystallite size (nm)	
Al phase	**Stress (MPa)**	**Perpendicular**	**Lateral**
As deposited	−120	50	47
165°C (105 d)	187	50	142
250°C (1 h)	181	50	72
Si phase			
165°C (105 d)	68	32	18
250°C (1 h)	47	18	11

3.4.2 Thermodynamics of ALILE

Upon annealing of a-Si/c-Al layered structures, an a-Si layer wets the (high-angle) Al GBs in the Al sublayer (Section 3.3.2.1) and subsequently this a-Si layer crystallizes there once the thickness of this wetting Si layer is beyond a critical value (Section 3.3.2.2). After the wetting Si layer at an Al GB crystallizes, the original Al GB has been replaced by two <Al>/<Si> interfaces. Upon further annealing, the weakly bonded, free Si atoms at the original a-Si/c-Al interface may continue to wet the newly developed <Al>/<Si> interfaces parallel to the original GB in the Al sublayer. The driving force for this process is given by $\gamma^{\text{interface}}_{\text{<Al>/<Si>}} - (\gamma^{\text{interface}}_{\text{<Al>/\{Si\}}} + \gamma^{\text{interface}}_{\text{<Si>/\{Si\}}})$. This driving force has been plotted as a function of temperature in Fig. 3.13 (in its inset the corresponding <Al>/<Si> interface energy and the sum of the <Al>/{Si} and <Si>/{Si} interface energies have been plotted) [71]. It follows that a positive driving force exists for a-Si to (continue to) wet the <Al>/<Si> interfaces formed within the Al sublayer.

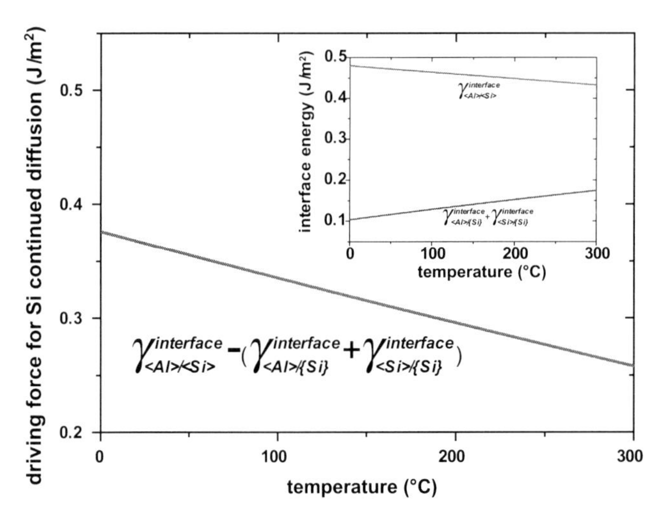

Figure 3.13 Driving force (as a function of temperature) for continued "wetting" by Si of the <Al>/<Si> interfaces formed in the original Al sublayer. The inset shows the <Al>/<Si> interface energy and the sum of the <Al>/{Si} and <Si>/{Si} interface energies. Reprinted with permission from Ref. [66], Copyright 2009, American Scientific Publishers.

For the a-Si wetting layer at the <Al>/<Si> interfaces parallel to the original GB in the Al sublayer, two cases are possible: (i) it may join the already existing c-Si grain or (ii) form a new c-Si nucleus at the <Al>/<Si> interface. In case (i) the existing c-Si grain grows laterally, and in case (ii) a new c-Si nucleus establishes two new interfaces (i.e., <Al>/<Si> and <Si>/<Si>), replacing the earlier formed <Al>/<Si> interface. The critical thickness for the c-Si grain growth in case (i) is given by

$$h^{\text{critical}}_{\text{Si grain growth}} = \frac{\gamma^{\text{interface}}_{\text{<Al>/<Si>}} - (\gamma^{\text{interface}}_{\text{<Al>/\{Si\}}} + \gamma^{\text{interface}}_{\text{<Si>/\{Si\}}})}{-\Delta G^{\text{crystallization}}_{\text{<Si>-\{Si\}}}} \tag{3.6}$$

The critical thickness for the formation of a new c-Si nucleus in case (ii) obeys

$$h_{\text{Si new nucleation}}^{\text{critical}} = \frac{\gamma_{\text{<Al>/<Si>}}^{\text{interface}} + \gamma_{\text{<Si>/<Si>}}^{\text{interface}} - (\gamma_{\text{<Al>/\{Si\}}}^{\text{interface}} + \gamma_{\text{<Si>/\{Si\}}}^{\text{interface}})}{-\Delta G_{\text{<Si>-\{Si\}}}^{\text{crystallization}}} \quad (3.7)$$

These two critical thickness values have been plotted as a function of temperature in Fig. 3.14 [71]. Again recognizing that the bond-weakening effect is restricted to about 2 ML of Si atoms at the <Al>/<Si> interface, it can be concluded that the Si wetting layer at the <Al>/<Si> interfaces will join the existing c-Si grain instead of forming a new c-Si grain at the <Al>/<Si> interface. Therefore, upon continued inward diffusion of Si atoms from the original a-Si sublayer into the <Al>/<Si> interfaces, the c-Si grains formed at the Al GBs in the Al sublayer grow laterally. This is in accordance with recent in situ heating VEFTEM observations (see Fig. 3.12a–d and the corresponding discussion in Section 3.4.1.3).

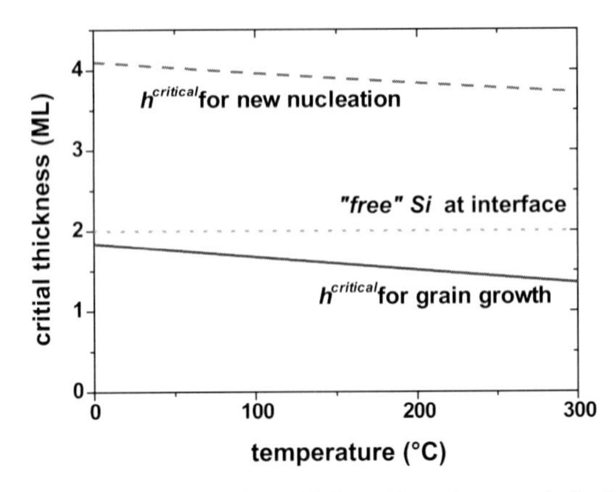

Figure 3.14 Critical thickness (in ML) for c-Si grain growth (solid line) and for formation of a new c-Si nuclei (dashed line) in the original Al sublayer (see Eqs. 3.6 and 3.7). The thickness of the "free" Si layers is about 2 ML at the interfaces with Al (dotted line). It follows that the c-Si formed originally at the Al GBs grows laterally and that the formation of new c-Si is impossible. Reprinted with permission from Ref. [66], Copyright 2009, American Scientific Publishers.

To arrive at a more or less complete description of the driving force for ALILE a number of microstructural processes involving energy change have to be considered. A sputter-deposited thin-film system is generally remote from an equilibrium state. Upon annealing, the energy stored in the thin-film system can be reduced via various mechanisms, such as mass transport (diffusion), grain growth or recrystallization of crystalline material, and crystallization of amorphous material. From a thermodynamic point of view, the total Gibbs energy change of the system upon annealing defines a driving force, which determines the overall direction of the kinetic process and can have a relation with the process rate (see Section 9.7 of Ref. [65]).

To assess the driving force for the ALILE process, for simplicity, the extreme case of "layer exchange" is considered: the sublayer sequence is completely reversed upon annealing, and the sublayers are composed of either pure crystalline Al or pure c-Si. The total Gibbs energy change ($\Delta G = G_{end} - G_{begin}$) of the system upon layer exchange consists of at least four contributions [67, 92]: 1) the energy change due to crystallization of the a-Si (ΔG_1), 2) the energy change due to the relaxation of macrostress and microstrain and the grain growth of the Al phase (ΔG_2), 3) the surface energy change (ΔG_3), and 4) the sublayer interface energy change (ΔG_4). The largest amount of energy released upon annealing is due to the crystallization of a-Si (ΔG_1); this energy release in principle is independent of the occurrence of layer exchange. Thus, a net driving force for the layer exchange itself is defined as $-(\Delta G_2 + \Delta G_3 + \Delta G_4)$, where the surface and interface energy changes, that is, ΔG_3 and ΔG_4, can be determined by applying the thermodynamic model [71, 72] and the energy change in the Al phase, that is, ΔG_2, can be obtained by determination of the microstructural changes of the Al phase. It should be mentioned that in the following a change in strain energy for the Si phase is not taken into account, because the measurement technique used (XRD) does not make it possible to extract such information from the initial *amorphous* phase.

By analyzing the macrostress and the crystallite size changes (see the discussion in Section 3.4.1.4) in the Al phase upon annealing of the a-Si (150 nm)/c-Al (50 nm) bilayer at various temperatures for one hour, the temperature dependence of the energy change in the Al phase, that is, ΔG_2, can be obtained [93] (the change in the microstrain of the Al phase is very small [67, 92], and therefore its contribution is not considered). The temperature dependences of the surface and the interface energy changes can be calculated using the methods as reviewed in Ref. [72]. As a result, the temperature dependence of the total driving force for layer exchange can be calculated: see the bold, dashed line in Fig. 3.15. The change in Gibbs energy, ΔG, is defined as $\Delta G = G_{\text{end}} - G_{\text{begin}}$, and thus the driving force equals $-\Delta G$. A negative value of the driving force (i.e., a positive value of the change in Gibbs energy) implies that layer exchange is unfavorable, whereas a positive value promotes layer exchange. It follows that (i) the driving force for layer exchange increases pronouncedly with temperature and (ii) the energy change in the Al phase promotes layer exchange, whereas the surface and interface energy changes oppose layer exchange. Hence, a critical temperature for layer exchange upon one-hour annealing can be indicated. For the current case this critical temperature is about 240°C, where $-(\Delta G_2 + \Delta G_3 + \Delta G_4) = 0$ J/m^2 (see Fig. 3.15). This prediction is consistent with the experimental observations shown in Fig. 3.11a, although it is recognized that the thermodynamic assessment only provides a rough estimate.

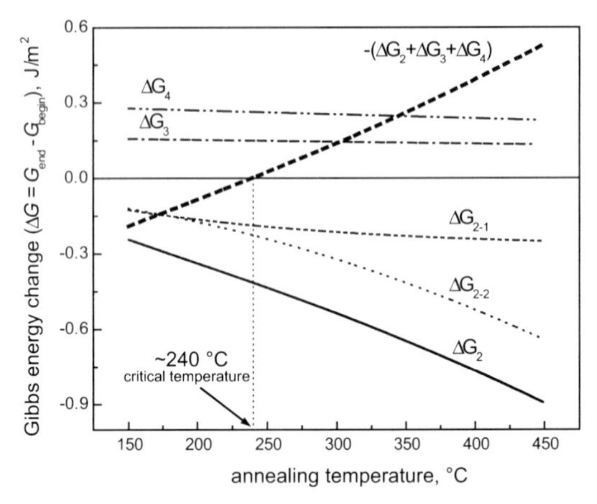

Figure 3.15 Gibbs energy changes ($\Delta G = G_{\text{end}} - G_{\text{begin}}$) upon annealing a-Si (150 nm)/c-Al (50 nm) bilayered specimens with high Al GB density in the Al phase at various temperatures for one hour. ΔG_2 is the Gibbs energy change in the Al phase due to (i) $\Delta G_{2\text{-}1}$ due to Al grain growth and (ii) $\Delta G_{2\text{-}2}$ corresponding to a change in strain/stress energy. ΔG_3 is a change in the total surface energy, and ΔG_4 is a change in the total interface energy. The term $-(\Delta G_2 + \Delta G_3 + \Delta G_4)$ thus represents the net, total driving force for layer exchange. Reprinted with permission from Ref. [93], Copyright 2006, Elsevier.

The negative value of the net driving force for $T < \sim 240°C$ is mainly because the (absolute value of the) energy change due to the Al grain growth is smaller for low-temperature annealing (e.g., annealing at 165°C for 1 hour or even for 10 days) than that for high-temperature annealing. If the value of the grain size of the Al phase is adopted as obtained for the annealing of the a-Si (150 nm)/c-Al (50 nm) bilayered structure at 165°C for 105 days (see Section 3.4.1.4 and Table 3.4), that is, further distinct grain growth of the Al grains has occurred, the net driving force becomes positive, that is, layer exchange now is possible energetically [91]. Indeed, practically complete Al→Si layer exchange has been observed for this annealing condition (see Fig. 3.16). On the same basis it was also predicted that the ALILE process would be faster for an a-Si/c-Al (i.e., a-Si on top of c-Al) bilayered structure than for a c-Al/a-Si (i.e., c-Al on top of a-Si) bilayered structure, which is consistent with experimental observations [84]. It is concluded that subtle microstructural effects can strongly influence the driving force and thus the occurrence, or not, of a macroscopic layer exchange process in an a-Si/c-Al or c-Al/a-Si bilayered or multilayered structure.

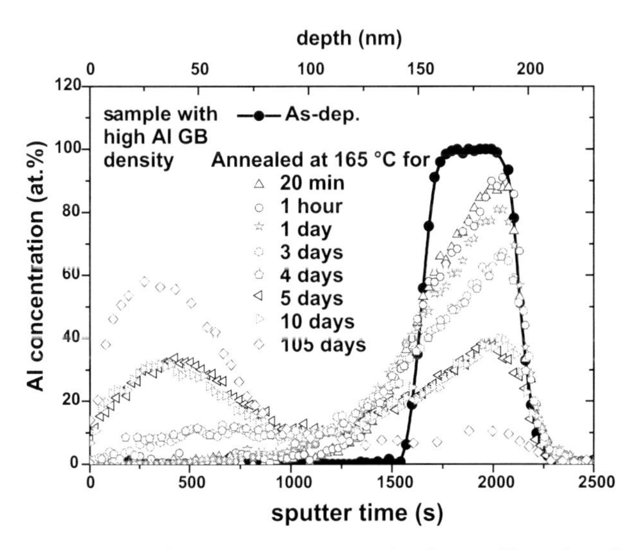

Figure 3.16 Measured Al concentration–depth profiles for the a-Si (150 nm)/c-Al (50 nm) bilayered specimens with high Al GB density in the Al phase as deposited and annealed at 165°C for various times. Reprinted with permission from Ref. [66], Copyright 2009, American Scientific Publishers.

3.4.3 Mechanism of ALILE

Considering the mass transport of Si into the original c-Al sublayer, and in view of the results presented in Section 3.4.2, the replacement of the c-Al sublayer by a Si sublayer (i.e., Al→Si layer exchange) can be ascribed straightforwardly to continued inward diffusion of Si initially along the original Al GBs and later along the produced boundaries between the c-Al and c-Si phases. Therefore, the "diffusion length" ($\sqrt{Dt}$) can be taken as a measure of the degree of Al layer→Si layer exchange.

Since the c-Al sublayer is laterally constrained by the thick, rigid substrate, the growth of c-Si grains at the original Al GBs by continuous mass transport of Si into the c-Al sublayer leads to the buildup of a large, local compressive stress, parallel to the original a-Si/c-Al interface, in the original c-Al sublayer. The compressive stress development in the c-Al bottom layer has been evidenced both by the in situ XRD analysis of the c-Al phase (overall result) and by HRTEM analyses of growing c-Si grains (local result) [85].

The fast diffusion of Si atoms along Al GBs, occurring via a vacancy exchange mechanism [94, 95], leads to the injection of free volumes in the a-Si sublayer, in particular adjacent to the Al GBs at which c-Si grains nucleate and grow. Since the a-Si matrix largely remains kinetically frozen at the processing temperature (~200°C), the free-volume concentration in a-Si would increase pronouncedly in the vicinity of the Al GBs at which c-Si grains nucleate and grow. The loss of Si atoms and the associated introduction of free volumes will not lead to a reduction of strived-for volume of the original a-Si layer. Hence a tensile stress parallel to the original a-Si/c-Al interface in the original a-Si layer develops because the lateral dimensions of the a-Si layer are dictated by the substrate (see above discussion for the originally pure c-Al layer).

This state of stress in the layer system (compressive in the c-Al sublayer and tensile in the a-Si sublayer) induces a stress gradient in the bilayer system. A diffusion potential gradient [96] thus exists for Al atoms in the Al sublayer such that they tend to move toward the adjacent, free-volume-rich a-Si sublayer. Thus, Al grains in the original c-Al sublayer are gradually dissolved. These dissolved Al atoms form Al plumes in the a-Si sublayer (see the in situ VEFTEM observations shown in Fig. 3.12). The Al plumes move (upward) within the solid a-Si layer and at the same time grow in size by collecting further Al atoms from the Al bottom layer (Fig. 3.12).

The *solid-state convection* process described before (see further Ref. [85]) eventually leads to the exchange of the original layer sequence. A continuous c-Si layer thus forms at the location of the original Al bottom layer by coalescence of separately developed Si grains, and the Al plumes developed in the original a-Si upper layer form a *near-continuous* Al upper layer, which, however, contains some "trapped" c-Si islands (see Fig. 3.12d; see also corresponding AES concentration–depth profiles of "layer-exchanged" a-Si/c-Al bilayers shown in Figs. 3.6a, 3.11b, and 3.16). As discussed in Section 3.3, Al GBs serve as the (only) favorable nucleation sites for low-temperature crystallization of a-Si. The Al plumes developed in the original a-Si upper layer are constituted of many Al nanocrystals and therefore contain newly formed Al GBs [85]. These Al GBs in the Al plumes provide additional sites for nucleation of c-Si but now at the location of the original a-Si upper layer. As a result, some a-Si crystallized at the original location of the a-Si upper layer, forming c-Si islands trapped in c-Al.

The mechanism of Al→Si layer exchange, as described before, on the basis of the real-time in situ-visualized Al→Si layer exchange process at the nanoscale (see also Section 3.4.1.3 and Ref. [85]), can be used to understand the observed evolution of Al concentration–depth profiles (as measured by AES depth profiling) during an Al→Si layer exchange process.

The Al concentration–depth profiles for the a-Si (150 nm)/c-Al (50 nm) bilayered specimens annealed at 165°C for various times have been gathered in Fig. 3.16 [91]. The evolution of the Al concentration–depth profiles exhibits three stages: (i) initial intermixing of Al and Si at both sides of the original interface, (ii) occurrence of a plateau (near the original a-Si/c-Al interface) and subsequently a maximum (closer to the surface of the original a-Si layer) in the Al concentration–depth profile at the location of the original a-Si sublayer, and (iii) establishment of Al→Si layer exchange, characterized by practically full replacement of the original Al sublayer by Si. These three stages can also be discerned in the results obtained upon annealing of the same bilayered specimen at 250°C for various times, as shown in Fig. 3.6a.

Upon inspection of the Al concentration–depth profiles presented in Fig. 3.16 (annealing at 165°C) and Fig. 3.6a (annealing 250°C), some detailed features can be detected additionally: (i) in a certain period of annealing time, the Al concentration–depth profile does not change significantly, that is, the progress of material redistribution appears to be retarded (e.g., see results shown in Fig. 3.16 for the anneals at 165°C from 20 minutes to 1 hour, 3 to 4 days, and 5 to 10 days); and (ii) with decreasing annealing temperature the height of the plateau region at the location of the original a-Si sublayer decreases and the position of the maximum of the Al concentration in the depth range of the original a-Si sublayer shifts to the surface.

The evolution of the Al concentration–depth profile may be interpreted as follows (utilizing the microstructural data presented in Section 3.4.1.3):

(i) **Initial "intermixing" of Al and Si**: This pertains to the initial nucleation and growth of c-Si at Al GBs and the initial development of small Al plumes in a-Si (see, e.g., Fig. 3.12a).

(ii) **Stagnation of material transport**: An initial relaxation of the stress gradient (developed due to Si transport into the Al sublayer and c-Si growth therein) decreases the driving force for Al transport. Continued (lateral) growth of the c-Si at the original Al GBs is required to generate a new stress gradient that drives the further growth of Al plumes.

(iii) **Occurrence of an Al plateau in a-Si (closer to the a-Si/c-Al interface)**: This corresponds with the development of Al plumes in a-Si directly above each original Al GB in which c-Si nucleates and grows (see, e.g., Fig. 3.12b and Fig. S3 of Ref. [85]).

(iv) **Occurrence of an Al maximum in a-Si (close to the surface of the a-Si layer)**: This observation reflects the mushroom shape of Al plumes at later stages of development in the a-Si layer (see Fig. 3.12c).

(v) **Temperature-dependence of the shape of the developed Al profiles in a-Si**: At a lower annealing temperature, the developed Al plateau in a-Si is lower because a smaller fraction of the Al GBs in the Al sublayer (i.e., those of sufficiently high energy) can be wetted by a-Si, allowing the occurrence of nucleation and growth of c-Si therein.

3.5 Main Conclusions

The operating diffusion mechanism in a-Si/c-Al layered systems at temperatures below 250°C is the diffusion of Si along GBs in the c-Al layer. Volume diffusion of Si (in c-Al) and of Al (in a-Si) can be neglected.

AIC is due to

(i) wetting of Al GBs by a-Si
(ii) crystallization of a-Si beyond a critical thickness of the a-Si wetting layer at Al GBs, at a temperature above about 140°C

The driving forces calculated for wetting of high-angle Al GBs by a-Si and crystallization of a-Si at Al GBs are indeed positive. No such driving force exists for occurrence of crystallization of a-Si at the a-Si/c-Al interface at temperatures below 400°C, and indeed crystallization of a-Si at the a-Si/c-Al interface does not occur at low temperatures.

ALILE can occur because

(i) a driving force exists for continuous wetting at the <Al>/<Si> interfaces along the original Al GBs after nucleation of c-Si has occurred there;
(ii) a driving force exists for lateral growth of the c-Si nucleated at the Al GBs; and
(iii) the unfavorable change in surface and interface energy upon layer exchange is overcompensated by the energy released due to the decrease of crystalline imperfections (as increase of the [c-Al] grain size and relaxation of macrostress).

The mechanism for ALILE derives from the development of a stress gradient in the bilayer system due to the inward Si transport and lateral c-Si growth at Al GBs in the Al sublayer. As a result, mushroom-shaped "plumes" of Al nanocrystals develop in the Si sublayer. The continuous growth (and eventually coalescence) of c-Si grains in the original Al sublayer and the concurrent development (and eventually coalescence) of Al plumes in the original a-Si sublayer eventually lead to the exchange of the original layer sequence.

Acknowledgments

We are particularly indebted to Dr. D. He and Prof. Y. H. Zhao for their contributions during their stay in the department of Prof. Mittemeijer at the Max Planck Institute for Intelligent Systems (formerly Max Planck Institute for Metals Research). We are grateful to Prof. Dr. F. Sommer for valuable discussions and to Dr. F. Phillipp and Prof. L. Gu (all at Max Planck Institute for Intelligent Systems) for their collaboration in the in situ TEM research performed in this project in the past years.

References

1. Bosnell, J. R., Voisey, U. C. (1970) The influence of contact materials on the conduction crystallization temperature and electrical properties of amorphous germanium, silicon and born films. *Thin Solid Films*, **6**, 161–166.
2. Boatright, R. L., McCaldin, J. O. (1976) Rapid growth of Si by solid-phase epitaxy, including comparisons to conventional Si crystal-growth. *Journal of Vacuum Science & Technology*, **13**, 938–939.
3. Boatright, R. L., McCaldin, J. O. (1976) Solid-state growth of Si to produce planar surfaces. *Journal of Applied Physics*, **47**, 2260–2262.
4. Nast, O., Puzzer, T., Koschier, L. M., Sproul, A. B., Wenham, S. R. (1998) Aluminum-induced crystallization of amorphous silicon on glass substrates above and below the eutectic temperature. *Applied Physics Letters*, **73**, 3214–3216.
5. Stradal, J., Scholma, G., Li, H., van der Werf, C. H. M., Rath, J. K., Widenborg, P. I., Campbell, P., Aberle, A. G., Schropp, R. E. I. (2006) Epitaxial thickening by hot wire chemical vapor deposition of polycrystalline silicon seed layers on glass. *Thin Solid Films*, **501**, 335–337.
6. Ekanayake, G., Quinn, T., Reehal, H. S., Rau, B., Gall, S. (2007) Large-grained polycrystalline silicon films on glass by argon-assisted ECRCVD epitaxial thickening of seed layers. *Journal of Crystal Growth*, **299**, 309–315.
7. Widenborg, P. I., Aberle, A. G. (2007) Hydrogen-induced dopant neutralisation in p-type AIC poly-Si seed layers functioning as buried emitters in ALICE thin-film solar cells on glass. *Journal of Crystal Growth*, **306**, 177–186.

8. Fuhs, W., Gall, S., Rau, B., Schmidt, M., Schneider, J. (2004) A novel route to a polycrystalline silicon thin-film solar cell. *Solar Energy*, **77**, 961–968.
9. Aberle, A. G. (2006) Progress with polycrystalline silicon thin-film solar cells on glass at UNSW. *Journal of Crystal Growth*, **287**, 386–390.
10. Gall, S., Schneider, J., Klein, J., Hubener, K., Muske, M., Rau, B., Conrad, E., Sieber, I., Petter, K., Lips, K., Stoger-Pollach, M., Schattschneider, P., Fuhs, W. (2006) Large-grained polycrystalline silicon on grlass for thin-film solar cells. *Thin Solid Films*, **511**, 7–14.
11. Wagner, T., Wang, J. Y., Hofmann, S. (2003) Sputter depth profiling in AES and XPS, in *Surface Analysis by Auger and X-Ray Photoelectron Spectroscopy* (eds. Briggs, D., Grant, J.) (IM Publications, West Sussex, UK, and SurfaceSpectra, Manchester, UK).
12. Wang, J. Y., Starke, U., Mittemeijer, E. J. (2009) Evaluation of the depth resolutions of Auger electron spectroscopic, x-ray photoelectron spectroscopic and time-of-flight secondary-ion mass spectrometric sputter depth profiling techniques. *Thin Solid Films*, **517**, 3402–3407.
13. Hwang, J. C. M., Ho, P. S., Lewis, J. E., Campbell, D. R. (1980) Grain-boundary diffusion of aluminum in polycrystalline silicon films. *Journal of Applied Physics*, **51**, 1576–1581.
14. Bukaluk, A. (1983) Analysis of diffusion mechanisms in thin polycrystalline Au-Ag films using Auger-electron spectroscopy. *Surface and Interface Analysis*, **5**, 20–27.
15. Hall, P. M., Morabito, J. M. (1976) Formalism for extracting diffusion-coefficients from concentration profiles. *Surface Science*, **54**, 79–90.
16. Highmore, R. J., Evetts, J. E., Greer, A. L., Somekh, R. E. (1987) Differential scanning calorimetry study of solid-state amorphization in multilayer thin-film Ni/Zr. *Applied Physics Letters*, **50**, 566–568.
17. Pamler, W. (1987) Application of Auger-electron depth profile analysis to thin-film interdiffusion studies. *Applied Physics A: Materials Science & Processing*, **42**, 219–226.
18. Hofmann, S. (1994) Atomic mixing, surface roughness and information depth in high-resolution AES depth profiling of a GaAs/AlAs superlattice structure. *Surface and Interface Analysis*, **21**, 673–678.
19. Wang, J. Y., Mittemeijer, E. J. (2004) A new method for the determination of the diffusion-induced concentration profile and the interdiffusion coefficient for thin film systems by Auger electron spectroscopical sputter depth profiling. *Journal of Materials Research*, **19**, 3389–3397.

20. Wang, J. Y., Zalar, A., Mittemeijer, E. J. (2004) Depth dependences of the ion bombardment induced roughness and of the interdiffusion coefficient for Si/Al multilayers. *Applied Surface Science*, **222**, 171–179.
21. Wang, J. Y., Hofmann, S., Zalar, A., Mittemeijer, E. J. (2003) Quantitative evaluation of sputtering induced surface roughness in depth profiling of polycrystalline multilayers using Auger electron spectroscopy. *Thin Solid Films*, **444**, 120–124.
22. Wang, J. Y., Zalar, A., Zhao, Y. H., Mittemeijer, E. J. (2003) Determination of the interdiffusion coefficient for Si/Al multilayers by Auger electron spectroscopical sputter depth profiling. *Thin Solid Films*, **433**, 92–96.
23. Wang, J. Y., He, D., Zalar, A., Mittemeijer, E. J. (2006) Interdiffusion in microstructurally different Si/Al multilayered structures. *Surface and Interface Analysis*, **38**, 773–776.
24. Murray, J. L., McAlister, A. J. (1984) *Bulletin of Alloy Phase Diagrams*, **5**, 74.
25. Wang, J. Y., He, D., Zhao, Y. H., Mittemeijer, E. J. (2006) Wetting and crystallization at grain boundaries: origin of aluminum-induced crystallization of amorphous silicon. *Applied Physics Letters*, **88**, 061910.
26. McCaldin, J. O., Sankur, H. (1972) Precipitation of Si from Al metallization of integrated-circuits. *Applied Physics Letters*, **20**, 171–172.
27. Ottavian.G, Sigurd, D., Marrello, V., McCaldin, J. O., Mayer, J. W. (1973) Crystal-growth of silicon and germanium in metal-films. *Science*, **180**, 948–949.
28. Sankur, H., McCaldin, J. O., Devaney, J. (1973) Solid-phase epitaxial-growth of Si mesas from Al metallization. *Applied Physics Letters*, **22**, 64–66.
29. Ottavian.G, Sigurd, D., Marrello, V., Mayer, J. W., McCaldin, J. O. (1974) Crystallization of Ge and Si in metal-films. 1. *Journal of Applied Physics*, **45**, 1730–1739.
30. Sigurd, D., Ottavian.G, Arnal, H. J., Mayer, J. W. (1974) Crystallization of Ge and Si in metal-films. 2. *Journal of Applied Physics*, **45**, 1740–1745.
31. Greene, J. E., Mei, L. (1976) Metal-induced crystallization of RF sputtered a-Si thin-films. *Thin Solid Films*, **34**, 27–30.
32. Hentzell, H. T. G., Robertsson, A., Hultman, L., Shaofang, G., Hornstrom, S. E., Psaras, P. A. (1987) Formation of aluminum silicide between 2 layers of amorphous-silicon. *Applied Physics Letters*, **50**, 933–934.
33. Radnoczi, G., Robertsson, A., Hentzell, H. T. G., Gong, S. F., Hasan, M. A. (1991) Al induced crystallization of a-Si. *Journal of Applied Physics*, **69**, 6394–6399.

34. Konno, T. J., Sinclair, R. (1992) Crystallization of silicon in aluminum amorphous-silicon multilayers. *Philosophical Magazine B: Physics of Condensed Matter Statistical Mechanics Electronic Optical and Magnetic Properties*, **66**, 749–765.
35. Konno, T. J., Sinclair, R. (1993) Crystallization and decomposition of co-sputtered amorphous silicon-aluminum thin-films. *Materials Chemistry and Physics*, **35**, 99–113.
36. Nakayama, T., Konno, T. J., Satoh, H., Sinclair, R. (1993) Structure and corrosion properties of Al/Si multilayers. *Journal of Magnetism and Magnetic Materials*, **126**, 167–169.
37. Haque, M. S., Naseem, H. A., Brown, W. D. (1994) Interaction of aluminum with hydrogenated amorphous-silicon at low-temperatures. *Journal of Applied Physics*, **75**, 3928–3935.
38. Konno, T. J., Sinclair, R. (1994) Metal-contact-induced crystallization of semiconductors. *Materials Science and Engineering: A*, **179**, 426–432.
39. Sinclair, R., Konno, T. J. (1994) In-situ HREM: application to metal-mediated crystallization. *Ultramicroscopy*, **56**, 225–232.
40. Ashtikar, M. S., Sharma, G. L. (1995) Silicide mediated low-temperature crystallization of hydrogenated amorphous-silicon in contact with aluminum. *Journal of Applied Physics*, **78**, 913–918.
41. Kim, J. H., Lee, J. Y. (1996) Al-induced crystallization of an amorphous Si thin film in a polycrystalline Al/native SiO2/amorphous Si structure. *Japanese Journal of Applied Physics Part 1: Regular Papers Short Notes & Review Papers*, **35**, 2052–2056.
42. Lee, S. W., Joo, S. K. (1996) Low temperature poly-Si thin-film transistor fabrication by metal-induced lateral crystallization. *IEEE Electron Device Letters*, **17**, 160–162.
43. Yoon, S. Y., Kim, K. H., Kim, C. O., Lee, J. H., Jang, J. (1997) Metal induced crystallization of hydrogenated amorphous silicon. *Journal of the Korean Physical Society*, **30**, S213–S216.
44. Drusedau, T. P., Blasing, J., Gnaser, H. (1998) Aluminum mediated low temperature growth of crystalline silicon by plasma-enhanced chemical vapor and sputter deposition. *Applied Physics Letters*, **72**, 1510–1512.
45. Yoon, S. Y., Oh, J. Y., Kim, C. O., Jang, J. (1998) Low temperature solid phase crystallization of amorphous silicon at 380 degrees C. *Journal of Applied Physics*, **84**, 6463–6465.
46. Drusedau, T. P., Diez, A., Blasing, J. (1999) Deposition of nanocrystalline silicon mediated by ultrathin aluminum underlayers by PCVD and sputter-deposition at 500 K. *Thin Solid Films*, **337**, 41–44.

47. Chambouleyron, I., Fajardo, F., Zanatta, A. R. (2001) Aluminum-induced crystallization of hydrogenated amorphous germanium thin films. *Applied Physics Letters*, **79**, 3233–3235.

48. Kishore, R., Hotz, C., Naseem, H. A., Brown, W. D. (2001) Aluminum-induced crystallization of amorphous silicon (alpha-Si : H) at 150 degrees C. *Electrochemical and Solid State Letters*, **4**, G14–G16.

49. Al-Dhafiri, A. M., Naseem, H. A., Kishore, R., Brown, W. D. (2002) Comparison of aluminum induced crystallization of sputtered amorphous silicon in Al/Si and Si/Al configurations. *Proceedings of the Eleventh International Workshop on the Physics of Semiconductor Devices, Vol. 1 & 2*, **4746**, 599–607.

50. Kishore, R., Sood, K. N., Naseem, H. A. (2002) Microstructural and analytical investigation of low temperature crystallized amorphous silicon/crystallized silicon interface using SEM and EDS. *Journal of Materials Science Letters*, **21**, 647–648.

51. Wang, R. C., Du, P. Y., Shen, G., Weng, W. J., Han, G. R. (2002) Study on Al-induced crystallization of Al/a-Si : H bilayer thin film. *Thin Solid Films*, **422**, 225–229.

52. Kishore, R., Hotz, C., Naseem, H. A., Brown, W. D. (2003) In situ x-ray diffraction studies of aluminium-induced crystallization of hydrogenated amorphous silicon. *Journal of Applied Crystallography*, **36**, 1236–1239.

53. Kishore, R., Hotz, K., Naseem, H. A., Brown, W. D. (2005) Transmission electron microscopy and x-ray diffraction analysis of aluminum-induced crystallization of amorphous silicon in alpha-Si : H/Al and Al/alpha-Si : H structures. *Microscopy and Microanalysis*, **11**, 133–137.

54. Pereira, L., Aguas, H., Vilarinho, P., Fortunato, E., Martins, R. (2005) Metal induced crystallization: gold versus aluminium. *Journal of Materials Science*, **40**, 1387–1391.

55. Zou, M., Cai, L., Brown, W. (2005) Nano-aluminum-induced low-temperature crystallization of PECVD amorphous silicon. *Electrochemical and Solid State Letters*, **8**, G103–G105.

56. Albarghouti, M., Naseem, H., Al-Jassim, M. (2006) TEM investigation of the role of a nano-oxide layer in aluminum-induced crystallization of a-Si : H. *Electrochemical and Solid State Letters*, **9**, G225–G227.

57. Srivastava, A. K., Sood, K. N., Kishore, R., Naseem, H. A. (2006) Interfacial diffusion effect on metal induced crystallization of an amorphous silicon: a microstructural pathway. *Electrochemical and Solid State Letters*, **9**, G219–G221.

58. Zou, M., Cai, L., Wang, H. Y., Xu, J. S. (2006) Silicon nanowires by aluminum-induced crystallization of amorphous silicon. *Electrochemical and Solid State Letters*, **9**, G31–G33.

59. Her, Y. C., Chen, C. W. (2007) Crystallization kinetics of ultrathin amorphous Si film induced by Al metal layer under thermal annealing and pulsed laser irradiation. *Journal of Applied Physics*, **101**, 043518.

60. Ikeda, K. I., Hirota, T., Fujimoto, K., Sugimoto, Y., Takata, N., Ii, S., Nakashima, H., Nakashima, H. (2007) In-situ heating observation for formation behavior of polycrystalline silicon thin films fabricated using aluminum induced crystallization. *Journal of the Japan Institute of Metals*, **71**, 158–163.

61. Gupta, S., Chelawat, H., Kumbhar, A. A., Adhikari, S., Dusane, R. O. (2008) Aluminum-induced in situ crystallization of HWCVD a-Si : H films. *Thin Solid Films*, **516**, 850–852.

62. Gong, S. F., Hentzell, H. T. G., Radnoczi, G., Charai, A. (1988) Solid-phase epitaxy and doping of Si through Sb-enhanced recrystallization of polycrystalline Si. *Applied Physics Letters*, **53**, 902–904.

63. Nast, O., Hartmann, A. J. (2000) Influence of interface and Al structure on layer exchange during aluminum-induced crystallization of amorphous silicon. *Journal of Applied Physics*, **88**, 716–724.

64. Mittemeijer, E. J. (1992) Analysis of the kinetics of phase-transformations. *Journal of Materials Science*, **27**, 3977–3987.

65. Mittemeijer, E. J. (2010) *Fundamentals of Materials Science: The Microstructure–Property Relationship Using Metals as Model Systems* (Springer, Berlin, Heidelberg).

66. Wang, J. Y., Wang, Z. M., Jeurgens, L. P. H., Mittemeijer, E. J. (2009) Mechanisms of aluminium-induced crystallization and layer exchange upon low-temperature annealing of amorphous Si/polycrystalline Al bilayers. *Journal of Nanoscience and Nanotechnology*, **9**, 3364–3371.

67. He, D., Wang, J. Y., Mittemeijer, E. J. (2005) The initial stage of the reaction between amorphous silicon and crystalline aluminum. *Journal of Applied Physics*, **97**, 093524.

68. Gall, S., Muske, M., Sieber, I., Nast, O., Fuhs, W. (2002) Aluminum-induced crystallization of amorphous silicon. *Journal of Non-Crystalline Solids*, **299**, 741–745.

69. Schneider, J., Klein, J., Muske, M., Gall, S., Fuhs, W. (2004) Aluminum-induced crystallization of amorphous silicon: preparation effect on growth kinetics. *Journal of Non-Crystalline Solids*, **338–340**, 127–130.

70. Wang, Z., Gu, L., Phillipp, F., Wang, J. Y., Jeurgens, L. P. H., Mittemeijer, E. J. (2011) Metal-catalyzed growth of semiconductor nanostructures without solubility and diffusivity constraints. *Advanced Materials*, **23**, 854–859.

71. Wang, Z. M., Wang, J. Y., Jeurgens, L. P. H., Mittemeijer, E. J. (2008) Thermodynamics and mechanism of metal-induced crystallization in immiscible alloy systems: experiments and calculations on Al/a-Ge and Al/a-Si bilayers. *Physical Review B*, **77**, 045424.

72. Jeurgens, L. P. H., Wang, Z., Mittemeijer, E. J. (2009) Thermodynamics of reactions and phase transformations at interfaces and surfaces. *International Journal of Materials Research*, **100**, 1281–1307.

73. Hiraki, A. (1986) Initial formation process of metal silicon interfaces. *Surface Science*, **168**, 74–99.

74. Wen, H. J., Dahneprietsch, M., Bauer, A., Cuberes, M. T., Manke, I., Kaindl, G. (1995) Thermal annealing of the epitaxial Al/Si(111)7x7 interface: Al clustering, interfacial reaction, and Al-induced p(+) doping. *Journal of Vacuum Science & Technology A: Vacuum Surfaces and Films*, **13**, 2399–2406.

75. Aberle, A. G., Harder, N. P., Oelting, S. (2001) Formation of large-grained uniform poly-Si films on glass at low temperature. *Journal of Crystal Growth*, **226**, 209–214.

76. Nast, O., Wenham, S. R. (2000) Elucidation of the layer exchange mechanism in the formation of polycrystalline silicon by aluminum-induced crystallization. *Journal of Applied Physics*, **88**, 124–132.

77. Widenborg, P. I., Aberle, A. G. (2002) Surface morphology of poly-Si films made by aluminium-induced crystallisation on glass substrates. *Journal of Crystal Growth*, **242**, 270–282.

78. Sieber, I., Schneider, R., Doerfel, I., Schubert-Bischoff, R., Gall, S., Fuhs, W. (2003) Preparation of thin polycrystalline silicon films on glass by aluminium-induced crystallisation: an electron microscopy study. *Thin Solid Films*, **427**, 298–302.

79. Doi, A. (2004) Fabrication of uniform poly-Si thin film on glass substrate by AlC. *Thin Solid Films*, **451–52**, 485–488.

80. Klein, J., Schneider, J., Muske, M., Gall, S., Fuhs, W. (2004) Aluminium-induced crystallisation of amorphous silicon: influence of the aluminium layer on the process. *Thin Solid Films*, **451–452**, 481–484.

81. Ekanayake, G., Quinn, T., Reehal, H. S. (2006) Large-grained poly-silicon thin films by aluminium-induced crystallisation of microcrystalline silicon. *Journal of Crystal Growth*, **293**, 351–358.

82. Sugimoto, Y., Takata, N., Hirota, T., Ikeda, K., Yoshida, F., Nakashima, H., Nakashima, H. (2005) Low-temperature fabrication of polyerystalline Si thin film using Al-induced crystallization without native Al oxide at amorphous Si/Al interface. *Japanese Journal of Applied Physics Part 1: Regular Papers Brief Communications & Review Papers*, **44**, 4770–4775.
83. Zalar, A., Wang, J. Y., Zhao, Y. H., Mittemeijer, E. J., Panjan, P. (2003) AES depth profiling of thermally treated Al/Si thin-film structures. *Vacuum*, **71**, 11–17.
84. He, D., Wang, J. Y., Mittemeijer, E. J. (2005) Reaction between amorphous Si and crystalline Al in Al/Si and Si/Al bilayers: microstructural and thermodynamic analysis of layer exchange. *Applied Physics A: Materials Science & Processing*, **80**, 501–509.
85. Wang, Z. M., Gu, L., Jeurgens, L. P. H., Phillipp, F., Mittemeijer, E. J. (2012) Real-time visualization of convective transportation of solid materials at nanoscale. *Nano Letters*, **12**, 6126–6132.
86. Welzel, U., Ligot, J., Lamparter, P., Vermeulen, A. C., Mittemeijer, E. J. (2005) Stress analysis of polycrystalline thin films and surface regions by x-ray diffraction. *Journal of Applied Crystallography*, **38**, 1–29.
87. de Keijser, T. H., Langford, J. I., Mittemeijer, E. J., Vogels, A. B. P. (1982) Use of the Voigt function in a single-line method for the analysis of x-ray-diffraction line broadening. *Journal of Applied Crystallography*, **15**, 308–314.
88. Delhez, R., de Keijser, T. H., Mittemeijer, E. J. (1982) Determination of crystallite size and lattice-distortions through x-ray-diffraction line-profile analysis: recipes, methods and comments. *Fresenius Zeitschrift fur Analytische Chemie*, **312**, 1–16.
89. Mittemeijer, E. J., Welzel, U. (2008) The "state of the art" of the diffraction analysis of crystallite size and lattice strain. *Zeitschrift fur Kristallographie*, **223**, 552–560.
90. Mittemeijer, E. J., Welzel, U. (2012) Diffraction line-profile analysis, in *Modern Diffraction Methods* (eds. Mittemeijer, E. J., Welzel, U.) (Wiley-VCH, Weinheim), 89.
91. Wang, J. Y., Wang, Z. M., Mittemeijer, E. J. (2007) Mechanism of aluminum-induced layer exchange upon low-temperature annealing of amorphous Si/polycrystalline Al bilayers. *Journal of Applied Physics*, **102**, 113523.
92. Zhao, Y. H., Wang, J. Y., Mittemeijer, E. J. (2004) Microstructural changes in amorphous Si/crystalline Al thin bilayer films upon annealing. *Applied Physics A: Materials Science & Processing*, **79**, 681–690.

93. He, D., Wang, J. Y., Mittemeijer, E. J. (2006) Thermodynamic and kinetic criteria for layer exchange in amorphous silicon/crystalline aluminium bilayers during annealing. *Scripta Materialia*, **54**, 559–561.
94. Mehrer, H. (2007) *Diffusion in Solids: Fundamentals, Methods, Materials, Diffusion-Controlled Processes* (Springer, Heidelberg).
95. Balluffi, R. W. (1984) Grain boundary diffusion mechanisms in metals, in *Diffusion in Crystalline Solids* (eds. Murch, G. E., Nowick, A. S.) (Academic Press, London), 319.
96. Larché, F. C., Cahn, J. W. (1985) The interactions of composition and stress in crystalline solids. *Acta Metallurgy*, **33**, 331–357.
97. Bergner, D., Cyrener, E. (1973) Diffusion von Fremdelementen in Aluminium-Mischkristallen. *Neue Hutte*, **18**, 356–361.
98. Fujikawa, S. I., Hirano, K. I., Fukushima, Y. (1978) Diffusion of silicon in aluminum. *Metallurgical Transactions A: Physical Metallurgy and Materials Science*, **9**, 1811–1815.
99. McCaldin, J. O., Sankur, H. (1971) Diffusivity and solubility of Si in Al metallization of integrated circuits. *Applied Physics Letters*, **19**, 524–526.
100. Fuller, C. S., Ditzenberger, J. A. (1956) Diffusion of donor and acceptor elements in silicon. *Journal of Applied Physics*, **27**, 544–553.
101. Kao, Y. C. (1967) On diffusion of aluminum into silicon. *Electrochemical Technology*, **5**, 90–93.
102. Ghoshtagore, R. N. (1971) Donor diffusion dynamics in silicon. *Physical Review B*, **3**, 2507–2514.
103. Rosnowski, W. (1978) Aluminum diffusion into silicon in an open tube high-vacuum system. *Journal of the Electrochemical Society*, **125**, 957–962.
104. Laferla, A., Torrisi, L., Galvagno, G., Rimini, E., Ciavola, G., Carnera, A., Gasparotto, A. (1993) Implants of aluminum in the 50-120 MeV energy-range into silicon. *Nuclear Instruments & Methods in Physics Research Section B: Beam Interactions with Materials and Atoms*, **73**, 9–13.
105. Galvagno, G., Lavia, F., Saggio, M. G., Lamantia, A., Rimini, E. (1995) 2-Dimensional aluminum diffusion in silicon: experimental results and simulations. *Journal of the Electrochemical Society*, **142**, 1585–1590.
106. Wang, Z. M., Wang, J. Y., Jeurgens, L. P. H., Mittemeijer, E. J. (2008) Tailoring the ultrathin Al-induced crystallization temperature of amorphous Si by application of interface thermodynamics. *Physical Review Letters*, **100**, 125503.

107. Fan, J. C. C., Anderson, C. H. (1981) Transition-temperatures and heats of crystallization of amorphous Ge, Si, and Ge1-xSix alloys determined by scanning calorimetry. *Journal of Applied Physics*, **52**, 4003–4006.

108. Donovan, E. P., Spaepen, F., Turnbull, D., Poate, J. M., Jacobson, D. C. (1985) Calorimetric studies of crystallization and relaxation of amorphous Si and Ge prepared by ion-implantation. *Journal of Applied Physics*, **57**, 1795–1804.

Chapter 4

Metal-Induced Crystallization by Homogeneous Insertion of Metallic Species in Amorphous Semiconductors

Antonio R. Zanatta and Fabio A. Ferri
Laboratório de Filmes Finos, Instituto de Física de São Carlos, Universidade de São Paulo, PO Box 369, São Carlos 13560-970, Brazil
zanatta@ifsc.usp.br

The present contribution aims to discuss some aspects related to the metal-induced crystallization of amorphous Si and Ge thin films prepared by cosputtering and probed by Raman scattering spectroscopy. As will be shown, the adopted experimental approach is suitable to investigate the crystallization phenomenon of amorphous thin films containing metallic species (controllably inserted and homogeneously distributed) at various concentration ranges. Additionally, the high sensitivity and spatial resolution provided by Raman spectroscopy are consistent with the detection of minute amounts of either crystalline or amorphous structures in a fast and nondestructive way. The whole process also involves the realization of cumulative thermal annealing treatments that give further insight into the crystallization kinetics of the systems

Metal-Induced Crystallization: Fundamentals and Applications
Edited by Zumin Wang, Lars P. H. Jeurgens, and Eric J. Mittemeijer

ISBN 978-981-4463-40-9 (Hardcover), 978-981-4463-41-6 (eBook)
www.panstanford.com

under investigation. To illustrate these features, films of amorphous Si and Ge containing manganese will be considered in detail. The main experimental findings involving amorphous Si and Ge films combined with other metals (Ni, Co, Fe, Gd, Er, and Al) will also be briefly discussed.

4.1 Introduction

Research involving amorphous silicon (a-Si) and amorphous germanium (a-Ge), and their alloys, has been carried out for more than five decades. Along all of these years, the interest in studying amorphous semiconductors (a-semiconductors) was driven by a mix of academic-technological features whose main purpose was to explore their properties, envisaging the development of efficient optical-electronic devices (see, e.g., [1]). Accordingly, considerable advances were obtained that greatly influenced both the microelectronics and photovoltaic industries. In fact, it became clear very soon that the use of a-semiconductors in technological applications has advantages over their crystalline counterparts such as [2, 3] (a) they can be synthesized following simple (low-budgetary) experimental methods; (b) they can be prepared over large areas and at relatively low temperatures; (c) because of their amorphous nature (or lack of structural constraints), a-semiconductor compounds can be produced at compositions ranging from the doping to the alloying regimes; and (d) their final optical-electronic properties can be adjusted by the insertion of foreign species and/or through the realization of postdeposition treatments. In spite of these desirable aspects, the disorder present in a-semiconductors severely limits their carrier mobility and, consequently, the performance of any electronic device based on these materials. A very convenient approach to minimize this drawback is to transform, after deposition, the amorphous matrix of either Si or Ge thin films to a polycrystalline material. This can be accomplished by submitting the films to energetic processes like, for example, conventional thermal annealing treatments [4], laser radiation (or laser-induced crystallization [LIC]) [5], and mechanical stress-strain (or pressure-induced crystallization [PIC]) [6], or by applying an electrical current (or current-induced crystallization [CIC]) [7].

In most (or all) of these methods, the temperature of crystallization T_c can be considerably decreased by the presence of metallic species in a process that is known by metal-induced crystallization, or simply MIC [8]. In such a case, Si- or Ge-based nano- or microcrystalline materials can be achieved from amorphous films deposited on inexpensive (and/or flexible) substrates, which, eventually, can lead to the development of devices with improved performance.

Figures 4.1a and 4.1b, for example, show the crystallization temperature of a-Si and a-Ge films according to the presence of different metals. As can be seen from the figures, all metals are responsible for some decrease in the crystallization temperature T_c of pure a-Si ($T_c \approx 800°C$) and pure a-Ge ($T_c \approx 600°C$). For most of the systems considered in Fig. 4.1, T_c is reduced by up to 25%: between ~800°C and 600°C for Si films and in the ~600°C–450°C range for Ge. For metals like Cu, Al, and Au such a reduction in the crystallization temperature can be of more than 50%, reaching values that can be as low as ~200°C.

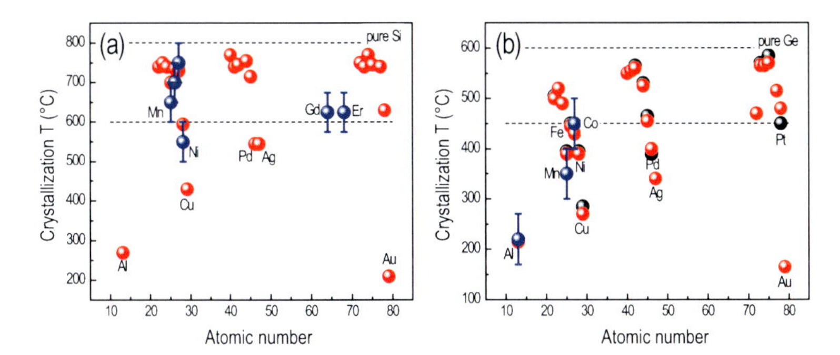

Figure 4.1 Crystallization temperature as a function of the atomic number of various metals when inserted into or in contact with (a) a-Si and (b) a-Ge thin films. The horizontal dashed lines indicate the crystallization temperature typically verified in pure a-Si (~800°C) and pure a-Ge (~600°C) films and 25% lower—a temperature range in which the majority of the a-semiconductor + metal systems crystallize. All information refers to experimental data as obtained by Knaepen et al. [11] (red spheres), Gaudet et al. [12] (black spheres), and our research group in Brazil (blue spheres).

Although it is clear that the crystallization temperature depends mostly on the characteristics of each metal, the mechanisms involved in the MIC of amorphous films are still under debate [9]. This is particularly true if one considers that any crystallization mechanism is highly influenced by the sample's details like [10] method of preparation and conditions, relative concentration of species, annealing procedure, and thickness, as well as the uniqueness of the techniques considered in their investigation.

Stimulated by these aspects, a few years ago, our group in Brazil started a systematic investigation of the MIC in a-Si and a-Ge films by adopting a different and complementary experimental approach. More specifically, whereas most of the works in this field considered the study of a-semiconductor–metal layers and X-ray diffraction (XRD) analyses, our investigations were focused on the controllable insertion of metal into the a-semiconductor matrix and their structural investigation by means of Raman scattering spectroscopy. This approach possesses some clear advantages when compared to the layered configuration, in the sense that it provides important physical-chemical information on the crystallization phenomenon and its dynamics with thermal annealing; it is not restricted to the a-semiconductor–metal interface; and consequently, the size or geometry of the samples does not influence the whole process. Furthermore, the adopted methodology is suitable to investigate changes in the optical bandgap of the films as well as to determine the development of any new phase (silicides, germanides, or even oxides) with high sensitivity and spatial resolution. Table 4.1 contains a summary of the work we conducted in the field of MIC by showing the systems we have studied so far. Table 4.1 also displays information regarding the metal concentration range, the sample annealing conditions, and the crystallization onset, for example. Part of the aspects involved in these investigations will be presented and discussed along the next pages, taking into consideration a-Si and a-Ge films containing different concentrations of manganese.

4.2 Sample Preparation and Characterization

The samples considered in Table 4.1 were prepared by radio frequency (13.56 MHz) sputtering a Si (or Ge) target in an atmosphere of pure argon. The only exception applies to a-Ge:H in which a mixture of Ar+H_2 gas was employed [9, 19]. Depending on

Table 4.1 Overview of a-Si and a-Ge films investigated at Laboratório de Filmes Finos (Brazil) by using the cosputtering method and Raman scattering spectroscopy. Information includes the metal concentration, annealing conditions (temperature range and temperature steps—given in parentheses), film thickness, temperature of crystallization or crystallization onset (as determined from Raman scattering analyses), optical bandgap, and deposition temperature T_d.

	Metal conc. range (at.%)	Annealing conditions*	Thickness (μm)	Crystallization onset (°C)	Optical bandgap (eV)**	T_d (°C)	Ref.
Crystalline Si	0	–	–	~1400 (melting)	2.5 ± 0.1	–	–
Pure a-Si	0	200–900 (100/150)	~0.5–2	750 ± 50	1.5 (1.7) ± 0.1	70–200	–
Si+Ni	0.1–10.0	200–800 (100)	~2	550 ± 50	1.4 (1.7) ± 0.1	200	[13]
Si+Mn	0.2–20.0	300–900 (150)	~1.7	650±50	2.0 (1.7) ± 0.1	150	[14]
Si+Co	2.8–10.3	300–900 (150)	~1	750 ± 50	1.4 (1.1) ± 0.1	150	[15]
Si+Fe	0.1–2.1	300–900 (150)	~0.5	700 ± 50	1.0 (1.4) ± 0.1	150	[16]
Si+Gd	0.1–16.8	300–800 (100)	~1	625 ± 50	–	70–150	[17]
Si+Er	0.1–1.2	300–800 (100)	~1	625 ±50	–	70–150	[17]
Crystalline Ge	0	–	–	~950 (melting)	1.0 ± 0.1	–	–
Pure a-Ge	0	200–600 (100)	~1–1.7	550 ± 50	0.9 (1.0) ± 0.1	150	–
Pure a-Ge:H	0	100–500 (100)	~1	500 ± 50	1.4 (1.0) ± 0.1	220	[9]
Ge+Al	0.005–2.5	100–500 (100)	~1	220 ± 25	1.3 (1.0) ± 0.1	220	[9]
Ge+Mn	0.1–24.0	200–600 (100)	~1.7	350 ± 50	0.8 (1.0) ± 0.1	150	[18]
Ge+Co	1.7–7.6	200–600 (100)	~1	450 ± 50	1.0 (1.1) ± 0.1	150	[15]

* All annealing treatments were 15 minutes long, cumulative, and carried out under a continuous flow of argon.

** Values correspond to the E_{04} optical bandgap (energy at which the absorption coefficient is 10^4 cm^{-1}) of films: as deposited and after crystallization (in parentheses). In the case of metal-containing films, the E_{04} values refer to samples with [metal] ~1 at.%.

the deposition time, target, and plasma characteristics, the thickness of the films can be varied from tens to thousands of nanometers on various substrates that were kept at a fixed temperature ($T_d \approx$ 70°C–220°C). The insertion of metallic species into the a-films was achieved by partially covering the Si (or Ge) target with small pieces of the desired metal. Such procedure allows control of the metal content in the films simply by varying the relative metal-to-Si (or metal-to-Ge) target area and taking into account their respective sputtering yield values [20].

For comparison reasons, each series of samples included one pure a-Si (or a-Ge) film that was prepared following exactly the same deposition conditions and subjected to the same annealing procedure and characterization analyses.

After deposition, and after each annealing step, the samples were investigated by different characterization techniques. The atomic composition of the films was determined by means of energy-dispersive X-ray (EDX) and, in certain cases, examined with auxiliary methods (Rutherford backscattering spectrometry or X-ray photoelectron spectroscopy [14]). This is an important practice since it confirms that the metallic species were effectively and homogeneously inserted in the a-Si (or a-Ge) films and, most importantly, it gives the exact atomic composition of the samples.

Figures 4.2a and 4.2b show the EDX spectra of some a-Si(Mn) and a-Ge(Mn) films as prepared by cosputtering. Along with the characteristic X-ray emission lines (due to Si, Ge, or Mn), it is clear the scaling of the Mn-related signal as the manganese target area was increased. This can be better seen in the insets of Figs. 4.2a and 4.2b that contain the [Mn] as a function of the absolute manganese target area employed during deposition.

Considering that all samples have been prepared under an atmosphere of argon, small amounts (~2 at.%) of this element can be found embedded in the a-Si and a-Ge hosts. Not only for the Si+Mn and Ge+Mn samples (Fig. 4.2) but for all systems considered in Table 4.1 it is important to notice that the metallic species were homogeneously distributed—even at concentrations as high as approximately 20 at.%. Likewise, no appreciable changes were observed in the metal concentration of the samples as-deposited and after thermal annealing.

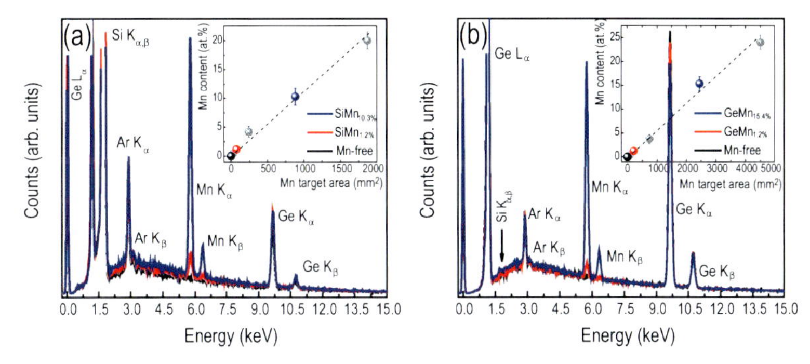

Figure 4.2 EDX spectra of (a) a-Si(Mn) and (b) a-Ge(Mn) films as-deposited. The main X-ray emission lines are indicated in the figures. To avoid inaccuracies in the determination of the atomic composition of the films, the spectra correspond to films deposited on c-Ge (a-Si films) and c-Si (a-Ge films) substrates. The insets show the [Mn], as determined from EDX, as a function of the manganese target area employed during deposition. The dashed straight lines in the insets are just guides to the eye. *Abbreviations*: c-Ge, crystalline Ge; c-Si, crystalline Si.

Postdeposition treatment of the Si(Mn) and Ge(Mn) films was performed under a continuous flow of argon in a temperature-controlled furnace. Each thermal anneal, in the 200°C–900°C temperature range, was 15 minutes long and performed in steps of 150°C (Si films) or 100°C (Ge films).

The treatments were cumulative and made on films deposited on different substrates to allow the investigation of their optical-structural properties. Films deposited on quartz, for example, were considered for optical transmission measurements in the ~500–2000 nm wavelength region (visible–near-infrared [VIS-NIR] range) rendering the thickness, index of refraction, and optical bandgap of the films [21]. In addition to this data, the optical measurements also indicate changes in the transmission profiles according to the metal concentration and as the thermal treatments advance [16]. In the former case, it is usual to observe a decrease in the optical transmission of the films because of extrinsic and/or free-carrier absorption processes associated with the presence of metallic species [22]. Thermal annealing of the films at increasing temperatures, on the contrary, induces some improvement of the optical transmission spectra, which can be attributed to the suppression of band-tail (or defect) states due to small changes in the atomic structure of the amorphous films.

As already said, the structural characterization of our samples was almost exclusively based on Raman scattering spectroscopy. The measurements were carried out in a commercial Raman microprobe system by means of a HeNe (632.8 nm) laser. At this specific wavelength the light penetration depth in a-Si and a-Ge films stays around 1–2 μm so that the laser energy is absorbed almost uniformly over the thickness of the samples. Since the Raman signal is influenced by temperature effects, one should be very careful in using small laser power densities in order to avoid unintentional sample heating or local crystallization [23].

All Raman spectra were taken, at room temperature, under the backscattering geometry and with a laser power of 0.2 mW (average spot size $\approx 1\ \mu m^2$).

Thorough analysis of the Raman spectra provides important morphological details (mainly related to the phonon frequencies and corresponding spectral linewidths) as well as a good estimate of the crystalline fraction of the samples [24, 25]. This is accomplished by fitting the Raman spectra of the films with *Gaussian* functions, each one associated with different contributions that can be due to either amorphous or crystalline nature [26]. In the case of Si films, these contributions take place, usually at ~470 ± 10 cm^{-1} (amorphous), at ~480–515 cm^{-1} (crystallites with different sizes and/or interface modes), and at ~520 ± 5 cm^{-1} (crystalline). For Ge films they take place at ~270 ± 5 cm^{-1} (amorphous), ~275–295 cm^{-1} (crystallites with different sizes and/or interface modes), and at ~300 ± 3 cm^{-1} (crystalline). On the basis of the relative amount of each of these contributions, the fraction of crystalline structures (X_F) present in Si or Ge films can be determined according to the following expression:

$$X_F = \frac{A_{xtal}}{A_{xtal} + \sigma_R A_{amorp}} \tag{4.1}$$

where A_{xtal} and A_{amorp} stand for the integrated areas due to the crystalline and amorphous phases, respectively. Differences due to the light-scattering cross sections of the amorphous (σ_{amorp}) and crystalline (σ_{xtal}) contributions are taken into account through the factor σ_R (= $\sigma_{xtal}/\sigma_{amorp}$), which is assumed to be equal to 0.8 for both Si and Ge films [25, 27].

At present, Raman scattering spectroscopy is extensively used in the characterization of various classes of materials. In fact, it represents one of the most convenient techniques to probe variations in the atomic structure of amorphous thin films as a result of compositional changes and/or due to postdeposition treatments [28].

Some of these aspects are illustrated in Fig. 4.3, which displays the Raman spectra of a-Si and a-Ge films, as-deposited, containing increasing concentrations of manganese. All features present in the spectra of Fig. 4.3 refer to the vibrational (transverse acoustical, TA, and transverse optical, TO) modes of Si-Si and Ge-Ge bonds. Actually, it is interesting to observe the effect of increasing amounts of manganese in the Raman spectra of the Si films, whereas no appreciable changes are noticeable in the spectra of the Ge films. This distinctive behavior has great influence on the MIC of the a-Si films, as will be shown in the following section.

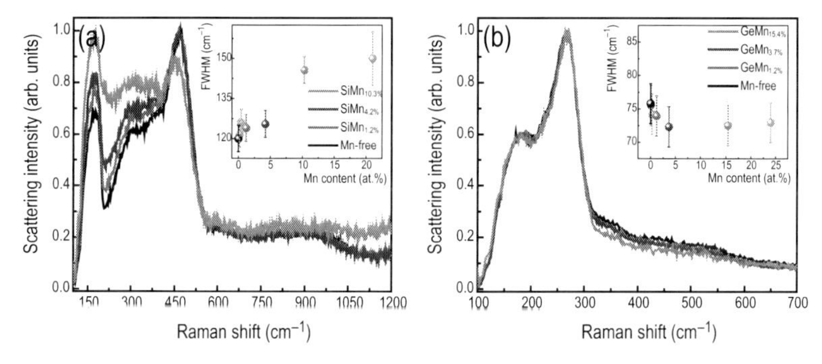

Figure 4.3 Raman scattering spectra of (a) a-Si and (b) a-Ge films containing different concentrations of manganese, as indicated in the figures. The spectra were normalized for comparison purposes and correspond to films as-deposited on crystalline quartz substrates. The main features present in the spectra of the Si films are related to the TA-like mode of Si-Si bonds at ~170 cm^{-1} and TO-like mode of Si-Si bonds at ~470 cm^{-1}. For the Ge films the most prominent vibrational mode occurs at ~270 cm^{-1} and refers to the TO-like mode of Ge-Ge bonds. The insets show the TO-like Raman linewidths (corresponding to the disorder present in the films) due to Si-Si and Ge-Ge bonds as a function of the manganese concentration.

4.3 Crystallization of Amorphous Si and Ge Films

Roughly, the Raman signal occurs when light is inelastically scattered by certain specific vibrational modes of atomic-molecular crystalline structures [29]. When disorder (or finite size effects) takes place, phonon selection rules break down and the Raman lines broaden and shift to smaller energies. The effect is so well established in the a-semiconductor literature that frequency shifts of the TO-like Raman mode (at ~470 cm^{-1} for Si and at ~270 cm^{-1} for Ge films) have direct association with changes in the dihedral bond angle between neighboring atoms [1, 3]. Correspondingly, the full width at half-maximum (FWHM) height, or linewidth, of the TO-like signal is a good qualitative indication of the degree of structural (dis)order present in a-semiconductor films. A similar reasoning applies to the ratio between the intensities of the TA-like and TO-like vibrational modes (I_{TA}/I_{TO}), which also has a close correspondence with the structural (dis)order present in a-semiconductors [30]. Considering that the phonon frequency is also susceptible to stress effects [31, 32], the structural (dis)order of a-semiconductors is more effectively described by variations in the TO-like Raman linewidth. In such a case, the higher the TO-like Raman linewidth (FHWM), the more disordered the atomic structure of the a-semiconductor.

Similarly to what happens in crystalline semiconductors, the presence of impurities in the matrix of both a-Si and a-Ge alters their atomic-electronic structure in a way that depends on the characteristics and relative amount of those foreign species. This is what can be seen in the inset of Fig. 4.3a in which the insertion of manganese is responsible for increased structural disorder, or broadening of the TO-like Raman mode of the a-Si(Mn) films. Interestingly, such a variation is not verified in the a-Ge(Mn) films (inset of Fig. 4.3b), even for manganese concentrations as high as ~25 at.%. Until this concentration range, and under the adopted experimental conditions, it seems that the atomic structure of the Ge films is not greatly influenced by the presence of manganese. Whereas the insertion of manganese into the a-Si host increases its disorder, the thermal annealing of both Si(Mn) and Ge(Mn) films causes structural changes that can lead, at high temperatures, to their crystallization. This can be seen in Fig. 4.4, which displays the Raman spectra of Si(Mn) films after thermal annealing at increasing temperatures. For comparison reasons, Fig. 4.4 exhibits the Raman spectra of a pure Si film (Mn-free) and those of a Si film containing 4.2 at.% of Mn. Part of the information contained in Fig. 4.4 is common to the Mn-free and $SiMn_{4.2\%}$ samples and can be summarized as follows: (1) They only crystallize after thermal annealing at ~750°C, as evidenced by the pronounced Raman signal at ~520 cm^{-1} (and its overtone at ~975 cm^{-1}); (2) close to the crystallization temperature, the samples present their narrowest Raman linewidths (see insets of Fig. 4.4), suggesting a smaller degree of structural disorder; and (3) after crystallization, the structural disorder tends to increase because of interface effects associated with the film–substrate, to the interaction involving the amorphous–crystalline phase, or to the advent of new crystal forms ($MnSi_{1.7}$ silicide phase in sample $SiMn_{4.2\%}$).

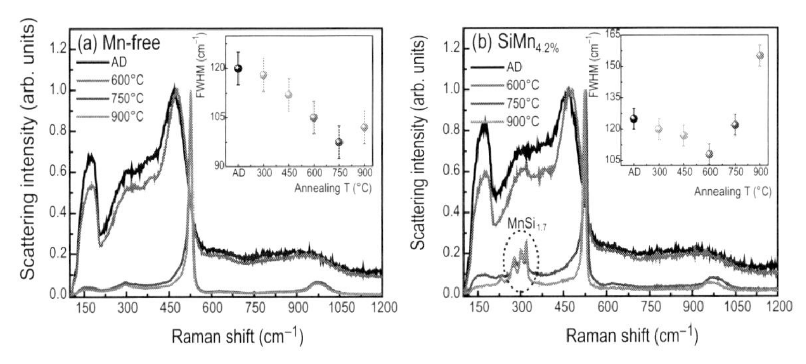

Figure 4.4 Raman scattering spectra of Si thin films, (a) Mn-free and (b) containing 4.2 at.% of Mn as-deposited (AD) and after thermal annealing at increasing temperatures. The main spectral features correspond to the TA-like (at ~170 cm^{-1}) and TO-like (at ~470 cm^{-1}) Raman modes due to Si-Si bonds. After crystallization, the TO-like Raman mode shifts to ~520 cm^{-1} and the presence of its overtone (at ~975 cm^{-1}) is evident. The insets show how the thermal annealing treatments affect the TO-like Raman linewidths (or disorder present in the films). The spectra were normalized for comparison purposes. Notice the development of the $MnSi_{1.7}$ phase in the $SiMn_{4.2\%}$ film after treatment at 750°C and 900°C.

In addition to the slightly different Raman linewidths presented by the Mn-free and $SiMn_{4.2\%}$ samples (particularly due to the presence of manganese species) the development of the $MnSi_{1.7}$ silicide phase in the $SiMn_{4.2\%}$ sample is noteworthy, which is at the origin of increased structural disorder and, consequently, a crystallization temperature of ~650 ± 50°C. Moreover, the presence of the $MnSi_{1.7}$ phase in the SiMn system is believed to limit the crystalline fraction of the SiMn samples, as will be shown in the following section.

Notwithstanding the relatively low efficiency of manganese in the MIC of a-Si, the SiMn system presents several interesting features such as [14, 33]: the structural-electronic properties of SiMn films are determined by the [Mn] and annealing temperature; once crystallized, SiMn films exhibit the $MnSi_{1.7}$ silicide phase, whose proportion scales with the [Mn]; the crystallization is accompanied by the development of submicrometer structures (essentially Mn-containing Si crystallites) on the surface of the samples; and, because of their shape and typical dimensions, the observed structures give rise to magnetic vortices.

Compared to SiMn samples, the analysis of Ge(Mn) films is not so exciting, but the MIC phenomenon is more remarkable. Figure 4.5 shows the Raman spectra of Ge films (Mn-free and containing 3.7 at.% of Mn) after annealing at increasing temperatures. As can be seen from the figure, as the thermal annealing advances, the Raman spectra indicate that some structural improvement is taking place in the Ge samples until they crystallize at ~500°C ($GeMn_{3.7\%}$) or at ~600°C (Mn-free). It is also important to mention that annealing treatments at temperatures higher than 600°C causes the destruction of the films and that just the amorphous and crystalline phases of Ge were observed in the Raman spectra of the present Ge(Mn) samples. The presence of the Mn_5Ge_3 germanide phase was detected only at the highest Mn concentration (~24 at.%) and by means of XRD measurements [18]. Furthermore, the present GeMn films exhibit magnetic properties that scale with the Mn concentration and tend to be more pronounced after crystallization [34]. Indeed, the present GeMn films were able to present either spin-glass behavior (when amorphous and with high [Mn]) or room-temperature ferromagnetism (at moderate [Mn] and after crystallization).

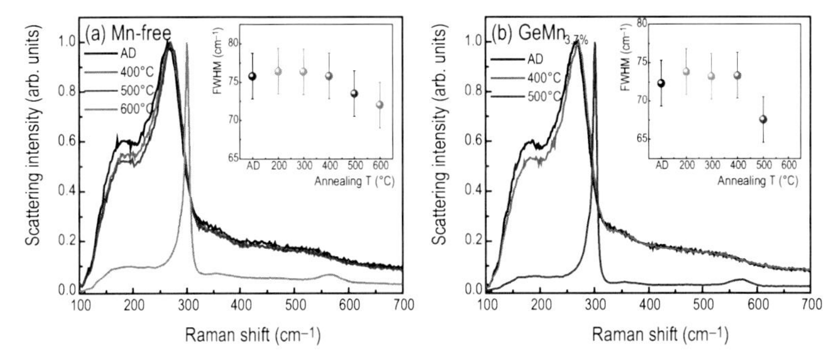

Figure 4.5 Raman scattering spectra of Ge thin films, (a) Mn-free and (b) containing 3.7 at.% of Mn as-deposited and after thermal annealing at different temperatures. The main spectral features are due to Ge-Ge bonds associated with the following TO-like Raman modes: amorphous (at ~270 cm^{-1}), crystalline (at ~300 cm^{-1}), and its overtone at ~570 cm^{-1}. The insets illustrate the corresponding TO-like Raman linewidths. The spectra were normalized for comparison purposes.

4.4 Structural Disorder and Crystallization

Hitherto, it is obvious the effect that manganese has on the atomic structure of the a-Si and a-Ge films, as well as on the temperatures at which they start to exhibit crystallization. The MIC effect is almost absent in the case of the SiMn samples but is concurrent with the increase of their structural disorder and with the development of $MnSi_{1.7}$ silicide crystals. In fact, the presence of manganese was able to reduce the temperature of crystallization by no more than 100°C: from 750 ± 50°C in pure a-Si to 650 ± 50°C when combined with Mn. Moreover, whereas the controllable insertion of manganese drastically increase the Raman linewidth (or disorder) of the SiMn films, no clear dependence has been observed in their respective temperatures of crystallization.

The structural changes experienced by the Si(Mn) films, as determined from their Raman linewidth, are shown in Fig. 4.6a—as a function of the annealing temperature and for a few manganese concentrations. According to the figure, thermal annealing at increasing temperatures causes some structural improvement (decrease of the FWHM value) until the crystallization takes place in the 600°C-750°C temperature range. This is consistent with the fact that increasing manganese concentrations will correspond to broader Raman linewidths and stronger Raman signals related to the presence of $MnSi_{1.7}$ crystals (Fig. 4.4b).

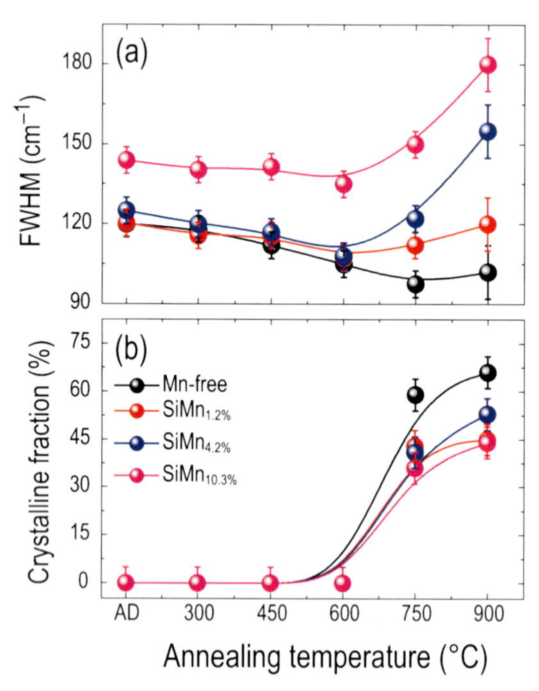

Figure 4.6 (a) FWHM height of the TO-like Raman signal, or simply Raman linewidth, as a function of annealing temperature. (b) Fraction of Si crystals, as determined from Raman analyses, as a function of the annealing temperature. As indicated in the figure, the data corresponds to a pure Si film (Mn-free) and containing manganese concentrations of 1.2, 4.2, and 10.3 at.%. Error bars take into account experimental resolution as well as uncertainties in data processing. The lines joining the experimental data points are just guides to the eye.

It is also worth to notice that increasing manganese concentrations seem to limit the fraction of c-Si in the samples—as determined from the Raman data. This can be seen in Fig. 4.6b, which shows the crystalline fraction X_F of the same samples considered in Fig. 4.6a, as a function of the annealing temperature. Such a behavior can be attributed to some experimental details (annealing conditions, nature of the substrate, etc. [10, 13, 32]) but, especially, to the fact that silicon crystals are present in the samples under two different forms, pure Si and pure $MnSi_{1.7}$ crystals.

On the basis of the present experimental findings, it is suggested that (allied to the distinctive chemical characteristics of the Si and Mn atoms) the structural disorder present in the Si network greatly inhibits the manifestation of the MIC effect in the SiMn system. It is evident that structural disorder is inherent to any amorphous network and its decrease requires energetic processing. However, the presence of manganese affects the whole process by introducing additional disorder before and after crystallization (with the growth of the $MnSi_{1.7}$ crystals).

The effect of manganese on the crystallization of the present GeMn films is more impressive, and the observed T_c values were either 350°C or 450°C—depending on the manganese concentration. These values should be compared with T_c = 550 ± 50°C, typically found in pure a-Ge films deposited and analyzed by the same methodology. Considering that Si and Ge share certain chemical characteristics and the fact that the present SiMn and GeMn samples possess a lot in common (prepared by cosputtering, similar Mn concentrations, identical annealing procedure, etc.), it is important to verify the low efficiency of manganese on the crystallization of the a-Si samples. Again, one possible explanation for such an aspect can be the different structural disorder experienced by the a-Si and a-Ge matrices under the influence of manganese. Analyses of the Raman spectra related to the GeMn samples indicate that within the experimental uncertainties, the a-Ge matrix is almost insensitive to the presence of manganese (Fig. 4.3b). In fact, it seems that the presence of manganese slightly decreases the structural disorder of the Ge matrix, as can be seen in the data of Fig. 4.7a. This is in contrast with the results obtained for the SiMn samples in which the presence of manganese always increased their structural disorder. The fraction of Ge crystals present in the Ge(Mn) samples is shown in Fig. 4.7b. The data was obtained from Raman analyses and corresponds to Ge films with different manganese concentrations and after thermal annealing at increasing temperatures.

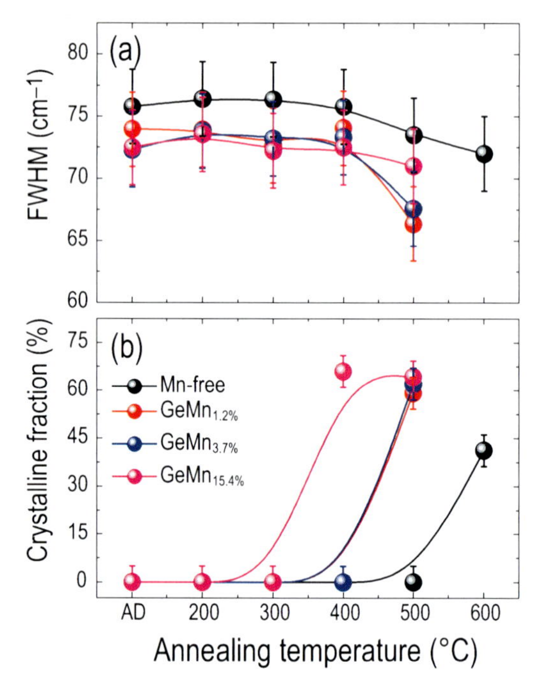

Figure 4.7 (a) FWHM height of the TO-like Raman signal and (b) a fraction of Ge crystals as a function of annealing temperature. The data corresponds to pure Ge (Mn-free) and Ge films containing manganese concentrations equal to 1.2, 3.7, and 15.4 at.%. Error bars take into account experimental resolution as well as uncertainties in data processing. The lines joining the experimental data points are just guides to the eye.

It is clear from Fig. 4.7b that depending on [Mn], crystallization takes place at different temperatures: at ~550°C for pure a-Ge (Mn-free), ~450°C for samples with [Mn] = 1.2 and 3.7 at.%, and ~350°C for the $GeMn_{15.4\%}$ sample. Moreover, the concentration of manganese seems to determine not only the crystallization temperature but also the total amount of Ge crystals, in perfect agreement with the concept of the MIC.

At this point the need for intermediate (or new) phases like silicides or germanides in the MIC of a-semiconductors is not unambiguous. In other words, while their role is well established in certain systems (see later) the present ensemble of SiMn and GeMn samples—prepared and analyzed following exactly the same experimental procedures—suggests that distinct systems behave differently. This concept is further supported by the results obtained in other metal-containing a-semiconductors presented in Table 4.1 and summarized as follows:

- *The SiNi system*: The MIC of the SiNi system takes place, typically at ~550 ± 50°C, and is influenced by factors like [10, 13] the Ni concentration, annealing temperature, nature of the substrate (c-quartz, c-Si, c-Ge, or glass), and sample thickness. The experimental results are consistent with the fact that $NiSi_2$ mediates the development of Si crystallites either by acting as crystallization seeds or by reducing the energy barrier that leads to the Si crystallization. Also, the Raman data indicates that only for the Ni-containing samples, the average size of Si nanocrystals is larger than in the Ni-free films and that thermal treatments at high temperatures cause a stress relieve due to the advent of nanostructures on the surface of the samples [32].
- *The SiCo and GeCo systems*: The presence of Co seems to have little (or no) effect on the crystallization behavior of the a-Si and a-Ge films prepared by cosputtering [15]. More specifically, whereas the GeCo films crystallize at ~450 ± 50°C (corresponding to a decrease of ~100°C when compared to pure Ge), the crystallization temperature of both SiCo and Si films is indistinguishable. Moreover, the crystallization process is insensitive to the Co concentration (in the ~0-10 at.% range), and the silicide or germanide phases were observed only by XRD measurements in films with the highest Co content.
- *The SiFe system*: The results provided by SiFe are very similar to those obtained from the SiMn system, that is, (a) they crystallize at slightly lower temperatures (~650°C–700°C in contrast with the ~750 ± 50°C of pure a-Si), independent of the Fe concentration, and (b) the development of the β-$FeSi_2$ phase is evident as soon as the crystallization process takes place, for Fe concentrations as low as ~0.8 at.% (see Fig. 4.8).
- *The SiGd and SiEr systems*: According to the data in Table 4.1, the presence of either Gd or Er metal can reduce the crystallization temperature of a-Si films by as much as ~125°C. A thorough inspection of the Raman data indicates

that the process is accompanied by some reduction of the structural disorder as the crystallization starts—in a way similar to that exhibited by the GeMn system (Fig. 4.7)—and that the crystallization temperature depends on the metal concentration. More precisely, the MIC phenomenon is effective only in samples containing Gd or Er concentrations higher than approximately 5 at.% or 1 at.%, respectively. Curiously, no silicide phases have been observed, even for films containing relatively large amounts of Gd [17].

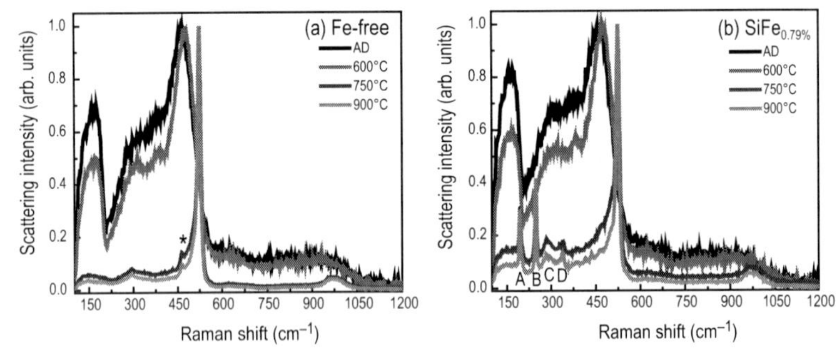

Figure 4.8 Raman scattering spectra illustrating the effect of thermal annealing treatments on a-Si films: (a) pure (Fe-free) and (b) containing ~0.8 at.% of Fe. The spectra were normalized for comparison purposes and correspond to films deposited on crystalline quartz substrates. The main features present in the spectra are related either to Si-Si bonds (at ~170, 300, 470, 520, and 975 cm^{-1}) or to the β-$FeSi_2$ phase (peaks A, B, C, and D at ~195, 250, 285, and 340 cm^{-1}) [16]. The signal indicated by a star comes from the substrate.

- *The GeAl system*: The Al-induced crystallization of a-Ge films prepared by sputtering can occur at temperatures of the order of 200°C. The process is influenced by the presence and relative concentration of both Al and hydrogen species, as well as by the nature of the substrates that support the films. On the basis of a number of experimental investigations [9, 19, 35], fourfold-coordinated Al (Al_4) atoms sitting at the center of tetrahedral Ge sites act as crystallization seeds, rendering a sort of either explosive or gradual crystallization, depending on the Al content. Within this context, mobile hydrogen atoms play a fundamental role in enabling the relaxation of the amorphous tissue surrounding the Al_4 crystal seeds. If the Al and/or H concentrations are appropriate the MIC of the GeAl system takes place at ~200°C, otherwise Ge crystals will appear only after annealing at ~500°C–550°C. Regardless of the situation, there is no evidence of secondary phases assisting or influencing the crystallization process.

4.5 Final Remarks

This chapter presents a summary of the main aspects related to the crystallization of a-Si and a-Ge thin films prepared by cosputtering. The deposition method is standard in the production of many different classes of materials (superconductors, metals, semiconductors, dielectrics, etc.) and perfectly matches the purpose of investigating the mechanisms behind the MIC of amorphous semiconductors. More specifically, the cosputtering method is consistent with the preparation of thin films of amorphous structure, and most importantly, it allows absolute control over their final atomic composition. In addition to the use of cosputtering for the preparation of films, the whole experimental procedure involving the MIC of a-semiconductors included the cumulative thermal annealing of the films at increasing temperatures and their posterior investigation by means of Raman scattering spectroscopy. Amongst the potential advantages of this experimental procedure one can mention: (a) the absence of a (macroscopic) a-semiconductor–metal interface, (b) the possibility of continuously changing the amount of metal species in the a-semiconductor matrix, (c) the access to particular physical-chemical information (like composition and structure) dynamically, that is, along the realization of thermal annealing treatments; and (d) the detection of new phases (crystalline, silicides, germanides, oxides, etc.) with great sensitivity and spatial resolution. Part of these aspects was discussed in detail by considering SiMn and GeMn systems.

Even though both SiMn and GeMn systems have presented the MIC phenomenon, the effect was more pronounced in the latter. According to our investigation, the SiMn films crystallize at 650 ± 50°C, regardless of the manganese concentration, whereas pure a-Si films become crystalline typically at 750 ± 50°C. These figures are in contrast with those obtained in GeMn films, which exhibited crystalline structures at 350 ± 50°C (considerably less that the 550 ± 50°C characteristic of pure a-Ge) and a clear dependence on the manganese concentration. The distinct effect that manganese has on the crystallization of the present a-Si and a-Ge films is somehow intriguing, especially if one considers that silicon and germanium atoms have a lot in common (belong to column IV of the periodic table) and form chemical bonds after hybridization of the *s* and *p* electronic orbitals, etc.) and that the films were prepared and analyzed following exactly the same procedures. Part of the differences verified in the MIC of the SiMn and GeMn systems, however, has to do with the structural disorder caused by the presence of Mn in the a-Si matrix as well as with the development of the $MnSi_{1.7}$ phase as the SiMn films are annealed at increasing temperatures. Put in other words, most of the energy provided during the thermal annealing of the SiMn films is expended trying to recover the increased atomic disorder caused by the presence of Mn in the a-Si host and in the growth of the $MnSi_{1.7}$ crystals. This is not the case of the GeMn system in which the insertion of manganese does not alter significantly the structural disorder of the films and where the occurrence of the Mn_5Ge_3 germanide is barely detected by Raman spectroscopy.

Along with the experimental results obtained in other metal-containing sputtered Si and Ge films, the present chapter emphasizes the role played by the chemical nature of the atomic species involved in the MIC of amorphous semiconductors. Because of its nonequilibrium nature, the crystallization mechanism of amorphous semiconductor thin films is not unique and depends very much on several experimental details (multilayered or doped systems, the presence of foreign atomic species, annealing conditions, characterization methods, etc.). Each system behaves in a different way, and a complete understanding of the MIC phenomenon will require further systematic experimental-theoretical work.

Acknowledgments

The present contribution was possible thanks to the involvement of a number of friends who enthusiastically worked in the search for answers to the MIC of a-semiconductors. To them, our special gratitude: Prof. I. Chambouleyron, Prof. F. Fajardo, Dr. C. T. M. Ribeiro, Dr. L. R. Muniz, and Dr. I. B. Gallo. This work was financially supported by the Brazilian agencies FAPESP and CNPq.

References

1. Street, R. A. (1991) *Hydrogenated Amorphous Silicon*, 1st ed. (Cambridge University Press, Cambridge, UK).
2. Street, R. A. (ed.) (2000) *Technology and Applications of Amorphous Silicon*, 1st ed. (Springer, Berlin).
3. Tanaka, K., Muruyama, E., Shimada, T., Okamoto, H. (1999) *Amorphous Silicon*, 1st ed. (John Wiley & Sons, Chichester, UK).
4. Blum, N. A., Feldman, C. (1972) The crystallization of amorphous silicon films. *Journal of Non-Crystalline Solids*, **11**, 242–246; Blum, N. A., Feldman, C. (1976) The crystallization of amorphous germanium films. *Journal of Non-Crystalline Solids*, **22**, 29–35.
5. Alferov, Z. I., Abakumov, V. N., Kovalvhuk, Y. V., Ostrovskaya, G. V., Portnoi, E. L., Smirnitskii, V. B. (1983) Interference laser annealing of semiconductors. *Soviet Physics Semiconductors-USSR*, **17**, 152–155.
6. Takamori, T., Messier, R., Roy, R. (1972) New noncrystalline germanium which crystallizes "explosively" at room temperature. *Applied Physics Letters*, **20**, 201–203.
7. Sameshima, T., Ozaki, K. (2000) Current-induced joule heating used to crystallize silicon thin films. *Japanese Journal of Applied Physics*, **39**, L651–L654.
8. Oki, F., Ogawa, Y., Fujiki, Y. (1969) Effect of deposited metals on the crystallization temperature of amorphous germanium films. *Japanese Journal of Applied Physics*, **8**, 1065.
9. Zanatta, A. R., Chambouleyron, I. (2005) Low-temperature Al-induced crystallization of amorphous Ge. *Journal of Applied Physics*, **97**, 094914-1–094914-11.
10. Ferri, F. A., Zanatta, A. R. (2008) Influence of film thickness on the crystallization of Ni-doped amorphous silicon samples. *Journal of Applied Physics*, 104, 013534-1–013534-5.
11. Knaepen, W., Detavernier, C., van Meirhaeghe, R. L., Sweet, J. J., Lavoie, C. (2008) In-situ x-ray diffraction study of metal induced crystallization of amorphous silicon. *Thin Solid Films*, **516**, 4946–4952; Knaepen, W., Gaudet, S., Detavernier, C., van Meirhaeghe, R. L., Sweet, J. J., Lavoie, C. (2009) In-situ x-ray diffraction study of metal induced crystallization

of amorphous germanium. *Journal of Applied Physics*, **105**, 083532-1–083532-7.

12. Gaudet, S., Detavemier, C., Kellock, A. J., Desjardins, P., Lavoie, C. (2006) Thin film reaction of transition metals with germanium. *Journal of Vacuum Science & Technology A: Vacuum, Surfaces, and Films*, **24**, 474–485.
13. Ferri, F. A., Zanatta, A. R., Chambouleyron, I. (2006) Metal-induced nanocrystalline structures in Ni-containing amorphous silicon thin films. *Journal of Applied Physics*, **100**, 094311-1–094311-7.
14. Ferri, F. A., Pereira-da-Silva, M. A., Zanatta, A. R. (2011) Development of the $MnSi_{1.7}$ phase in Mn-containing Si films. *Materials Chemistry and Physics*, **129**, 148–153.
15. Ferri, F. A., Pereira-da-Silva, M. A. (2012) The composition, structure and optical properties of weakly magnetic Co-containing amorphous Si and Ge films. *Materials Chemistry and Physics*, **134**, 153–157.
16. Gallo, I. B., Zanatta, A. R. (2011) Structural-electronic aspects related to the near-infrared light emission of Fe-doped silicon films. *Solid State Communications*, **151**, 587–590.
17. Zanatta, A. R. (unpublished data).
18. Ferri, F. A., Zanatta, A. R. (2009) Structural, optical and morphological characterization of amorphous $Ge_{100-x}Mn_x$ films deposited by sputtering. *Journal of Physics D: Applied Physics*, **42**, 035055-1–035055-6.
19. Chambouleyron, I., Fajardo, F., Zanatta, A. R. (2001) Aluminum-induced crystallization of hydrogenated amorphous germanium thin films. *Applied Physics Letters*, **79**, 3233–3235.
20. Vossen, J. L., Kern, W. (eds.) (1978) *Thin Film Processes*, 1st ed. (Academic Press, New York, USA).
21. Swanepoel, R. (1983) Determination of the thickness and optical constants of amorphous silicon. *Journal of Physics E: Scientific Instruments*, **16**, 1214–1222.
22. Pankove, J. I. (1971) *Optical Process in Semiconductors*, 1st ed. (Dover, New York, USA).
23. Zanatta, A. R., Ribeiro, C. T. M. (2004) Laser-induced generation of micrometer-sized luminescent patterns on rare-earth-doped amorphous films. *Journal of Applied Physics*, **96**, 5977–5981.
24. Lannin, J. S. (1988) Local structural order in amorphous semiconductors. *Physics Today* (July), 28–35.

25. Tsu, R., Gonzalez-Hernandez, J., Chao, S. S., Lee, S. C., Tanaka, K. (1982) Critical volume fraction of crystallinity for conductivity percolation in phosphorus-doped Si:F:H alloys. *Applied Physics Letters*, **40**, 534–535.

26. Smit, C., van Swaaij, R. A. C. M. M., Donker, H., Petit, A. M. H. N., Kessels, W. M. M., van de Sanden, M. C. M. (2003) Determining the material structure of microcrystalline silicon from Raman spectra. *Journal of Applied Physics*, **94**, 3582–3588.

27. Fujii, M., Hayashi, S., Yamamoto, K. (1991) Growth of Ge microcrystals in SiO_2 thin film matrices: a Raman and electron microscopy study. *Japanese Journal of Applied Physics (Part 1)*, **30**, 687–694.

28. Gwenael, G., Colomban, P. (2007) Raman spectroscopy of nanomaterials: how spectra relate to disorder, particle size and mechanical properties. *Progress in Crystal Growth and Characterization of Materials*, **53**, 1–56.

29. See, for example, Lewis, I. R., Edwards, H. G. M. (2001) *Handbook of Raman Spectroscopy*, 1st ed. (Marcel Dekker, New York, USA).

30. Pilione, L. J., Maley, N., Lustig, N., Lannin, J. S. (1983) Modifications of intrinsic disorder in amorphous silicon. *Journal of Vacuum Science & Technology A: Vacuum, Surfaces, and Films*, **1**, 388–391.

31. de Wolf, I., Maes, H. E., Jones, S. K. (1996) Stress measurements in silicon devices through Raman spectroscopy: bridging the gap between theory and experiment. *Journal of Applied Physics*, **79**, 7148–7156.

32. Zanatta, A. R., Ferri, F. A. (2007) Crystallization, stress, and stress-relieve due to nickel in amorphous silicon thin films. *Journal of Applied Physics*, **102**, 043509-1–043509-5.

33. Ferri, F. A., Pereira-da-Silva, M. A., Zanatta, A. R. (2009) Evidence of magnetic vortices formation in Mn-based sub-micrometer structures embedded in Si-Mn films. *Journal of Physics D: Applied Physics*, **42**, 132005-1–132005-5.

34. Ferri, F. A., Pereira-da-Silva, M. A., Zanatta, A. R., Varella, A. L. S., de Oliveira, A. J. A. (2010) Effect of Mn concentration and atomic structure on the magnetic properties of Ge thin films. *Journal of Applied Physics*, **108**, 113922-1–113922-5.

35. Muniz, L. R., Ribeiro, C. T. M., Zanatta, A. R., Chambouleyron, I. (2007) Aluminium-induced nanocrystalline Ge formation at low temperatures, *Journal of Physics: Condensed Matter*, **19**, 076206-1–076206-16.

Chapter 5

Aluminum-Induced Crystallization: Applications in Photovoltaic Technologies

Abdelilah Slaoui,[a] Prathap Pathi,[b] and Özge Tuzun[c]
[a]*ICUBE, UMR 7357 UdS-CNRS, 23 rue du Loess, 67037, Strasbourg Cedex 2, France*
[b]*National Physical Laboratory, K.S. Krishnan Road, New Delhi 110012, India*
[c]*Department of Physics, Düzce University, Düzce, Turkey*
abdelilah.slaoui@unistra.fr, pathiprathap@gmail.com, tuzun.ozge@yahoo.com

Metal-induced crystallization (MIC), particularly aluminum-induced crystallization (AIC), has attracted great attention as a technique to prepare high-quality, large-grain silicon films from an amorphous silicon (a-Si) film precursor at a lower annealing temperature below the eutectic temperature of the Al/a-Si system (<577°C) on low-cost foreign substrates like glass or metallic foils. In this chapter, a brief explanation on the fundamental physics behind MIC, typically AIC of a-Si as precursor films, has been presented. Recent findings concerning a-Si films crystallized by the AIC process and their ultimate utilization for solar cell fabrication as a seed layer are also described.

Metal-Induced Crystallization: Fundamentals and Applications
Edited by Zumin Wang, Lars P. H. Jeurgens, and Eric J. Mittemeijer

ISBN 978-981-4463-40-9 (Hardcover), 978-981-4463-41-6 (eBook)
www.panstanford.com

5.1 Introduction

To make photovoltaic energy competitive with other electricity sources, the cost per watt produced has to be lowered to €0.5 per watt peak [1]. Consequently, reducing production costs through materials' consumption has become an important challenge while maintaining efficiency. In this regard, fabrication of very thin, crystalline Si (c-Si) solar cells on foreign substrates is being intensively investigated worldwide to produce efficient photovoltaic devices at lower cost [2]. The high conversion efficiencies that have now come close to 25% [3, 4] on thick silicon stimulate to put extensive efforts on c-Si films to approach this value, provided appropriate optical confinement and surface passivation are applied. Indeed, although c-Si is an indirect bandgap semiconductor and therefore a relatively poor light absorber, efficient solar cells with reasonable currents can be achieved even in thin layers only a few microns thick, provided efficient light confinement is applied to obtain high absorbance of the incident light [5, 6]. The light management can be realized by texturing the front side of the active Si layer to obliquely couple the light into the layer, combined with a back reflector. The optical confinement is particularly important for wavelengths larger than 800 nm, which are weakly absorbed in silicon.

On the other hand, high cell efficiencies are only achievable using high-quality Si material and complex processing. Therefore, it would be highly desirable to be able to form large-area, single-crystal films on a supporting structure and in an economical manner. This is why the combination of c-Si thin films (<10 μm) with foreign substrate materials such as ceramics, glass, or even plastics has invoked a lot of research activity this last decade.

As discussed thin-film silicon devices can be realized by depositing the active material on a supporting material of a nonsemiconductor, non-single-crystal materials, which in turn, however, results in polycrystalline silicon (poly-Si) material rather than in a single-crystalline film. Yet, this thin-film poly-Si with high crystallinity is an attractive material for high-efficiency solar cells, in terms of less consumption of material than bulk c-Si and longer carrier diffusion length than that of amorphous silicon (a-Si)-containing thin films due to higher carrier mobility, in addition to no light-induced degradation [7, 8] effect.

The polysilicon film is defined as having a grain size between 1 μm and 1 mm. Such a film is composed of many silicon crystalline grains of varied crystallographic orientation. The main drawback of these polysilicon films for solar cell application is the presence of a high density of grain boundaries, which play the role of recombination centers for the generated carriers. As a consequence the cell performances are strongly affected. Figure 5.1a gives the relationship between grain size and open circuit voltage (V_{oc}) as summarized by Werner [9]. V_{oc} is considered to be a parameter reflecting the cell characteristics and crystalline properties. Dependence of the open circuit voltage as a function of grain size indicated a voltage drop proportional to $\ln(g^*/g)$ with the critical grain size $g^* = 400$ μm [2]. This grain size denotes the transition to a constant voltage for $g > g^*$. For pn-junction solar cells, a decrease of the open circuit voltage from 600 mV to 400 mV was noted for a decrease of grain size from 1 mm to 1 μm [10]. Accordingly, superior characteristics are obtained with a grain size of 100 μm, but the characteristics are worse with a grain size of a few tens of micrometers. In other words, the grain size of an active layer has strong influence on cell efficiency, as shown in Fig. 5.1b [11].

Figure 5.1b shows that the cells fabricated from poly-Si of a grain size $g >> 1$ mm showed conversion efficiencies exceeding 15%, while poly-Si-based thin-film cells significantly drops for $g << 1$ μm as a result of recombination of minority carriers at grain boundaries, which in turn causes the drop in the open circuit voltage with grain size. Therefore, high-quality poly-Si films in terms of grain size (greater than the film thickness) with an intragrain quality comparable to wafer-based silicon seem to be a potential material candidate for the fabrication of high-efficiency thin-film silicon solar cells on foreign substrates.

To produce such large-grain silicon films, high-temperature deposition methods need to be employed (>600°C), which poses several constraints on the substrates as well as electronic quality of the deposited Si material [12, 13]. Most of the groups are combining high-temperature processes, including chemical vapor deposition (CVD) of silicon and its crystallization by zone melting. Mitsubishi Corp. reported a solar cell efficiency of 16.5% with a Si film epitaxially deposited on a zone-recrystallized Si film on oxidized Si wafers [14]. Fraunhofer ISE reaches efficiencies of 11% for thin-film solar cells on graphite and 9.3% for cells on SiSiC-ceramics using a zone-melting process [15].

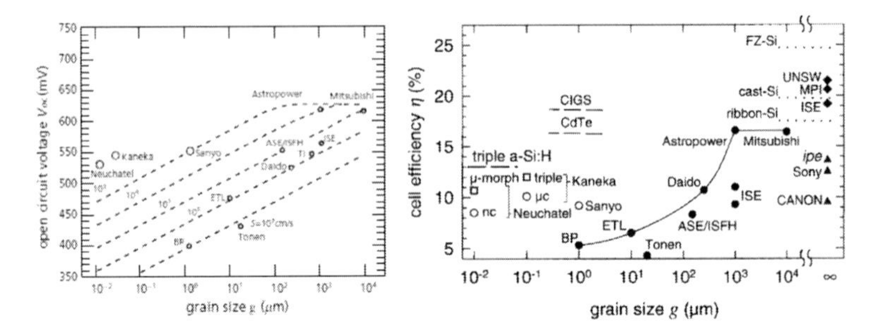

Figure 5.1 (a) The relationship between grain size and open circuit voltage (V_{oc}). Reprinted with permission from Ref. [9], Copyright 2001, Elsevier. (b) The relationship between grain size and solar cell efficiency are correlated to the carrier lifetime (diffusion length). In the figure, S indicates the recombination velocity at the grain boundaries. Reprinted with permission from Ref. [11], Copyright 1999, Springer.

Another way to produce directly polysilicon films is low-temperature solid-phase crystallization (SPC) at around 600°C of a stack of amorphous layers deposited at 300°C. This results in a grain size of the order of 1–2 μm. This approach requires a very long time (>12 hours) but allowed production of single-junction cells at efficiencies just under 10% [16, 17].

To deposit large-grained poly-Si at low temperatures, T < 650°C, several methods were explored, like laser-crystallized Si films as epitaxial seedling layers on glass substrates [18] or crystallization induced by aluminum [19], enabling the formation of grains with a size of around 10 μm. To improve the throughput of the crystallization process, rapid crystallization techniques are expected to be applied instead of conventional furnace annealing. Flashlamp annealing (FLA) is a millisecond-order rapid annealing technique, which is expected to be a high-throughput crystallization process to form poly-Si films [20]. However, metal-induced crystallization (MIC), particularly aluminum-induced crystallization (AIC), appears to be the preferable approach to form c-Si films with a grain size substantially larger than 25 μm on glass substrates [21].

The MIC of a-Si can be used to produce poly-Si with grains larger than those achievable either by thermal annealing of a-Si or by direct deposition of poly-Si by CVD. If a-Si is deposited at low temperatures on substrates coated with certain metals, and then heated to a temperature >300°C, the a-Si film can be converted to poly-Si. Alternatively, if the deposition of a-Si can be carried out at such higher temperatures on these metals, one can obtain large-grain poly-Si films directly. Here the metal acts as a catalyst to induce crystallization. MIC involves intermixing of metal with Si and the formation of a high concentration of metal alloy in the amorphous/crystalline interface. Furthermore, it was found that the growth of the crystalline phase would stop when no more metal is available. It must be noticed that using gold and aluminum metals makes it possible to crystallize a-Si with very low temperatures (<150°C). The speed of crystallization from the amorphous phase to the crystalline phase can be increased up to a factor of about 50 for gold and 66 for aluminum [22]. Additionally, the metal concentrations are very important for the speed of MIC. In these cases, the segregation of metal is observed between a-Si and the crystal. However, crystallization of metal/a-Si systems at lower temperatures makes it compatible with the use of conventional, low-cost soda-lime glass substrates.

5.2 Metal-Induced Crystallization of Amorphous Silicon

It is well known that a-Si has a higher free energy than c-Si. As a result, a-Si will be converted into more stable c-Si or poly-Si upon annealing, which is said to be the driving force behind the crystallization of a-Si. Upon annealing of a-Si, small, organized clusters of atoms, known as nuclei, start forming. This process is known as nucleation, and the nuclei continue to grow into crystals upon further annealing. The activation energy for nucleation is larger than that for crystallization. Therefore, higher-temperature annealing steps are needed for the crystallization of a-Si. However, for metals that form a eutectic with Si, the mechanism of crystallization involves lowering the nucleation activation energy. The metal atoms weaken the bonds in Si and make a-Si more unstable [23]. As a result, its conversion to poly-Si or c-Si is more favorable. Therefore, with the use of metals, heterogeneous nuclei can be formed at low temperature where the homogeneous nucleation rate is small. These nuclei can then be grown into crystals at a low temperature without a significant increase in nucleation. The nucleation rate can be kept low, and crystal growth can be started from the existing nuclei when the crystal growth is carried out at low-temperature annealing. Since the density of nuclei is low, large grains of c-Si can be obtained. In addition to increasing grain size, it is also important to start nucleation at controlled locations.

Metals like Ni, Pd, Ti, W, and Cu in contact with silicon react to form a thin silicide layer (reactive metal alloys) at certain characteristic temperatures with lattice constants close to that of Si. The silicide layer reduces the activation energy for a-Si crystallization, and a-Si under the silicide is thermally crystallized into poly-Si before other parts of the a-Si. Since the crystallization is initiated by metal, the process is called "metal-induced crystallization." The metal atoms in the MIC region diffuse to the grain boundaries, where they are trapped. The trapped metal atoms react with silicon to form a thin layer of silicide at the grain boundaries. At the region of the MIC–a-Si interface, the metal silicide boundaries exist as a continuous sandwich layer between MIC polysilicon and a-Si and act as a reactive layer that is responsible for grain growth. Such a layer is also called a reactive grain boundary (RGB) [24]. The silicide, RGB, a few nanometers in thickness, moves into the a-Si and lowers the activation energy of a-Si crystallization, leaving the poly-Si behind. As the diffusivity of the metal in crystallized silicon is relatively faster, the metal atoms in the polysilicon diffuse quickly to the new silicon grain boundary. This increases the concentration of metal at the RGB and subsequently pushes the metal atoms to the a-Si region continuously. As a result, the RGB propagates in the lateral direction—hence the crystallization. Therefore, the process is called "metal-induced lateral crystallization (MILC)." MILC can be distinguished from MIC in that MILC occurs through a lateral phase transformation from the MIC region, and this lateral crystallization is mediated by the silicide formed in the MIC region, as discussed before [25].

For such a catalytic phase transformation to occur, three different atomic fluxes are required in the system, as shown in Fig. 5.2:

1. Firstly, the bond breaking of a-Si atoms and migration of each atom toward the interface between a-Si and the silicide and adsorption of migrated silicon atoms at the silicide surface to create metal vacancies.

2. Secondly, the hopping of the created metal vacancies inside the silicide to reach the interface between the silicide and polysilicon. Hopping of the metal vacancies is coupled with the metal ions in the silicide.
3. Finally, rearrangement of the dissociated silicon atoms to be attached to the dangling bonds of the polycrystal.

Phase transformation of one atomic layer can be completed by the rearrangement of dissociated silicon atoms [26].

In the case of MILC, the polysilicon growth rate depends on the thickness of the RGB, the temperature, and the catalysts [27]. When Ni is used as a catalyst for MILC, the growth rate is about 5 μm/hour at a temperature of 550°C, while for Pd, the growth rate is 2 μm/hour at 500°C [28, 29]

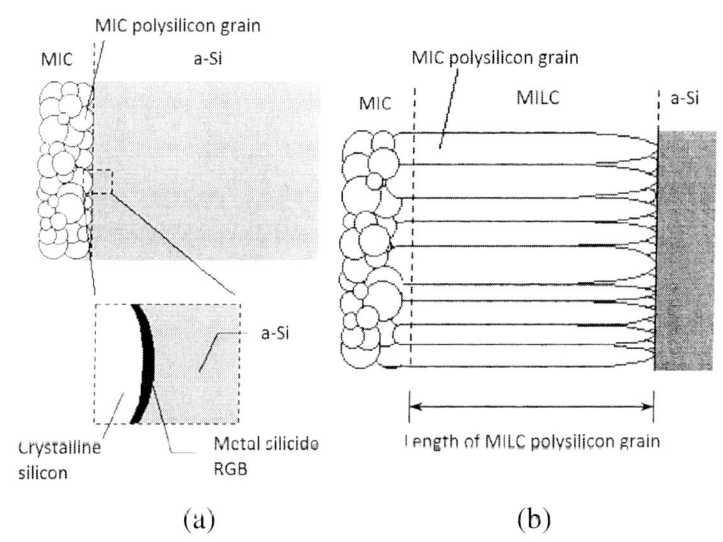

Figure 5.2 MILC polysilicon formation during the annealing process. (a) At the beginning of the annealing process, many metal atoms are trapped and a metal silicide is formed at the grain boundaries of the MILC polysilicon region. These metal silicide grain boundaries at the MIC–a-Si interfaces, which are reactive regions, are responsible for MILC formation. (b) During the annealing process, the metal silicide RGB absorbs silicon atoms from the a-Si region and rejects them to the MIC polysilicon region. As a result, the polysilicon grain grows up in the lateral direction. Reprinted with permission from Ref. [24], Copyright 2003, IEEE.

In the interaction of the metal/a-Si system during annealing, the crystallization of silicon is mediated by forming a compound system, usually stable metal silicide phases in thermodynamic equilibrium (e.g., Ni/Si, Pd/Si, Pt/Si), by forming simple eutectic systems that do not form stable metal silicides (e.g., Al/Si, Ag/Si, Au/Si), by forming a eutectic alloy, or by migration of metal silicide, or by combining some of these phenomena [30]. The stable metal silicide phases existing in thermal equilibrium play an important role during the crystallization process. For simple eutectic systems, the MIC process takes place below the eutectic temperature as a solid phase transition, which involves the dissociation of Si atoms from a-Si and the diffusion of Si atoms through the metal, followed by nucleation and incorporation of Si atoms into already existing Si crystals [31]. The crystallization temperature of silicon in contact with some selected metals and the eutectic temperature of metal/Si bilayer systems are given in Table. 5.1.

Table 5.1 Crystallization temperature and eutectic temperature of various metal/Si bilayer systems

Metal/a-Si	T_{eutec} (°C)	T_{crist} (°C)	Ref.
Al/a-Si	577	150	[32]
Ag/a-Si	830	200	[33]
Au/a-Si	360	130	[34]
Sb/a-Si	630	430	[35]
Cu/a-Si	802	485	[36]
Ni/a-Si	964	485	[29]
Pd/a-Si	760	500	[37, 38]
Co/a-Si	1204	700	[39, 40]
Pt/a-Si	830	–	[41, 42]

Though the exact driving force behind MILC is not clear, Hayzelden et al. [43] suggested a model, for instance, for nickel-induced lateral crystallization using a free-energy diagram for $NiSi_2$, that the chemical potential of Ni is lower at the $NiSi_2$/a-Si interface, whereas the chemical potential of Si atoms is lower at the $NiSi_2$/c-Si interface. So there is a driving force for Ni to move toward a-Si to reduce its free energy. The Ni moving forward in turn reacts with a-Si to form new $NiSi_2$, and the process repeats. The Si atoms remaining behind attach to the $NiSi_2$ template to form c-Si since their chemical potential is lower at the $NiSi_2$/c-Si interface. The nickel-induced crystallization (NIC) process is shown to occur below 600°C with much faster growth rates [44]. In addition to the fast growth rate, NIC leads to the formation of a <110>-textured poly-Si, which is favorable for better passivation in order to minimize grain boundary recombination velocity. NIC-derived films have a lower density of dangling bonds due to the better coordination of <110>-oriented grains. The solubility of Ni in Si at 600°C is higher than at room temperature. This leads to the formation of Ni precipitates along the grain boundaries, which act as recombination centers.

Other metals like Co or Pd also cause a reduction of the crystallization temperature. The silicide acts as a medium for the transport of atoms. Using a system of a-Si/Pd2Si/c-Si, Tu et al. [45] showed that dissociation of Pd_2Si occurs at its interface with c-Si and Pd diffuses through Pd_2Si toward a-Si. A similar mechanism was suggested by Hayzelden et al. [43] for $NiSi_2$-mediated growth. Growth of c-Si can occur using the silicide as a template. $NiSi_2$ has a 0.4% and $CoSi_2$ has a 1.2% lattice mismatch compared to c-Si. So $NiSi_2$ will be the best template for MILC and will yield the best quality of c-Si.

Non-silicide-forming metals like Al function merely as catalysts. The metal and silicon are dissolved up to the solubility limit. The metal phase can easily be removed. Additionally, Ni creates deep-level acceptor levels at 230 meV above the valence band and at 350 meV below the conduction band in c-Si, while Al is a shallow acceptor—only 67 meV above the valence band [46]. Thus Ni enhances carrier recombination and thus strongly decreases the minority carrier lifetime. In photovoltaic applications the photogenerated minorities make the minority carrier lifetime crucial for good solar cells. Hence, Al is favored over Ni in photovoltaic applications.

5.3 Aluminum-Induced Crystallization

The a-Si in contact with aluminum transforms into c-Si at temperatures below the eutectic temperature of the Al/Si system, which is much lower than the temperature required for solid-phase crystallization of a-Si, resulting in an exchange of the position of the bilayer system. The process of crystallization of a-Si is known as "aluminum-induced crystallization." The thermodynamics for bulk materials of Si and Al reveals that the Si/Al system does not form compounds and is rather immiscible [47]. However, depending on the annealing temperature, the AIC process could be distinguished in two modes due to the differences in the interaction of Al and Si. AIC can be observed in a wide temperature range, typically between 200°C and 577°C, the eutectic temperature of the AlSi alloy. The maximum solubility of Si in Al is about 0.005 at.% at 200°C but increases up to 0.75 at.% at 500°C. Therefore, the effects of dissolution and volume diffusion of Si into and through Al grains can be neglected at temperatures below 200°C. However, it affects the exchange process considerably at a high temperature around 500°C.

Initially, the Si layer is amorphous and the Al layer is crystalline, which are subjected to a compressive macrostress and exhibit lattice distortions. As the annealing progressed from low temperatures, the Al and Si layers exchanged their locations. Thereafter a-Si transformed into c-Si and the Al grains grew laterally, leading to the relaxation of the macrostress and microstrain in the Al layer. In other words, the transformation of a-Si into c-Si is driven by the relaxation of elastic strain energy associated with the macrostress and microstrain due to the drastic atomic rearrangement within the bilayer by the layer exchange. Thus the misfit experienced in the initial (unannealed) condition can be relieved. The c-Si nucleates within the Al layers and penetrates the Al as the c-Si grows. The continued diffusion of Si atoms makes the initial c-Si grains at the Al grain boundaries to grow laterally. The degree of mass transport depends on the diffusion length of Si along Al grain boundaries.

The Al/Si layer stack is started to be annealed at higher temperatures (<577°C), where the solubility of Si into Al becomes considerable, and the interaction of amorphous silicon with Al makes the dissociation of the amorphous phase and Si atoms diffuse across the Al/Si interface. Once Si is dissolved in the metal, the Si atoms diffuse within the Al layer and Si atoms nucleate within the Al layer to form a small c-Si cluster. This cluster grows until it is vertically confined within the Al layer by the continuous incorporation of further Si atoms. The grain continues to grow in the lateral direction only until the entire Al is replaced, which results in the formation of a continuous poly-Si layer on the substrate with the exchange of the entire layer. Silicon is also crystallized within the top layer in the form of small nanosized Si (nc-Si) crystallites that agglomerate as more compact Si islands in spite of the fact that the actual layer exchange is completed. These islands can be selectively removed to reach the crystallized poly-Si layer.

Since the dissolution and diffusion of Si into Al is prominent at higher temperatures, an interfacial oxide layer is essential at the interface between the initial Al and a-Si layers to control the exchange rate and hence the microstructural quality of the resulting crystallized layer. Usually, this interfacial oxide layer would be AlO_x because of the simplicity in forming such layers, which can be made by simple exposure of the Al to ambient air. This oxide layer functions as a permeable membrane separating the top and bottom layers throughout the process [48].

Although the actual mechanisms of the interdiffusion and the associated crystallization in the Al/a-Si system and the driving force for the layer exchange are not well understood, some studies have been made to explain the crystallization process. The physics behind the AIC process involving the Al/Si system is explained in the following sections.

5.3.1 Solid Solubility of the Al/Si System

If the Al/Si system is under thermal annealing, the thermal equilibrium between pure Al and a-Si breaks, resulting in a solid Al solution with a Si solute. The Al/Si system is a simple binary eutectic with limited solubility of aluminum in silicon and limited solubility of silicon in aluminum. Figure 5.3 depicts the binary Al/Si equilibrium phase diagram that shows the relationships between the various phases that appear within the Al/Si system under equilibrium conditions [49]. There is an equilibrium eutectic reaction at approximately 12.2 at.% silicon at the eutectic temperature (577°C). At this particular temperature, the solubilities of silicon in aluminum and aluminum in silicon are 1.65 at.% and 0.5 at.%, respectively. Otherwise, the solubility of silicon decreases to 0.05 at.%, 0.2 at.%, and 0.8 at.% at 300°C, 400°C, and 500°C, respectively, as shown in Fig. 5.3 [50]. When the temperature is such that the solubility approaches that of the eutectic temperature (577°C), rapid solid-state diffusion occurs, with the silicon segregating from the diffusion interface and migrating through the aluminum. There is only one invariant reaction in the Al/Si phase diagram, namely

$$L \rightarrow \alpha + \beta \text{ (eutectic)} \tag{5.1}$$

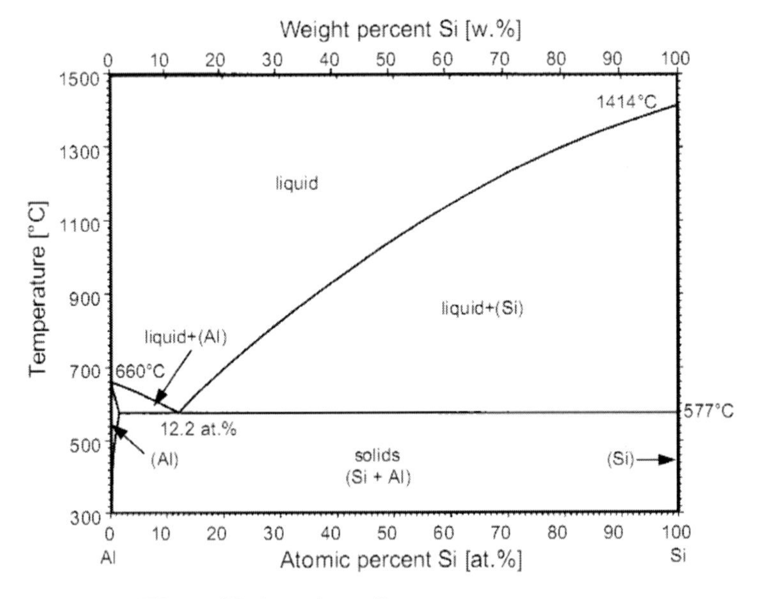

Figure 5.3 AlSi equilibrium phase diagram.

In Eq. 5.1, L is the liquid phase, α is predominantly aluminum, and β is predominantly silicon. Generally phase diagrams represent the relationship between temperature and composition under atmospheric pressure. From the phase diagram, it can be understood that the Al/Si eutectic can form in three ways:

(i) Directly from the liquid, in the case of a silicon concentration of 12.2% (i.e., for a eutectic aluminum–silicon alloy)

(ii) In the presence of primary aluminum, if silicon contents < 12.2% (i.e., for hypoeutectic aluminum–silicon alloys)

(iii) In the presence of primary silicon crystals, if silicon contents > 12.2% (for hypereutectic aluminum–silicon alloys)

5.3.2 Diffusion of Silicon in Aluminum

At annealing temperatures below the eutectic temperature, solid Al dissolves the silicon and permits it to diffuse more rapidly than in wrought (shaped) Al, as shown in Fig. 5.4 [50]. The diffusion coefficient (D) of Si in solid aluminum becomes enhanced as much as 1.5 orders of magnitude for evaporated aluminum. Specimens annealed at the same temperature but for different times show that the enhancement in D decreases for longer annealing duration due to imperfections in the Al film. The amount of grain boundary in the film must be decreased; however, other imperfections such as dislocations remain at a high concentration that cause lower diffusivity for higher temperatures. The activation energy (E_A) is also quite different for evaporated- and wrought-Al systems. The E_A is 0.79 eV and 1.36 eV for evaporated Al and conventional Al, respectively. This difference can be explained by the accelerated diffusion at the diffusion levels in aluminum (such as boundaries and dislocations). The diffusion coefficient of about 5×10^{-7}–2×10^{-8} cm^{2}s^{-1} in the temperature range of 400°C–500°C for solid Al is lower than that of liquid-phase Al (>10^{-4} cm^{2}s^{-1}) [51].

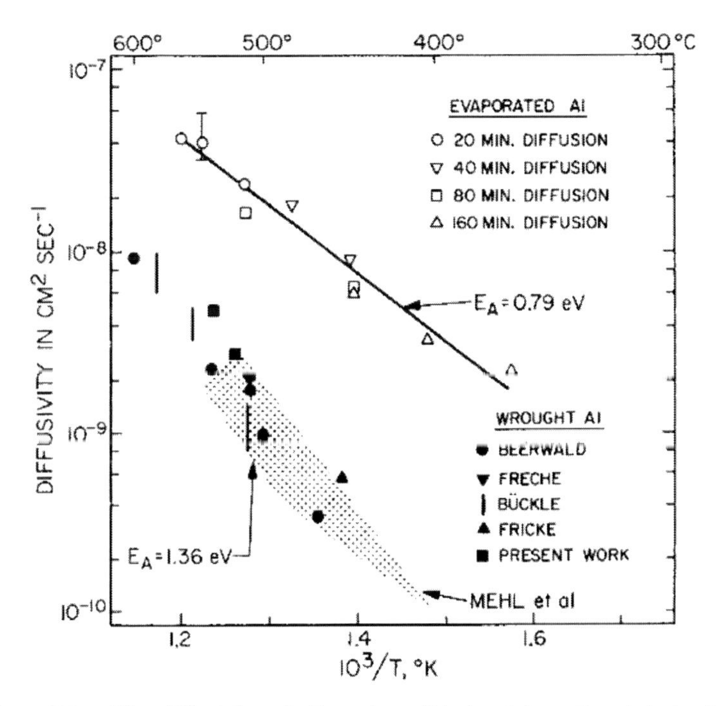

Figure 5.4 The diffusivity of silicon in solid aluminium. Reprinted with permission from Ref. [50], Copyright 1971, American Institute of Physics.

5.3.3 Thermodynamics and Kinetics of Al/Si Bilayers

Gibbs's classification serves as the fundamental basis for division of phase transformation processes. For this reason, to understand the thermodynamics and kinetics of crystallization of a-Si in contact with Al, Gibbs energy changes caused by this layer exchange process in Al/Si bilayers must be considered. Although at the beginning of the process, sublayers contains either pure Al or pure Si, the sublayer sequence is completely reversed upon annealing. Gibbs energy changes (ΔG) of the bilayer system, that is, the Gibbs energy after layer exchange minus the Gibbs energy before layer exchange, are calculated per unit area parallel to the surface. The Gibbs free-energy diagram for the amorphous–crystalline system is shown in Fig. 5.5.

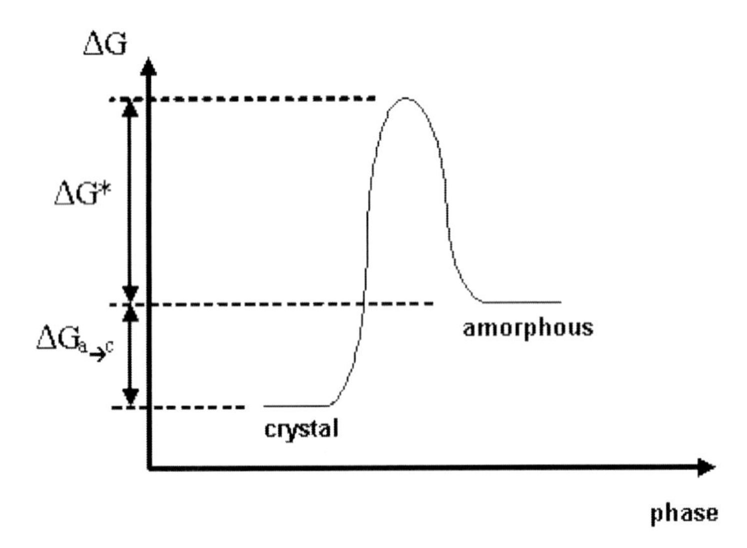

Figure 5.5 Gibbs free-energy diagram for the amorphous–crystalline system.

The driving force of dissolution of silicon in contact with aluminum at constant temperature is the difference of Gibbs energy ($\Delta G_{a\to c}$) between the metastable a-Si phase (G_a) and the stable c-Si phase (G_c):

$$\Delta G_{a\to c} = G_a - G_c \tag{5.2}$$

The variation of Gibbs free energy depends on the silicon phase. The Gibbs free energy of a system at any moment in time is defined as the enthalpy of the system minus the product of the temperature times the entropy of the system.

$$G = H - TS \tag{5.3}$$

Here H is the enthalpy, S the entropy, and T the absolute temperature. The Gibbs free energy of the system is a state function because it is defined in terms of thermodynamic properties that are state functions. The change in the Gibbs free energy of a system that occurs during a reaction is therefore equal to the change in the enthalpy of the system minus the change in the product of the temperature times the entropy of the system.

$$\Delta G = \Delta H - \Delta(TS) \tag{5.4}$$

If the reaction is run at constant temperature, this equation can be written as follows:

$$\Delta G = \Delta H - T\,\Delta S \tag{5.5}$$

Here ΔH and ΔS are the corresponding enthalpy and entropy changes. A consequence of the second law of thermodynamics is that spontaneous reactions occur at constant temperature and pressure when ΔG is negative, that is, $\Delta G < 0$. This condition implies that a system will naturally tend to minimize its free energy by successively proceeding from a value G_i to a still lower, more negative value G_f until it is no longer possible to further reduce G, at which $\Delta G = 0$. The system has achieved equilibrium, and there is no longer a driving force for change. In other words, for a process that cannot occur, $\Delta G > 0$. It is important to note that neither the sign of ΔH nor that of ΔS taken individually is sufficient to determine the reaction direction; rather, it is the sign of the combined function ΔG that is crucial in this regard. The concept of minimization of free energy is a criterion for stability in a system and forward change in a reaction or process. The balance between the contributions from the enthalpy and entropy terms to the free energy of a reaction depends on the temperature at which the reaction takes place.

The mechanism of dissolution and diffusion of silicon (a-Si) into Al and growth to the crystalline phase by annealing the system still remains to be understood. This transformation can be understood by the change of Gibbs free energy. It is widely accepted that the general driving force behind the crystallization of silicon is obviously the dominant reduction of the Gibbs energy of the system upon silicon crystallization [48, 52–54], but this does not explain the occurrence of layer exchange. The Gibbs energy change in Si/Al bilayers suggested that the rearrangement of the Si and Al phases is promoted by the release of elastic strain/stress energy and grain growth for the Al phase [45, 46]. It is believed that there is a particular effect of annealing temperature on the kinetics of the process of layer exchange upon annealing a-Si /crystalline Al (c-Al) bilayers.

The peculiar exchange observation of the locations of Al (bottom to top) and Si (top to bottom) sublayers may have a kinetic origin, a thermodynamic origin, or both. In addition, the formation of amorphous reaction layers at interfaces and the formation of amorphous oxide layers on metal surfaces can have a thermodynamic origin different from the kinetic origin thought [55, 56].

It is important to know that the exchange can occur in a wide temperature range, <200°C to below the eutectic temperature, 577°C. As suggested by Wang et al. [57], the temperature at which the reaction takes place has a considerable effect on the layer exchange process. The layer exchange process can also occur upon annealing pure a-Si/c-Al bilayers at a temperature as low as 165°C. According to the SiAl phase diagram, the maximum solubility of Si in Al is only about 0.75 at.% at 500°C and drops to 0.005 at.% at 200°C [58]. Therefore, substantial dissolution and volume diffusion of Si into and through Al grains can be neglected at temperatures below 200°C, on the contrary, which would affect the exchange process considerably at a high temperature around 500°C. Accordingly, in the AIC of a-Si, there exist two temperature regimes, a low-temperature regime (T < 300°C) and a high-temperature regime (T>400°C). Since, both the solubility and the diffusivity of Si in Al grains are negligible for the low-temperature AIC process, the nucleation and growth of c-Si are exclusively controlled by diffusion of Si at Al grain boundaries. Extensive efforts have been made to understand the low-temperature exchange process at Max Planck Institute for Metals Research, Germany. For instance, the transformation of the Al (50 nm)/a-Si (150 nm) bilayer, where the aluminum has a columnar structure, annealed at 250°C is explained by Zhao et al. [52], as shown in Fig. 5.6. At the beginning of the annealing at 250°C, the a-Si layer and the Al layer are subject to a compressive macrostress and microstrain. The stress in the sublayer is composed of the stress after layer deposition at room temperature and the thermal stress due to heating up to the annealing temperature. The thermal stress is caused by the difference in thermal expansion of the substrate and Al layer. At the end of the annealing, the Al and Si layers exchange their locations and a-Si crystallizes into poly-Si. The grain size of Al increases laterally, and the macrostress and microstrain of the Al layer are relaxed. According to results, the stress in the Al layer has relaxed entirely during the annealing. Upon annealing, the Gibbs energy of the thin-film system can be reduced, which is the driving force for layer exchange. The driving force determines not only the overall direction of the kinetic process, that is, change from a nonequilibrium state to an equilibrium state, but also the process rate, as explained next.

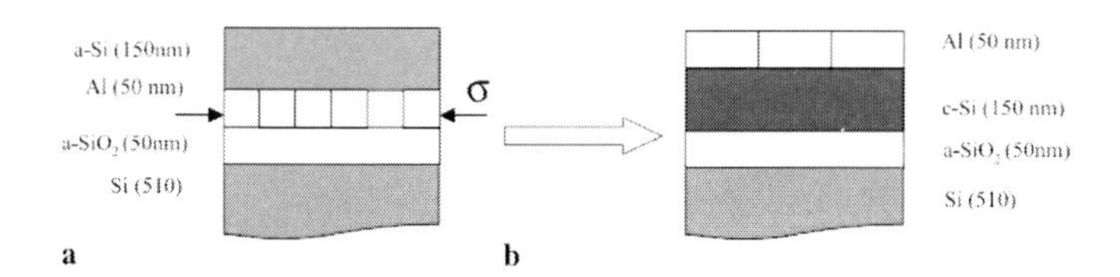

Figure 5.6 Schematic representation of the layer exchange for the Al (50 nm)/a-Si (150 nm) bilayer annealed at 250°C. Reprinted with permission from Ref. [52], Copyright 2004, Springer.

The first step is inward diffusion of Si along grain boundaries in the Al layer. As discussed in the previous section, when the a-Si layer is in contact with the Al layer, the free electrons of the Al layer near the interface have a Coulomb screening effect on the adjacent Si atoms, which leads to a weakening of the covalent bonds of the adjacent Si atoms [59, 60]. Upon annealing, free Si atoms diffuse into the Al layer along the grain boundaries of the Al phase because of which a decrease of energy is realized [57].

The second step is the nucleation of Si at grain boundaries of the Al layer. Usually, nucleation takes place in a short time period, t_N, called the nucleation phase. After the nucleation phase, no new stable nuclei are formed. The crystalline form of a-Si in contact with Al at relatively low temperature is due to the presence of free Si atoms in contact with Al. The silicon layer keeps its amorphous state until attaining a critical thickness. Beyond this critical thickness, nucleation of Si crystallization can start. The critical thickness for nucleation for Si crystallization at the Al grain boundary is ~4 monolayers (ML) (0.88 nm) of Si atoms.

The final process step is the lateral growth of existing Si grains in the Al layer. Once crystallization of the Si layer at an Al grain boundary has occurred, the original Al grain boundary has been replaced by two Al/Si interfaces. Further annealing makes the weakly bonded Si atoms at the original c-Al/a-Si interface to diffuse into the newly developed Al/Si interfaces because of the decrease of energy. Hence, a positive driving force exists for the diffusion of Si atoms into the Al/Si interfaces. Upon the diffusion of Si atoms into the Al/Si interface, the original Al grain and the new c-Si grain are separated by those Si atoms (a-Si layer). Then, one of two cases is possible: (1) coalescence of the existing c-Si grain or (2) formation of a new c-Si nucleus at the Al/Si interface after reaching a critical thickness upon further annealing. In case 1, the existing c-Si grain grows laterally, and in case 2, a new c-Si nucleus establishes two new interfaces. However, the Si atoms diffusing into the Al/Si interfaces would prefer to coalesce the existing c-Si grain instead of forming a new c-Si grain at the Al/Si interface in order to minimize the surface energy. Therefore, upon continued diffusion of Si atoms from the original a-Si layer into the Al/Si interfaces, the c-Si grains formed originally at the Al grain boundaries grow laterally. The degree of mass transport, in other words the exchange process from Al to the Si layer, depends on the diffusion length of Si along Al grain boundaries [61].

For high-temperature AIC, the solubility and diffusivity of Si in Al grains become large enough and start to play considerable roles in the AIC process. It was claimed that an interfacial oxide layer between the a-Si and c-Al sublayers would be necessary to control the dissolution and diffusion of silicon in aluminum and hence the exchange process [62, 63]. The kinetics of the exchange process in this mode (high temperature > 300°C) involves dissolution of Si atoms in Al and diffusion within the Al sublayer; c-Si nucleation takes place when a certain concentration of Si in the Al grains of the Al sublayer is exceeded; and Si grain growth occurs in lateral directions due to the confinement by the interfacial oxide layer and the substrate until the Al sublayer is eventually replaced by the growing Si grains. The schematic representation of the growth process at higher temperatures is shown in Fig. 5.7.

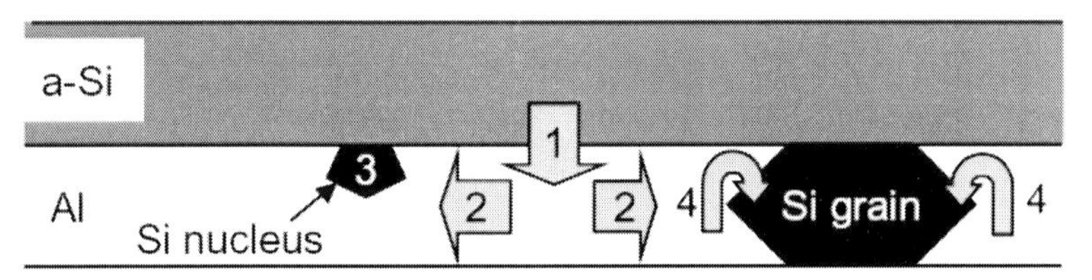

Figure 5.7 ALILE growth model suggested by Nast [19]. Step 1: Si dissolution and transport into the Al layer. Step 2: Si diffusion in the Al layer. Step 3: Si nucleation. Step 4: Growth of the Si grain. *Abbreviation:* ALILE, aluminium induced layer exchange.

Nucleation of Si grains occurs when the Si concentration in the Al layer exceeds the solid solubility limit for c-Si, C_S, and is related to the change of Gibbs energy of the Al/Si system. During the nucleation, Gibbs energy increases due to the formation of interfaces between Al and c-Si and decrease due to the Si bulk formation [64]. The critical radius of the crystal silicon cluster, R_c, is referred to as the radius at which the Gibbs energy of the cluster has its maximum value. The value of the critical radius depends on the supersaturation ($S = C_{Si}/C_S$, where C_{Si} is the Si concentration and C_S is the saturation concentration) and the annealing temperature (T_A). According to this model, the critical radius of the cylindrical grain is

$$R = \frac{\sigma\Omega}{kT_A \ln S}, S \rangle 1, \tag{5.7}$$

This explanation can be correlated to the change in Gibbs energy as

$$\Delta G_c = \frac{\pi\sigma^2 d_{Al}\Omega}{kT_A \ln S} \tag{5.8}$$

Here σ is Gibbs energy per unit area of the interface between the Si grain and Al, $\Omega = 2 \times 10^{-29}$ m^3 is the atomic volume of Si, and $k = 8.6 \times 10^{-5}$ eV/K is the Boltzmann constant.

$$\frac{dR}{dt} = \frac{D_{Si}\Omega}{R}(C_{Si} - C_S).$$

$$\frac{dC_{Si}}{dt} = -\frac{\pi R_C^2 J}{\Omega} - 2\pi D_{Si}(C_{Si} - C_S)\sum_i n_i(R_i)$$

The existing grains were assumed to grow according to the diffusion-limited mechanism. In this case, the evolution of the grain radius is described by Ref. [65].

Consuming the available Si atoms the grain nucleation and growth lead to decrease of C_{Si} as Ref. [64]. The summation in the second term of this expression was carried over all the Si grains of different radii, R_i.

For the diffusion-limited growth, the growth rate is proportional to the difference of silicon concentration far away from the grain-to-silicon concentration just in front of the grain. The growth rate is given by $J_c \alpha C_{Si} - C_S = C_S(S-1)$and is, therefore, proportional to the supersaturation. This implies that a stable nucleus cannot be formed at $C_{Si} = C_S$. Formation of a stable nucleus occurs if C_{Si} exceeds the critical concentration C^*. In other words, the supersturation, S, has to be above the critical supersaturation S^*. Therefore, grains grow when the silicon concentration exceeds the saturation concentration [66].

The evolution of C_{Si} in Al with respect to time at a particular temperature is evaluated by Sarikov et al. [64], as shown in Fig. 5.8 using the above model. The C_{Si} remains zero (region 1) prior to reaching the membrane/Al interface. The concentration of Si in Al continuously increases, but no nucleation of Si grains yet occurs since $C_{Si} < C_S$ (region 2). The nucleation of Si grains starts when the Si concentration in the Al layer exceeds C_S ($Cs < Csi \leq C_m$) (region 3), and the C_{Si} continues to grow due to the diffusion of Si from the a-Si layer. Equilibrium would be established by nucleation and growth of Si. Thus the consumption of Si balances the increase of C_{Si} by the increase of grain density. Thus a balance between the supply and consumption of Si is established. The C_{Si} decreases and tends to saturate; however, $C_{Si} > C_S$ (region 4). In this case, the existing grains continue to grow but nucleation is suppressed because C_{Si} stays between C_S and C^* ($Cs < Csi \leq C^*$), which is a characteristic feature of the AIC process.

Nast et al. [67] explained this kind of self-regulated suppression of nucleation as overlapping of Si depletion regions around existing grains. As shown in Fig. 5.9, the Si growth process is constrained by the Al/a-Si interface. Si atoms dissolved in the Al layer have high mobility and hence diffuse quickly within the film and/or along the interfaces toward the growing Si grains [50]. The newly formed Si grains within the Al matrix are far apart and do not influence each other at an early stage of the crystallization process. The concentration, C_i, of Si atoms in the immediate vicinity of the interface between the Si grain and the Al matrix is assumed to be equal to the equilibrium concentration, established by the incorporation and dissolution of Si atoms into and from the grain, respectively, at a particular annealing temperature. The concentration C_n describes the amount of Si solute in the Al matrix at which nucleation is likely to occur. As mentioned before, the nucleation rate strongly depends on the defects and grain boundaries in the Al layer as well as the temperature and corresponding silicon concentration. The concentration difference between C_i and C_n establishes a concentration gradient and, hence, diffusive flow of Si atoms in the direction toward the grain. Consequently, Si solute depletion occurs up to the effective diffusion distance, *d*, in the vicinity of the growing grain [68]. In this region, the possibility of new nucleation decreases with increasing depletion toward the advancing Al/Si grain interface. When the effective diffusion distances of adjacent grains begin to overlap, competition for the available Si atoms dissolved in the Al layer occurs. At this point the concentration decreases below C_n and the probability for the formation of new nuclei becomes less.

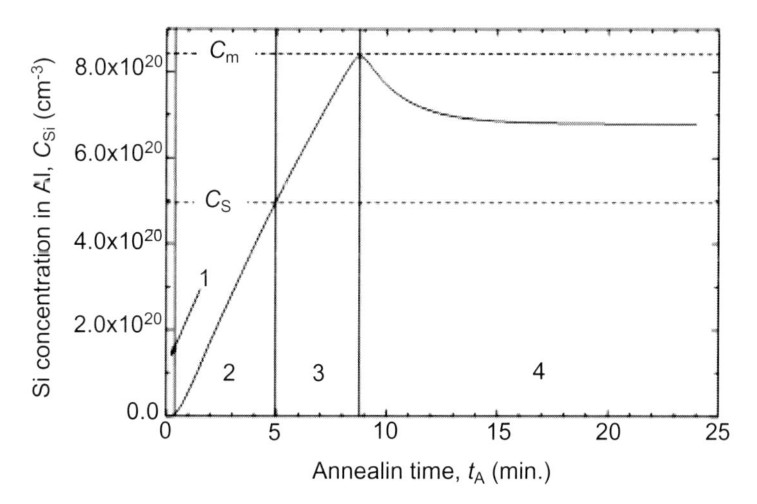

Figure 5.8 Dependence of Si concentration in Al, C_{Si}, on the time of annealing, t_A, at T_A = 500 °C. Reprinted with permission from Ref. [64], Copyright 2006, Elsevier.

5.3.4 Formation of AIC Layers

It is attractive to use a foreign substrate as a supporting material for the formation of poly-Si layers by the AIC process at high temperatures (~500°C). The following discussion concerns the high-temperature growth of polysilicon layers. Different kinds of substrates like transparent glass-ceramic (GC) and ceramic alumina, borosilicate glass, etc., were used by different researchers.

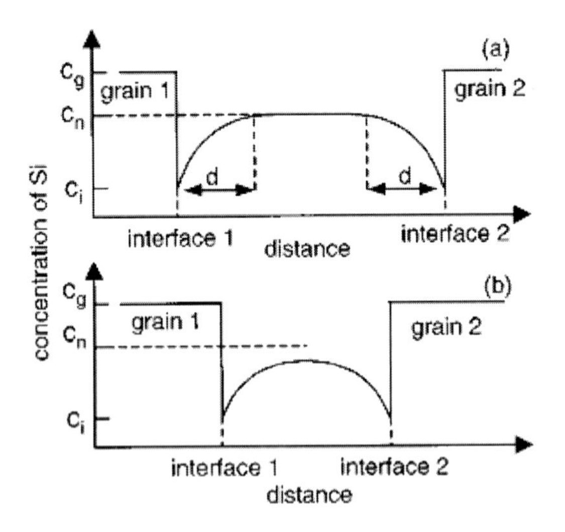

Figure 5.9 Schematic presentation of the spatial concentration of a Si solute within the Al layer. (a) When neighboring grains are far apart and (b) when the interference in the form of an overlap of the Si solute depletion areas occurs. C_g is the concentration of Si inside the growing grain. Reprinted with permission from Ref. [67], Copyright 2000, American Institute of Physics.

Usually, the substrates can be cleaned ultrasonically in acetone followed by alcohol. The precursors for the AIC process are aluminum (Al) and a-Si, which can be deposited by electron beam evaporation or sputtering. Plasma-enhanced chemical vapor deposition (PECVD) can also be used to deposit a-Si films. In general, excessive thickness of the a-Si film than metal is required in order to form continuous crystalline films on the substrates. As required, a thin, permeable membrane of AlO_x at the interface of Al/Si is formed by exposure of the Al film to air or thermal oxidation prior to the deposition of a-Si. The isothermal annealing of the Al/a-Si bilayer structure below the eutectic temperature of the Al/Si system (T_{eu} = 577°C) leads to the exchange of layers and concurrent formation of c-Si. The AlO_x membrane remains at the same position during the layer exchange, as shown in Fig. 5.10 [69].

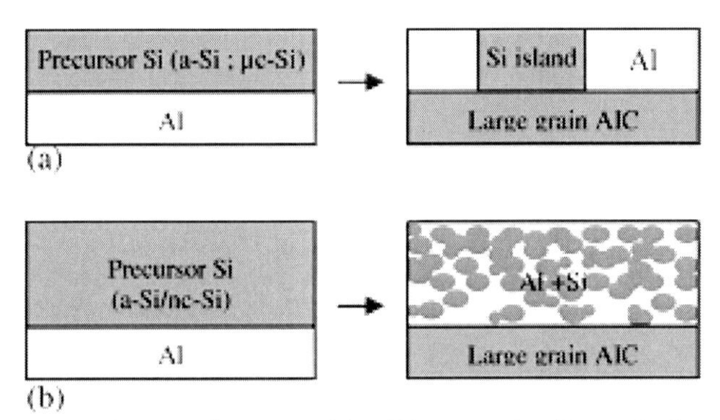

Figure 5.10 Schematic diagrams of the AIC process leading to the formation of (a) Si islands and (b) an Al/Si network. Reprinted with permission from Ref. [69], Copyright 2004, Elsevier.

The layer exchange is mediated by different steps like 1) dissolution of silicon in aluminum, 2) formation of nuclei, 3) growth of nuclei, and 4) coalescence. The exchange process starts with the dissociation of a-Si and the subsequent diffusion of Si atoms through the permeable membrane into the initial Al layer. Nucleation of Si grains in the Al layer starts when the Si concentration reaches the solubility limit for c-Si. Then, Si nuclei are formed locally within the initial Al layer and continue to grow in all directions, until they are confined vertically between the glass substrate and the permeable membrane. The kinetics of the Si transport into the Al layer, that is, silicon concentration in aluminum and the density of nuclei, are controlled by annealing temperature and membrane thickness. Both thinner membranes and higher annealing temperatures result in more rapid increase of Si concentration in Al and vice versa. Raising the annealing temperature leads to a decrease of the nucleation time; on the other hand, increasing the thickness of the membrane results in an increase of the nucleation time. By varying the annealing temperature and membrane thickness, one may influence the kinetics of Si diffusion through the membrane in a controlled way and hence the nucleation kinetics, as explained by Sarikov et al. [70].

In the later stage, Si growth is fed by lateral diffusion of Si atoms within the initial Al layer toward existing grains and the growth continues only laterally until adjacent grains coalesce. Finally, a continuous poly-Si film can be formed on the substrate. During the course of the growth of Si grains within the initial Al layer, the Al is displaced to the initial a-Si layer, resulting in an Al(+Si) layer on top of the poly-Si film. The Al layer reaching the top of the poly-Si film contains silicon inclusions, which are referred to as Si "islands." The residual Al+Si layer formed on top of the structure can be removed by two different ways depending on the subsequent processes: mechanical polishing with colloidal silica or wet chemical etching. For the chemical treatment two steps are used. First, the aluminum excess is etched by the following selective composition: 16 cc phosphoric acid (H_3PO_4, 85%), 2 cc deionized water (DI-H_2O), 1 cc nitric acid (HNO_3, 69%), and 1 cc acetic acid (CH_3COOH, 100%) at 40°C. The etching rate of the selective composition for aluminum is 600 nm/minute at 40°C. Then the Si islands are etched by the second chemical etchant that is nonselective: it is composed of 72.5 cc nitric acid (HNO_3, 69%), 1.5 cc hydrofluoric acid (HF, 49%), and 28 cc DI-H_2O. The etching rates for Si and Al are 0.8 μm/minute and 1.1 μm/minute, respectively, at room temperature.

The layer exchange can also occur in the "inverse structure" for which the a-Si is deposited on the substrate, followed by Al deposition [71]. The permeable membrane required for the "inverse structure" can also be prepared by exposure to air. However, the membrane is a SiO_x layer in this case. The formation of SiO_x is relatively a slower process compared to the formation of AlO_x. The "inverse structure" is advantageous compared to the "normal structure" in terms of 1) a smoother poly-Si surface and directly access for subsequent process steps without chemical or mechanical treatments and 2) an Al layer between the poly-Si film and the glass substrate serving as a contact layer in device configurations, provided that the following processing steps are limited to the eutectic temperature of the Al/Si system (577 °C). This is because of the fact that the Al/Si system forms a liquid phase above 577 °C.

The resulting poly-Si seed layers were p+-type due to Al doping during the AIC process. This is due to the fact that Al is a shallow acceptor in Si with an energy level of 67 mV above the valence band edge [72]. Therefore, the AIC of a-Si always results in a p-type conducting material.

5.3.5 Effect of Growth Parameters on the AIC Growth Kinetics and the Resulting c-Si Microstructure

Many physical parameters have an important influence on the reaction of aluminum and a-Si for the formation of a continuous poly-Si film. Identified important parameters are schematically summarized in Fig. 5.11.

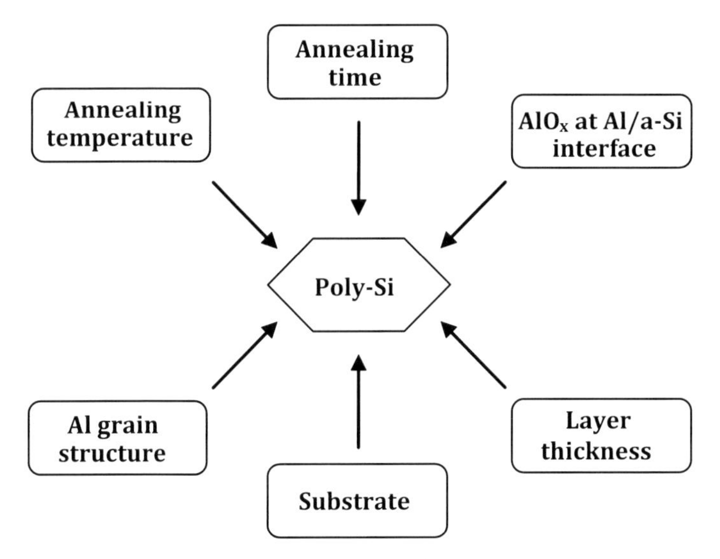

Figure 5.11 Illustration of the parameters that influence the AIC process.

The effects of each parameter on the AIC process for the formation of a poly-Si seed film is described in the following sections. Especially the Al structure, the Al/Si interface, the substrate, and the exchange annealing conditions are investigated in terms of the crystallographic quality of the poly-Si film.

5.3.5.1 Effect of thermal budget (temperature and time)

The annealing temperature and time are crucial parameters of the Al/Si layer exchange process [21, 73]. The crystallization and growth kinetics can be described through the analysis of surface coverage and density of the grains that could be evaluated from optical microscope images. For instance, the growth kinetics of the Si grains from the Al/Si bilayer system annealed at 450°C for different times ranging from 40 minutes to 90 minutes on a flowable oxide

(FOx)-coated alumina substrate was presented [73] and shown in Fig. 5.12. The dendritic shape of the Si grains is observed at all annealing temperatures, *T*, for a short time annealing, *t*. The shape of the Si grains is strongly depending on the crystallization time: as the annealing time lasts, the Si grains grow uniformly in all lateral directions until their coalescence and form a continuous film. The crystallized fraction F_c, defined as *the silicon surface area divided by the total area*, can be deduced for each T/t couple.

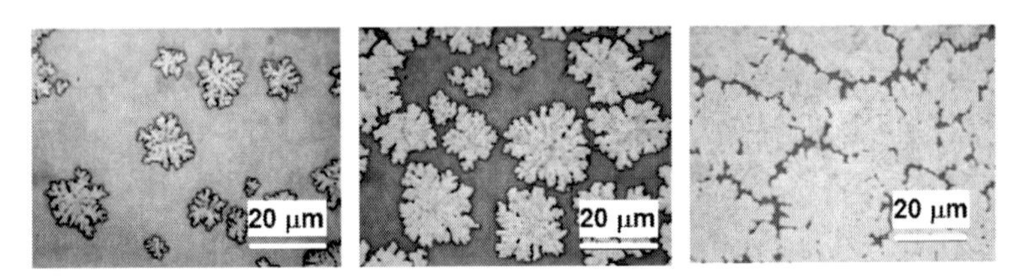

Figure 5.12 Optical microscope images of AIC dendritic grains grown on a thermally oxidized silicon substrate (t-SiO_2) at 450°C for (a) 40 minutes, (b) 60 minutes, and (c) 90 minutes. Reprinted with permission from Ref. [73], Copyright 2007, Elsevier.

The crystallized fraction, F_c, as a function of annealing time for different temperatures (400°C–550°C) after AIC on FOx-coated alumina substrates is shown in Fig. 5.13. It can be understood from the figure that an incubation time (minimum time needed to form the initial nuclei, t_N) is necessary before the crystallization started and a particular time (t_G) is needed to grow the initial nuclei to form a continuous film. Both t_N and t_G decrease with an increase of the AIC annealing temperature. Also, the time necessary to complete the layer exchange, called crystallization time t_A, is found to depend strongly on the annealing temperature. At high temperatures, F_c increases very rapidly and then saturates while at low temperature and then F_c increases moderately and saturates after a long time. An activation energy for the time corresponding to F_c = 50 is deduced from this data: 1.447 eV in the temperature range 425°C–550°C. The activation energy again varies with the method of deposition and thickness of AlO_x membrane that controls the diffusion during the exchange process [74, 75].

The electron backscattered diffraction (EBSD) measurements performed on poly-Si films formed by AIC on FOx-coated alumina substrates as a function of annealing temperature are shown in Fig. 5.14. The color of the mapping images of Fig. 5.14 corresponds to a particular crystalline orientation: the graded red color highlights Si grains with less than 15° from the (100) plane parallel to the surface, while the blue color corresponds to an orientation deviation of less than 15° from the (111) plane. The other orientations are indicated in yellow. Inverse pole figures shows the color coordinate from which the orientation of a grain can be identified. This AIC poly-Si film is strongly textured and preferentially oriented within the (100) plane parallel to the surface.

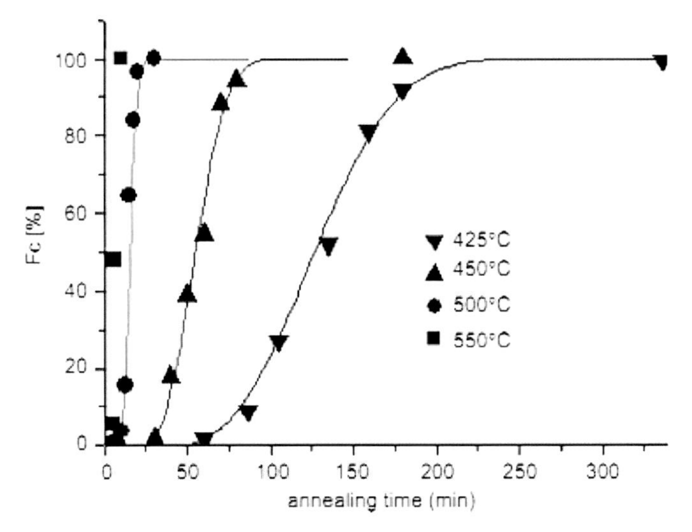

Figure 5.13 Crystallized fraction of polysilicon films formed at 425°C, 450°C, 500°C, and 550°C on an alumina substrate as a function of annealing time. Reprinted with permission from Ref. [73], Copyright 2007, Elsevier.

The distribution of analyzed pixels as a function of deviation angle θ is plotted in Fig. 5.15 for all annealed samples. The deviation is accumulated by segments of 5° compared to the <100> orientation. In this graph, the minimum deviation (5°) means a fully <100>-oriented layer, while the maximum deviation of 54.7° corresponds to the <111> plane. As an example 33% of the poly-Si AIC film formed at 425°C has an orientation deviation between 10° and 15° from the (100) orientation. A majority of grains (50%–80%) have an orientation deviation between 5° and 20° relative to (100) and increasing with reducing annealing temperature. Given the fact that no Si (110) is expected in the layers, (111)-oriented Si grains are witnessed by the presence of deviation angles in the range 45°–55°. It is said that most of the Si grains grown by AIC have a deviation angle from the (100) orientation between 5° and 25° for all investigated temperatures. However, increasing the annealing temperature tends to decrease the (100) preferred orientation.

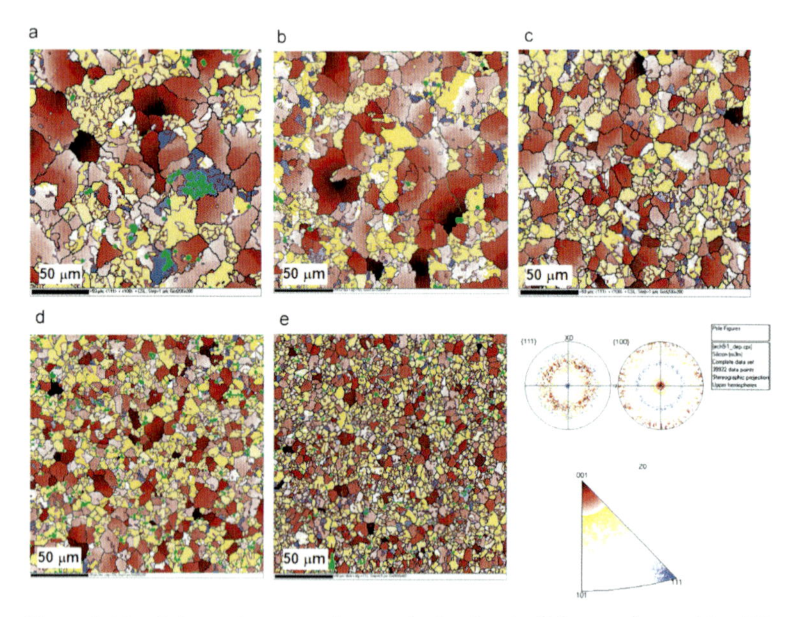

Figure 5.14 Orientation-mapping analysis of poly-Si layers formed by AIC on FOx-coated alumina substrates as a function of annealing temperature: (a) 400°C, (b) 425°C, (c) 450°C, (d) 500°C, and (e) 550°C. The pole figure and inverse pole (f) correspond to the 450°C sample. Here, the layer thicknesses of Al and Si layers are 270 nm and 500 nm. Al was exposed to air about one week prior to Si deposition, which gives the thickness of AlO_x ≈ 4.5 nm. The surface was prepared by chemical mechanical polishing for EBSD measurements. Reprinted with permission from Ref. [73], Copyright 2007, Elsevier.

At high annealing temperatures (>450°C) the competition between growing grains is high. The oversaturation decreases between grains, and consequently all growth directions are possible. In contrast, at low annealing temperatures there is no competition between grains and the orientations with a high growth rate parallel to the surface are preferred. This rapid lateral growth is in competition with the (100) vertical growth and most probably leads to the formation of twins in the growing grain. Such a phenomenological model might explain the increasing trend of twins at low annealing temperatures, and (100)-oriented grains are twin free.

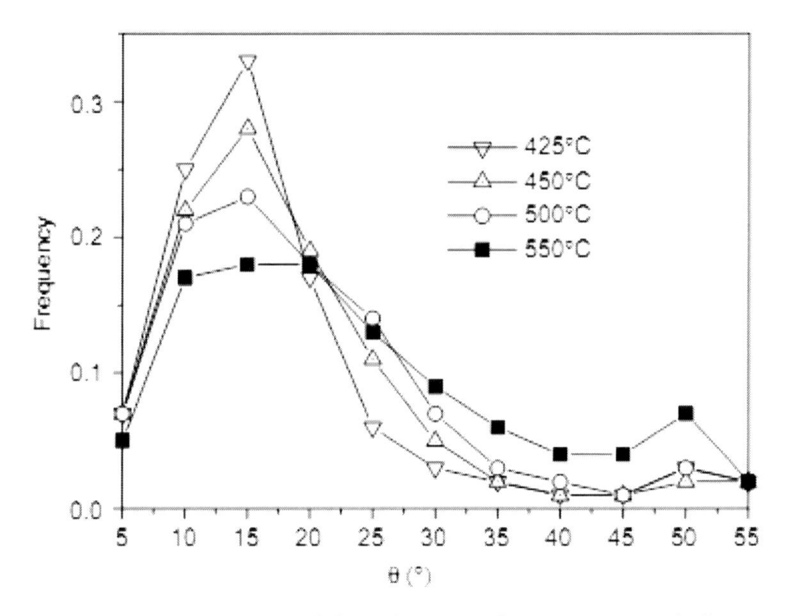

Figure 5.15 (100) texture of the poly-Si samples on FOx-coated alumina formed by AIC at different temperatures. The deviation is considered by segments of 5°. The lines are here as a guide to the eye. Reprinted with permission from Ref. [73], Copyright 2007, Elsevier.

The average grain size (D_g) decreases from about a 23 μm diameter at 400°C down to 7 μm at 550°C. The grain size distribution becomes less homogenous for a lower annealing temperature, but the grain size is much larger (refer to Ref. [73] for more details). Such a shape of distribution means that the process is diffusion controlled instead of interface controlled [76]. The difference in the average grain size can be attributed to the growth morphology of the crystals, which depends on the annealing temperature. A correlation between the exchange annealing temperature and the grain size can be found in such a way that the grain size increases when the nucleation rate decreases by reducing the annealing temperature. Since the grain size distribution is dependent on the ratio of the grain growth to the nucleation rate, the ratio increases with decreasing annealing temperature. The decrease in grain growth due to a lower annealing temperature is less than the decrease in the nucleation rate. The effective diffusion distance is longer at lower annealing temperatures, leading to larger depletion areas around the growing grains, preventing further nucleation. Therefore, the grains grow to a larger size before impingement occurs. It has to be taken into account that nucleation as well as diffusion-controlled growth are thermally activated processes. Hence, the overall crystallization process takes longer at lower annealing temperatures [48].

The inter- and intragrain defects in these AIC polysilicon films contains low-angle grain boundaries (LAGB) (angle $< 2°$) and coincident site lattice (CSL) boundaries consisting of twin boundaries of the first order ($\Sigma 3$), second order ($\Sigma 9$), and third order ($\Sigma 27$). The AIC layers formed on GC substrates at 475°C contain about 38.6% of $\Sigma 3$ twin boundaries, lower than that found at 500°C/6 hours (46.5%) [21]. The density of twin $\Sigma 9$ (~5%) and $\Sigma 27$ (~2%) boundaries is almost the same for both annealing processes. Thus, the larger the grains, the less defective they are. It is clearly seen that the AIC layer on GC substrates formed at 475°C/8 hours has optimal structural properties in terms of grain size and defects.

5.3.5.2 Effect of the interfacial oxide layer (AlO_x)

Generally, diffusion barriers of thin films are used to separate materials from direct contact in order to prevent them from reacting. Diffusion barriers are used in thin-film crystallization systems, an example being silicides to prevent direct AlSi contact. Ideally, a barrier layer X sandwiched between Al and a-Si layers

(i) should constitute a kinetic barrier to the traffic of Al and a-Si across it. In other words, the diffusivity of Al and a-Si in the layer X should be small;
(ii) should be thermodynamically stable with respect to Al and a-Si at the highest temperature of use; further, the solubility of X in Al and Si should be small;
(iii) should be easy to form, adhere to the involved films, possess low stress, and be compatible with other processing [77].

The oxide layer is formed on top of the aluminum layer before a-Si deposition influences the exchange process between Al and a-Si. The AlO_x layer behaves as a membrane by allowing the diffusion throughout the process without participating in the layer exchange process. Hence, this native oxide layer remains in position and separates the top and bottom layers during the exchange process of layers. For this reason, an aluminum oxide interfacial layer is necessary during the exchange process in order to form a continuous poly-Si film. The purpose of AlO_x layer formation was explained in detail [78]. As shown in Fig. 5.16a, the AlO_x interfacial layer leads to a continuous poly-Si layer, while the porous silicon (Fig. 5.16b) is formed when the AlO_x layer is not present [79]. Silicon islands are formed on top of the poly-Si surface, which are originating from the secondary crystallization.

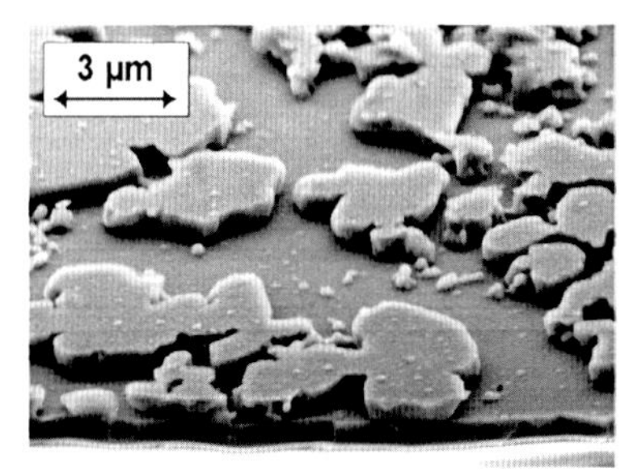

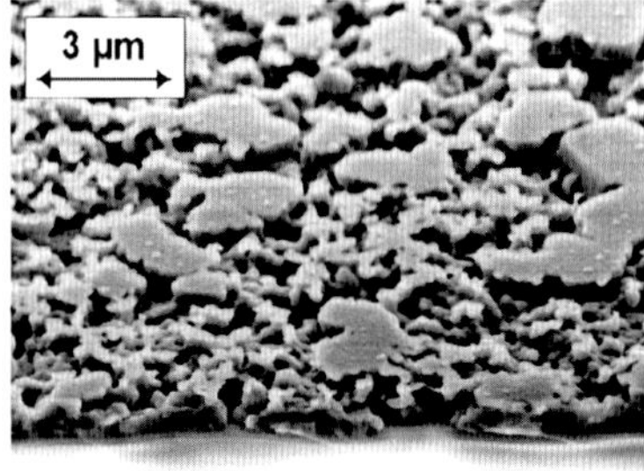

Figure 5.16 SEM images of crystallized silicon after the removal of residual aluminum by wet-chemical etching. (a) With the AlO_x interfacial layer obtained by oxidation for two hours' exposure to air and (b) without the AlO_x layer, that is, the precursor layer was deposited without breaking the vacuum of the deposition chamber. The thickness of Al and a-Si layers was about 300 nm and 375 nm, respectively, by DC-magnetron sputtering. Reprinted with permission from Ref. [79], Copyright 2002, WIP Renewable Energies. *Abbreviation*: SEM, scanning electron microscopy.

The formation of aluminum oxide was studied by L. P. H. Jeurgens [80]. It has been demonstrated that an amorphous oxide film of relatively uniform thickness is formed when the sample was exposed to oxygen at low temperatures (≤100°C). At higher temperatures (≥200°C), an amorphous oxide film is formed initially, which gradually becomes crystalline γ-Al_2O_3. Both lateral diffusion of the oxygen species and the rate of oxide formation increase with increasing temperature. The thickness is in the range of 2–4 nm for a growth temperature higher than 200°C. At exposure at room temperature (~20°C), the oxide layer has an amorphous structure.

Hence, the amorphous nature can possibly change toward the crystalline phase during crystallization annealing. The thickness of AlO_x changes with the duration of air exposure. An exposure of three minutes, one day, and one week resulted in a thickness of 50 Å, 64 Å, and 125 Å, respectively.

The effects of oxidation duration on the crystallography of the final poly-Si film formed by the AIC method were studied using EBSD. For instance, evaporated aluminum layers of 300 nm were exposed to ambient air for one day and one week prior to the deposition of a 550 nm thick a-Si layer. The crystalline orientation maps for poly-Si layers formed on FOx-coated alumina at 500°C for five hours as a function of aluminum oxidation duration are shown in Fig. 5.17,a,b. The colors of orientation mapping show the crystallographic orientation of grains. Other orientations are showed by gray color. Black points appear as defects in the AIC layer [81].

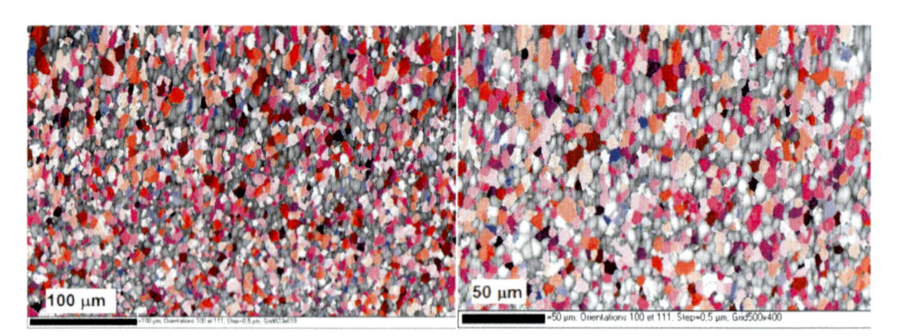

Figure 5.17 Crystalline orientation map of grains for a fully crystallized poly-Si layer at 500°C for five hours on FOx-co ated alumina. (a) One-day aluminum oxidation, (b) one-week aluminum oxidation (colors: red = vertical axis//(100) ± 9°, blue = vertical axis//(111) ± 9°, violet = vertical axis//(103) ± 9°). The thickness of Al/a-Si is 300/500 nm, and Al layers were exposed to ambient air for one day and one week prior to a-Si layer deposition. Reprinted with permission from Ref. [81], Copyright 2009, Université de Strasbourg, France.

The continuous poly-Si film was obtained with the average and maximum grain size of 10 μm and 18 μm, respectively, for one-week oxidation. Even if the oxide layer for one-week exposures becomes thicker, which influences the crystallization time by increasing the crystallization duration [82], the crystallization was fully completed by annealing at 500°C for five hours. Furthermore, the AIC films grown on alumina substrates are <103> preferentially oriented instead of <100> as commonly found. The <100> orientation is free of twins, whereas the other orientations present twins. The average grain size becomes almost half of that of one-week-oxide films by the value of 5 μm; the maximum grain size is 14 μm, while sustaining a similar crystallographic orientation. The grain size tends to decrease by reducing the thickness of the aluminum oxide layer due to a faster nucleation rate. It can be concluded that the oxidation thickness affects the nucleation and its growth velocity.

5.3.5.3 Effect of the aluminum layer

The crystallization of a-Si occurs in the aluminum layer, and the structure of aluminum has a key role in the AIC process and the quality of the resultant poly-Si film. There are two parameters concerning the aluminum layer that are influencing the nucleation rate of Si, namely, the oxygen concentration and the grain size of aluminum.

Klein et al. [62] showed that the exchange process time and grain size of the final poly-Si are affected by the presence of oxygen in the Al layer. Figure 5.18 depicts the influence of oxygen flow (fO2) during aluminum deposition on the annealing time. The layer exchange could be completed in 90 minutes without oxygen flow, while the exchange is completed in 17 minutes of annealing time in the presence of a 2 sccm oxygen flow during aluminum deposition. The reason for the dramatic decrease of the annealing time with the aluminum deposition under O2 can be correlated to the faster diffusion of Si atoms through the Al/a-Si interface and within the aluminum layer. The aluminum layer containing a higher oxygen concentration includes more defects that causes faster diffusion and thereby results in reduced process time.

The effect of oxygen within the aluminum layer was studied by changing the vacuum level of the electron beam evaporation system during Al deposition on FOx-coated alumina substrates. In the present case, the samples, A and B, were prepared with different vacuum levels during the aluminum deposition. Sample A was prepared with a low vacuum of 1.2×10^{-2}–1×10^{-4} mbar, while sample B was prepared with a high vacuum of 2×10^{-6} mbar.

The EBSD images of samples A and B are shown in Fig. 5.19a,b, and the inverse pole figures that identify the orientation by color codes are shown in Fig. 5.19c,d, respectively. The average grain size is 10 μm and 6 μm for samples A and B, respectively. Many grains are <100> oriented for sample A, as identified from Fig. 5.19c. In contrast, there is no significant preferred orientation for sample B. This implies that the higher oxygen compound in Al results in poor crystalline quality. The defects resulting in Al due to increased oxygen content enhance the nucleation rate and cause smaller average grain size.

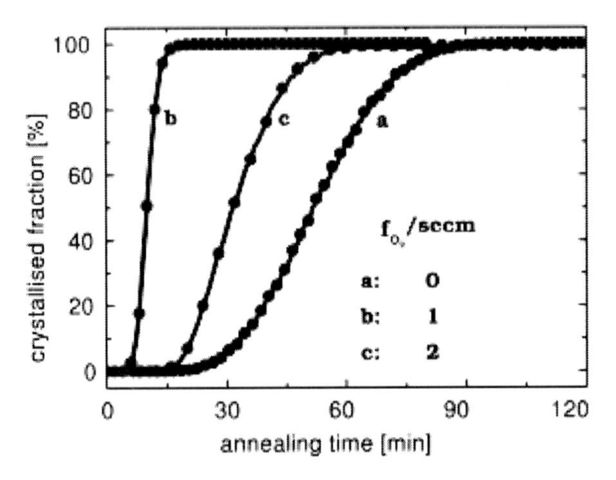

Figure 5.18 Crystallized fraction as a function of annealing time for samples with different oxygen flows (f_{O2}) during aluminium deposition. (a) f_{O2} = 0 sccm, (b) f_{O2} = 1 sccm, and (c) f_{O2} = 2 sccm. Reprinted with permission from Ref. [62], Copyright 2004, Elsevier.

The second parameter that strongly affects the nucleation of Si is the grain size of the aluminum layer. Indeed, the final grain size of the poly-Si film depends on the crystallization rate and thereby on the size of Al grains. The exchange process is faster with a smaller Al grain size that gives rise to a higher nucleation rate and a smaller final Si grain size [82]. The grain size of polycrystalline aluminum depends on its deposition rate: the higher the deposition rate, the larger the grain size [83].

The Al films, named C and D, grown at deposition rates such as 16 Å/s and 25 Å/s, respectively, were used for the AIC process on an alumina substrate [84]. These characteristics most probably affect the morphology and structure of the polycrystalline aluminum film and therefore the AIC layer exchange. In Table 5.2, an overview of the deposition rate of aluminum and its mean and maximum grain sizes as deduced for samples C and D is given.

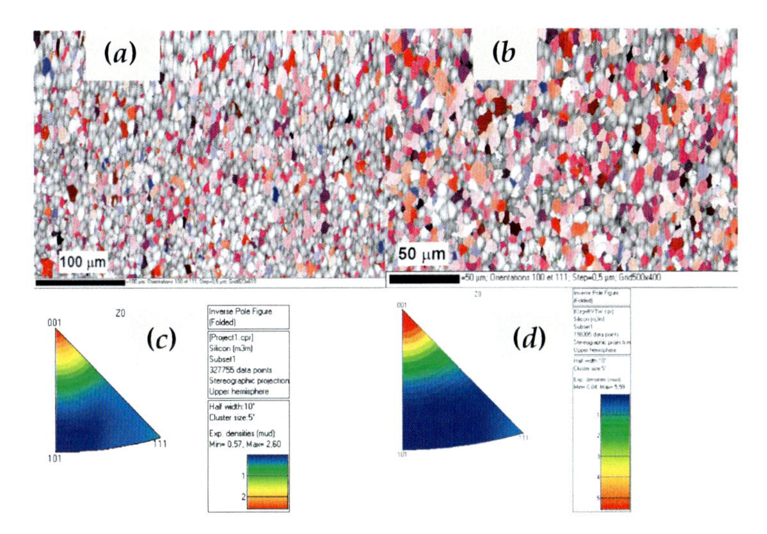

Figure 5.19 EBSD analysis of polysilicon films on FOx-coated alumina substrates for (a) sample A and (b) sample B, giving the crystalline orientation map of the grains (colors: red = vertical axis//(100) ± 9°, blue = vertical axis//(111) ± 9°, violet = vertical axis//(103) ± 9°). (c and d) Inverse pole figures for samples A and B. The thickness of Al/a-Si is 300/500 nm, and Al layers were exposed to ambient air for one week prior to a-Si layer deposition. Here, sample A was prepared with high concentration of oxygen and sample b with low oxygen concentration. Reprinted with permission from Ref. [81], Copyright 2009, Université de Strasbourg, France.

Table 5.2 Details of the deposition rate of aluminum and its mean and maximum grain sizes as deduced for samples C and D

Name	Al deposition rate (Å/s)	Mean grain size (μm)	Maximum grain size (μm)
C	16	15	65
D	25	7	24

The EBSD analyses of the AIC samples C and D are shown in Figs. 5.20a and 5.20b, respectively. The average grain diameters are about 15 μm and 7 μm, respectively, for samples C and D. On the other hand, the AIC films grown on alumina substrates are <103> preferentially oriented for C and D (Fig. 5.20c,d). Furthermore, no twins are observed for the <103>-oriented grains for C and D, whereas the other orientations present twins. It is deduced that lowering the Al deposition rate increases the Al grain size, which results in a poly-Si film with larger grain sizes.

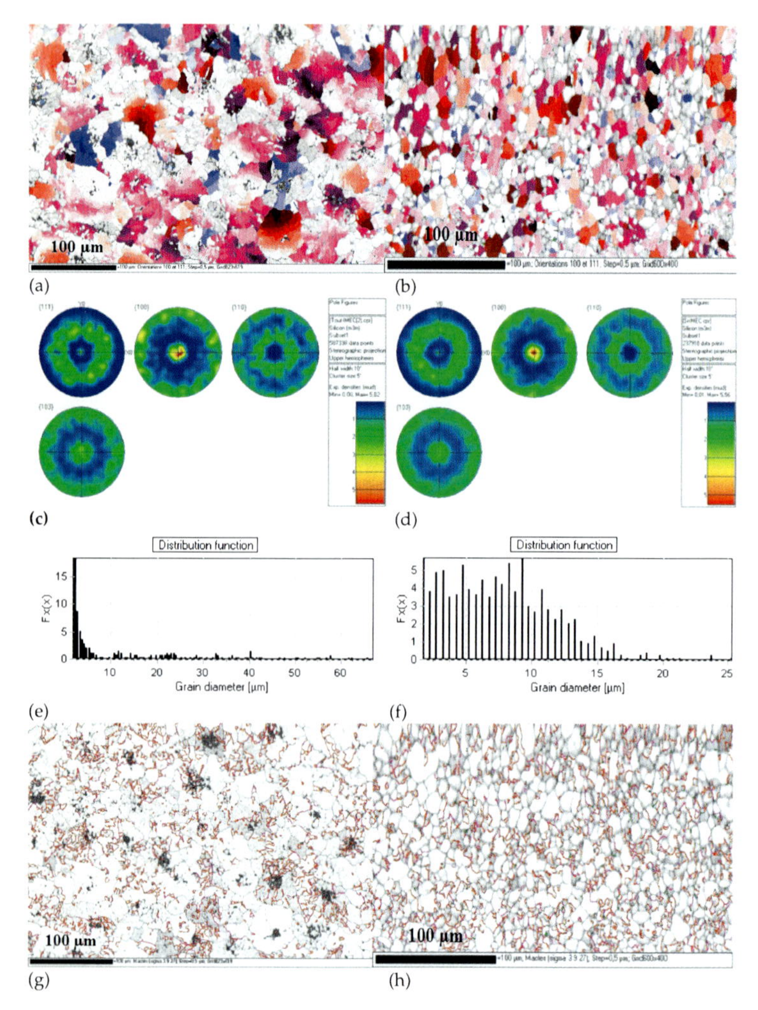

Figure 5.20 EBSD analysis of the grains for samples C and D. (a and b) Crystalline orientation map of the grains (colors: red = <100> ± 9°, blue = <111> ± 9°, violet = <103> ± 9°), (c and d) pole figures, (e and f) grain size distribution, and (g and h) map of the twins (red = $\Sigma 3$, purple = $\Sigma 9$, yellow = $\Sigma 27$). The experiments were carried out on FOx-coated alumina substrates, and the thickness of Al/a-Si is 300/500 nm. Reprinted with permission from Ref. [84], Copyright 2008, Elsevier.

The crystallographic defects and the comparison of defect distributions for samples C and D are shown in Fig. 5.21. Reactive grain boundaries (RGBs), LAGB, and twin boundaries of Σ3 (59°–60°), Σ9 (38°–40°), and Σ27 (35°–37°) are observed. The distribution of RGB is almost the same for both samples, while a very high distribution of Σ3 twin boundaries is observed for sample C. If the twin boundaries of sample D are compared with those of sample C, the distribution of high-energy grain boundaries (Σ9 and Σ27) increases when that of Σ3 (low-energy grain boundary) is decreasing. Furthermore, the distribution of the low-energy CSL boundary (Σ3) is significantly higher than that of high-energy grain boundaries. This distribution difference of CSL boundaries can be attributed to the fact that low-energy CSL boundaries are more likely to be growing in a larger area. During grain growth, some grain boundaries are eliminated when two grains separate or a tetrahedral grain disappears, while others are created when two grains impinge. Because grain growth leads to a reduction in the total interfacial area, more grain boundaries must be eliminated than created. However, higher-energy grain boundaries are preferentially eliminated from the network during grain growth, which leads to a higher population of low-energy grain boundaries [85]. The lower-energy boundaries, which are growing, have larger areas than the higher-energy boundaries. In analogy, the bigger grain size and higher Σ3 distribution can be explained for sample C.

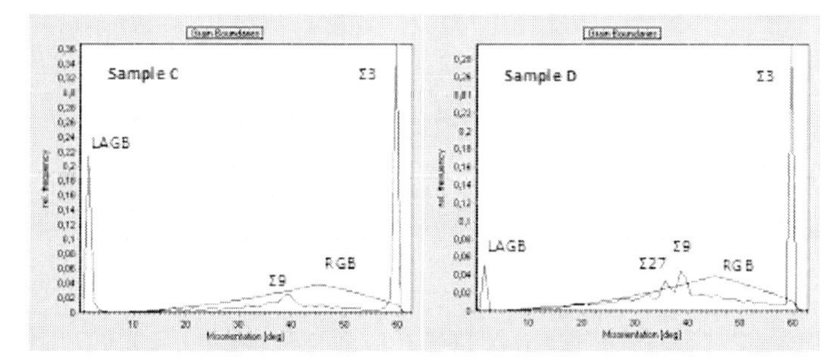

Figure 5.21 Distribution of CSL boundaries and random boundaries for (a) sample C and (b) sample D. Reprinted with permission from Ref. [84], Copyright 2008, Elsevier.

In conclusion, it is proved that the amount of oxygen as a pollutant during aluminum deposition and the grain size of the aluminum layer strongly affect the morphology of grains and the kinetics of growth. Aluminum layer containing a higher oxygen concentration gives more defective Al layer; and this causes faster diffusion in grains. Hence, enhancement in nucleation and reduction in grain size would be the result of increased oxygen content during Al deposition. Finally, poly-Si films with larger grain size are formed using Al films deposited at a lower rate.

5.3.5.4 Al/Si thickness ratio

The effect of Al layer thickness on morphology was studied by Nast et al. [67]. In the AIC sample for which the a-Si layer was only half the thickness of the Al layer showed a pore structure as shown in Fig. 5.22a. As the Si grains grow to a thickness equal to that of the Al film, further growth is confined to the plane bounded by the substrate and the interface between the two layers. Then, the lateral growth progressed with the lack of a Si source and the Si source was fully consumed before the neighboring grains met, which led to pores in the layers [74]. Therefore, shortage of Si prevents the formation of a continuous poly-Si layer prior to depletion of the silicon. For the Al layer thinner than a-Si (a-Si layer was three times as thick as the Al layer), a continuous poly-Si layer can be formed on the substrate and the layer could be as thick as the original Al film after the crystallization. A surplus of Si material resulted in the formation of a Si network, visible after Al etching, as depicted in Fig. 5.22b. In comparison, a thicker a-Si layer facilitates the formation of a continuous poly-Si layer but with a thickness determined by the original Al layer thickness. Thus, it is concluded that to form continuous poly-Si films under steady temperature conditions, the a-Si layer must be at least as thick as the Al layer.

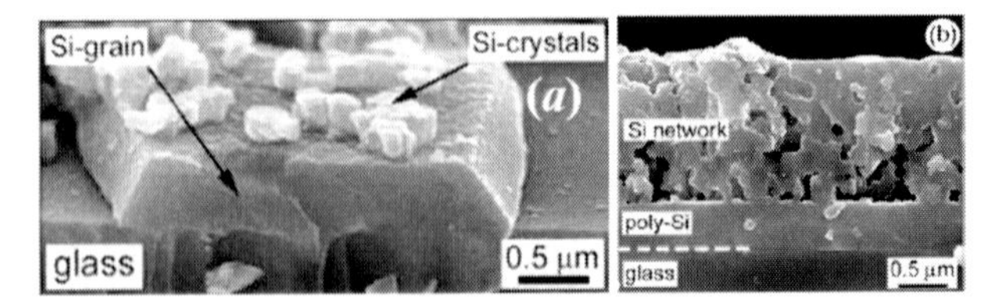

Figure 5.22 Cross-sectional SEM micrograph of a specimen after crystallization and Al etching. (a) The original a-Si layer was only half the thickness of the Al layer, and (b) the a-Si layer was thicker than the Al layer. Reprinted with permission from Ref. [67], Copyright 2000, American Institute of Physics.

5.3.5.5 Effect of hydrogen in a-Si

As discussed, the microstructural quality of AIC-grown films like crystallographic orientation and grain size distribution are known to be changed according to the exchange conditions as well as the stress state and degree of disorder of the precursor films. The case of a-Si films deposited by PECVD results in varied content of hydrogen, which is either in atomic form or in molecular form. In this connection, hydrogen content in the precursor a-Si films, which changes the state of a-Si films [86], is expected to have considerable effect on the resultant films. A comprehensive study of the influence of hydrogen content in a-Si:H deposited by electron cyclotron resonance (ECR)-PECVD on the crystallization temperature and defect distribution in an AIC-grown poly-Si layer is made by Prathap et al. [87].

The hydrogen concentration, which was estimated from ERDA measurements with respect to H_2 dilution of silane, showed that the concentration of hydrogen in a-Si:H films varied from 14.1 at.% to 19.2 at.% with an increase of H_2 dilution in the precursor gas mixture ($H_2/(H_2+SiH_4)$ = 0%–85%). The microstructure ratio of as-deposited a-Si:H films increased from 56.1% to 66.8% with an increase of the hydrogen dilution ratio from 0% to 85%. Incorporation of hydrogen atoms into the a-Si network modifies the structure and internal energy of the network. This might induce a larger free-energy difference between aluminum and a-Si networks.

The crystallographic orientation and defect analysis of AIC-grown films with respect to the hydrogen content were studied using EBSD measurements. The predominant orientation of the grains is found to be <100> and maintained for all hydrogen dilution ratios rather than defect distribution. The grains are found to have crystallographic defects of low angle (<15°), high angle (>15°), and CSL boundaries of Σ3, Σ9, and Σ27, as shown in Fig. 5.23. The analysis indicated that the frequency of low- and high-angle grain boundaries relatively increased significantly at higher dilution ratios/hydrogen content. The frequencies of different types of CSL boundaries, where the predominant CSL boundaries are, seem to be Σ3 followed by Σ9 and Σ27. However, the CSL grain boundaries seem to be less sensitive to the hydrogen content in the a-Si layers.

Generally, the exchange can be promoted by the release of elastic strain/stress energy and grain growth for the Al phase [53], and the microstructural quality of AIC-grown poly-Si films mainly depends on the rate of dissolution of silicon atoms in Al. On the other hand it is known that the local environment of a-Si influences the nucleation and hence the crystallization kinetics enormously.

The a-Si:H films were deposited over Al films at 250°C, and annealed above 425°C, which is greater than the deposition temperature of a-Si:H, at which evolution of hydrogen starts from the film surface. The evolution/migration of hydrogen leaves dangling bonds in the film. The increasing density of dangling bonds generates more structural disorder, in addition to the existing disorder in a-Si:H as hydrogen evolves. This phenomenon modifies the structure and internal energy of the a-Si:H system and hence the activation energy for the diffusion of silicon atoms into aluminum. The frequency of defects, mainly low-angle grain boundaries, also increases with hydrogen content, as shown in Fig. 5.23. The generation of defects in the AIC process is more closely related to the

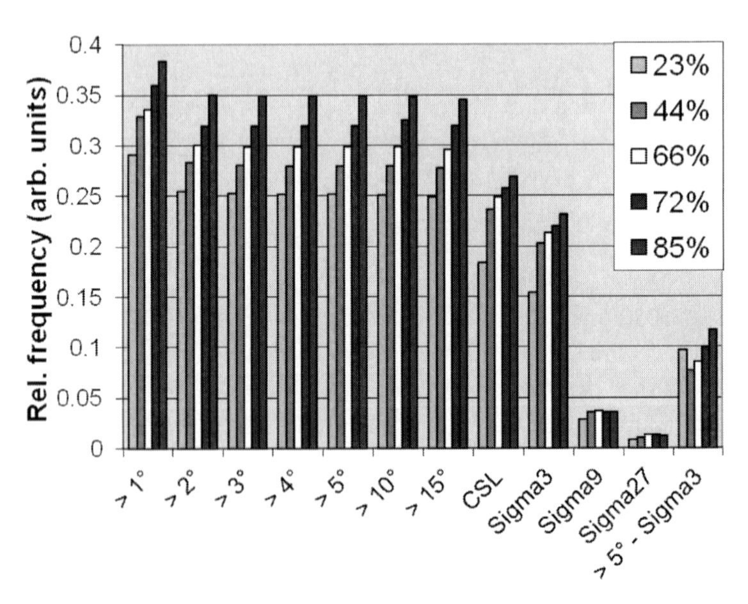

Figure 5.23 Defect distribution in AIC-grown poly-Si films (450°C for 8 hours) as a function of the hydrogen dilution ratio used for a-Si:H film deposition. Reprinted with permission from Ref. [88], Copyright 2011, Elsevier.

dissolution rate of Si atoms in Al. As with the increase of hydrogen content, the rate of dissolution is expected to increase with the aid of additional stress/strain energy generated at the interface of Al/a-Si:H due to the evolution of hydrogen when the precursor films have started to be annealed. This causes the generation of a pronounced density of defects at higher hydrogen dilution ratios/concentrations.

5.3.5.6 Effect of substrate

The surface roughness of the substrate controls the nucleation rate and therefore the final crystallographic quality of the formed poly-Si seed layer by AIC. This effect was verified using two kinds of substrates like FOx-coated alumina and GC substrates.

The alumina substrates were spun with FOx-25 (Dow Corning) solution to reduce the surface roughness of the alumina substrate [94], and GC is coated with SiN. The final surface roughness of the alumina and GC is found to be 160 nm and 8.2 nm, respectively, as shown in Fig. 5.24.

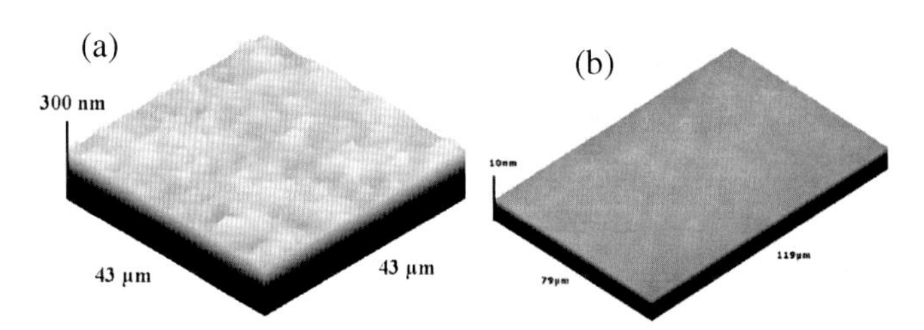

Figure 5.24 The surface morphologies of (a) FOx-coated alumina and (b) SiN-coated GC.

The EBSD results of the continuous poly-Si films fabricated at similar annealing conditions (475°C for 6 hours) on alumina and GC are shown in Fig. 5.25 [88]. The difference in grain size for both samples is obvious. The grain size distribution as extracted from the EBSD map is shown in Fig. 5.26. The shape of the distribution profile is significantly different from one to another substrate. The distribution of the grain size is more homogeneous, which implies that the nucleation rate during the crystallization process is constant in the case of GC substrates [76]. The average grain size of the AIC layer formed on GC reached 25.2 μm, while it is close to 8 μm on alumina substrates. The small average grain size on alumina can be correlated to the roughness of the surface prior to the a-Si deposition and the exchange process. Increased surface roughness favors enhanced nucleation rates, resulting in competing growth of grains, and finally leads to the formation of small grains.

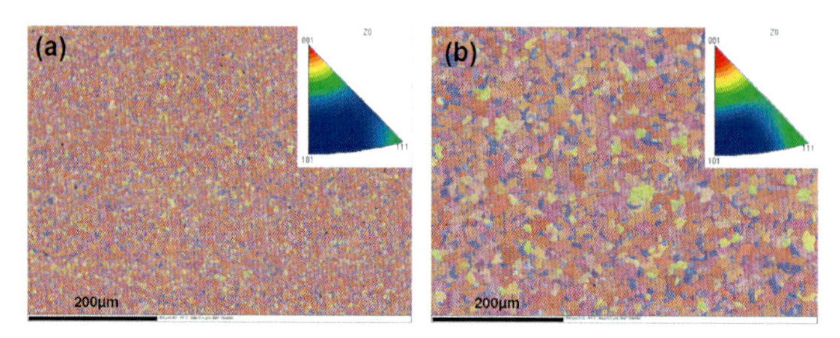

Figure 5.25 Orientation-mapping analysis of poly-Si layers formed by AIC at 475°C on (a) FOx-coated alumina and (b) SiN_x-coated GC. Inverse pole figures in the insets correspond to the (a) alumina sample and (b) GC sample (colors: red = <100> ± 15°, blue = <111> ± 15°, green = <101> ± 15°). The thickness of Al/a-Si is 200/370 nm, and Al layers were exposed to ambient air for one week prior to a-Si layer deposition that gives an AlO_x of a thickness of 4.5 nm. The surface was prepared by chemical mechanical polishing for EBSD measurements. Reprinted with permission from Ref. [88], Copyright 2011, Elsevier.

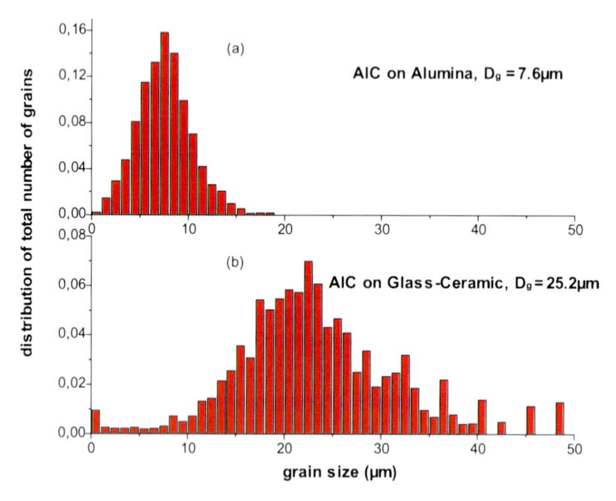

Figure 5.26 Average grain size (D_g) and grain size as a function of the substrate for the poly-Si films formed by AIC at 475°C on alumina and GC substrates. Reprinted with permission from Ref. [88], Copyright 2011, Elsevier.

With more detailed analysis on the <100> fiber texture deduced from these EBSD maps, the distribution of the angular deviation (ϑ) from the <100> orientation was made. The resulting frequency distribution as a function of ϑ is given in Fig. 5.27. The crystallographic orientations are comparable for AIC layers formed on alumina and GC. In this graph, the minimum deviation (5°) means a fully <100>-oriented layer, while the maximum deviation of 55° corresponds to the <111> plane. The majority of grains (60%–70%) in these poly-Si films has an orientation deviation between 5° and 25° relative to the <100> orientation. The occurrence of the <100> orientation is almost the same for SiN_x-coated GC and FOx-coated alumina used samples. The <110>-orientated silicon is indicated by deviation angles in the range of 30°–45°, while that is in the range of 45°–55° for <111>-orientated silicon. Either <101> or <111> shows lower distribution for both cases of samples than the <100> orientation. A 74% of the total surface was covered by <100> for the alumina used sample, while surface coverage of <101>- and <111>-orientated grains is only 18% and 8%, respectively. The surface fractions of <100>, <101>, and <111> are 72%, 20%, and 8%, respectively, when considering the GC substrate used poly-Si film. This results show that <100> is the preferred orientation for both types of samples. No twins are observed for the <100>-oriented grains, whereas the other orientations present a layer density of twins. The reason for this behavior is the orientation of the initial nucleus close to the <100> plane [88].

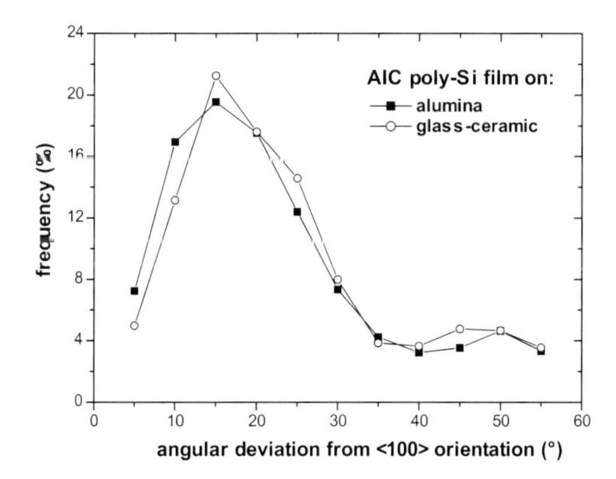

Figure 5.27 The distribution of misorientation angles as deduced from EBSD analysis for the poly-Si formed at 475°C by AIC on FOx-coated alumina and SiN_x-coated GC substrates. Reprinted with permission from Ref. [81], Copyright 2009, Université de Strasbourg, France.

The main crystallographic defects present in the continuous poly-Si layers are the LAGB (angle < 2°) and the CSL boundaries consisting of twin boundaries of the first order (Σ3), second order (Σ9), and third order (Σ27). Such defects are present in both films irrespective of the substrates, as shown in Fig. 5.28. This implies that the type of defects present in the grains does not depend on the substrate used, while the grain size does.

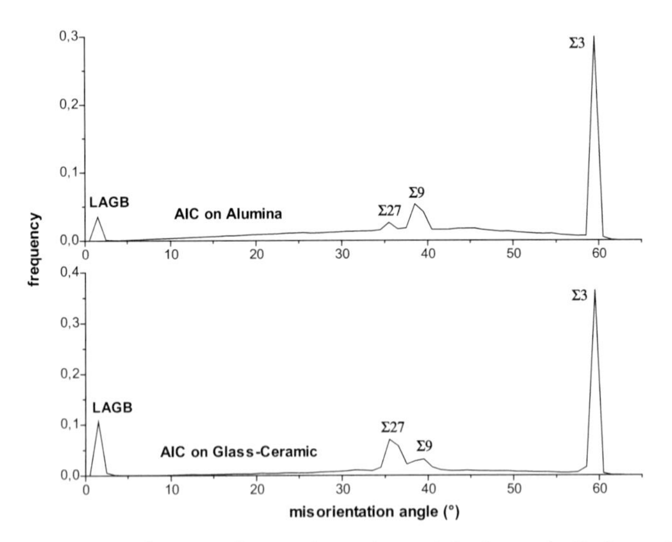

Figure 5.28 Distribution of misorientation angle for poly-Si formed by AIC on FOx-coated alumina and SiN_x-coated GC substrates at 475°C. Reprinted with permission from Ref. [81], Copyright 2009, Université de Strasbourg, France.

5.4 Preferential (100) Crystal Orientation of AIC Layers

In most of the experimental conditions the crystallites in AIC-grown films were grown in the (100) orientation. Schneider et al. [90] and Sarikov et al. [91] proposed a model to explain the observed experimental results. The model depends on the double-pyramid (octahedral) structure of {111} planes, which presents the lowest specific surface energy (σ). According to this model, the specific interface energy depends on the orientation of the crystal surface. In c-Si, the {111} faces have the lowest σ and are preferentially formed.

The result is an octahedral or double-pyramid structure of Si clusters with {111} faces, <110> ledges, and <100> tips, when the Si clusters are formed completely within the Al bulk.

The interface energy can also be reduced by nucleation at preferential sites, which is referred to as heterogeneous nucleation. In the Al layer, the grain boundaries and interfaces to the glass and the Al oxide layer are suitable places for heterogeneous nucleation. The formation of Si nuclei at the Al/AlO_x interface was demonstrated experimentally. In such a case, it is possible that the Si nuclei with minimum surface energy might be deviated from a perfect octahedral shape with a cutoff by the nucleation plane. The double-pyramidal Si nucleus shaped by the {111} Si planes and cut off by the interface plane with the AlO_x membrane is shown in Fig. 5.29. The position and orientation of the cutoff plane have been determined by the directing normal vector to this plane, **n**, tilted by angle θ ($0 \leq \theta \leq \pi$) from the Z axis and rotated by angle φ ($0 \leq \varphi \leq 2\pi$) relative to the X axis.

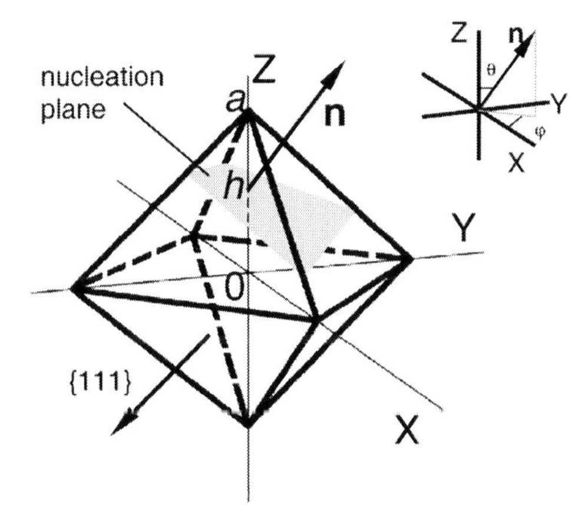

Figure 5.29 The sketch of the double-pyramid model. The part of the pyramid above the nucleation plane is cut off. Reprinted with permission from Ref. [91], Copyright 2007, Elsevier.

Monte-Carlo calculations of the change of Gibbs energy as a function of orientation of the nucleus at the interface showed that a nucleus aligned with the <100> tip perpendicular to the interface is energetically favorable. The degree of preferential orientation, R (100), can be controlled with a suitable barrier, a decrease of the annealing temperature, and/or the supersaturation of Al with Si.

5.5 Electrical Characteristics of AIC Poly-Si Layers

It was shown that the Al concentration within the crystallized poly-Si film after AIC evaluated from chemical analytic methods like energy dispersive spectroscopy (EDS)/secondary ion mass spectrometry (SIMS) [19, 92] was less than that of the concentration evaluated from the Hall effect method. Therefore, it was suggested that part of the incorporated Al is electrically inactive. The Al incorporated on interstitial and substitutional sites in the Si lattice can be thermally activated during heat treatments at 900°C so that it ends up on substitutional sites [93]. The AIC-grown poly-Si films, which are p-type conducting, can be converted to n-type poly-Si by overdoping, via diffusion carried out using spin-on dopant solutions above 900°C [94]. This enables the fabrication of n-type solar cells [95].

5.6 Applications of AIC for Solar Cell Fabrication

Usually the poly-Si films grown on foreign substrates using the AIC process are used as a seed layer, which serves as a template, for subsequent homoepitaxial growth of an absorber layer of a poly-Si thin-film solar cell. The highest-recorded photovoltaic conversion efficiency with an AIC-grown seed layer on an alumina substrate is 8% [96]. The epitaxy can be made using different ways like direct deposition at elevated temperatures using CVD, named as the high-temperature method, or deposition of a-Si at lower temperatures using evaporation or physical vapor deposition (PVD) and annealing above 600°C, named as the low-temperature method. The inherent characteristic of the (100) orientation of AIC films favorable for epitaxial thickening even at lower deposition temperatures below 600°C [97] also leads to lower defect density in comparison to the epitaxy made on other seed layers with different orientations. However, CVD methods are the most commonly used methods for high-quality epitaxial layers.

5.6.1 Epitaxy and Its Characteristics

The epitaxial thickening of AIC-grown poly-Si films on glass substrates was made using electron–cyclotron resonance chemical vapor deposition (ECRCVD) at temperatures below 600°C [98]. The cross-sectional image of such layers is shown in Fig. 5.30. It was found that the quality of the ECRCVD-grown films strongly depends on the orientation of the underlying seed layer grains. The epitaxial layers showed a (100) preferential orientation due to a mainly favorable orientation of the seed layers.

Figure 5.30 TEM cross section of an ECRCVD-grown film deposited on an AIC seed layer on glass. Reprinted with permission from Ref. [98], Copyright 2004, Elsevier. *Abbreviation*: TEM, transmission electron microscopy.

For the AIC layers thickened by vapor-phase epitaxy at 1200°C, the crystallographic orientations of the AIC seed layer were not completely reproduced in the epitaxial Si layer [89]. The changes in crystallographic orientation of the epilayer compared to its AIC seed are probably due to a faster growth rate of the highly deviated grains compared to (100) ones. In addition, the frequency of the coincidence index $\Sigma3$ was almost twice as high in the epi-Si layer as in the seed film. This behavior was attributed to defects induced by the high-temperature epitaxy process and seems to depend on the orientation of the seed grains.

The intragrain defect analysis was made on the epitaxial layers formed on AIC poly-Si layers by low (electron beam evaporation at 600°C) and high-temperature epitaxy (thermal CVD at 1130°C) [99]. It was found that the AIC seed layers themselves were found to be major sources of intragrain defects. The AIC seed layers prepared under different conditions can lead to different intragrain defect densities in the epitaxial layers, as shown in Figs. 5.31a and 5.31c, for which the epitaxy is made by the low-temperature method. For the same seed layer, the degree of defects and their evolution are similar irrespective of the method of epitaxy formation, as shown in Figs. 5.31b and 5.31c, for which the epitaxy is made by low-temperature and high-temperature methods, respectively. Therefore, the microstructural quality of the AIC seed layer defines the ultimate quality of the epitaxial layer and hence the photovoltaic conversion efficiency [100]. It was also proved that defect formation could be influenced by the difference in the coefficient of thermal expansion between the substrate and silicon during crystallization [101].

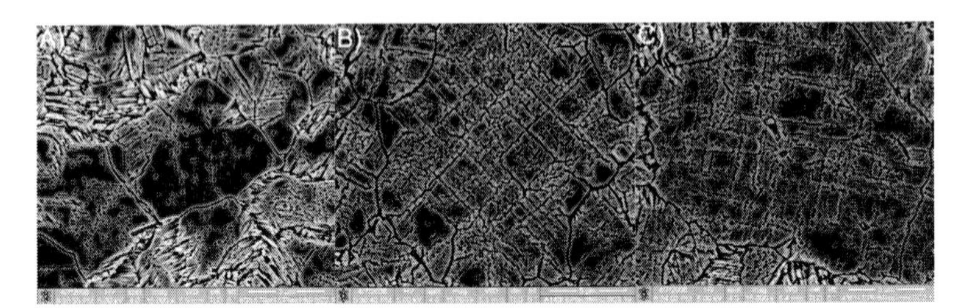

Figure 5.31 Top-view SEM images after defect etching of epitaxial samples. Reprinted with permission from Ref. [99], Copyright 2009, Elsevier.

The crystalline orientation-mapping image for the epitaxial poly-Si films formed by LPCVD at 1000°C on the AIC seed layer on FOx-coated alumina is shown in Fig. 5.32a. The grain size can be estimated by summing over the number of data points in each grain. The distribution of the grain size, Fig. 5.32b, is relatively homogenous (σ = 3.7 μm), which means diffusion-limited grain growth [76] for epitaxy of poly-Si on the AIC seed layer during crystallization. The poly-Si absorber layer exhibits a ~6.9 μm average grain size. This result is consistent with that of the underlying AIC poly-Si seed layer.

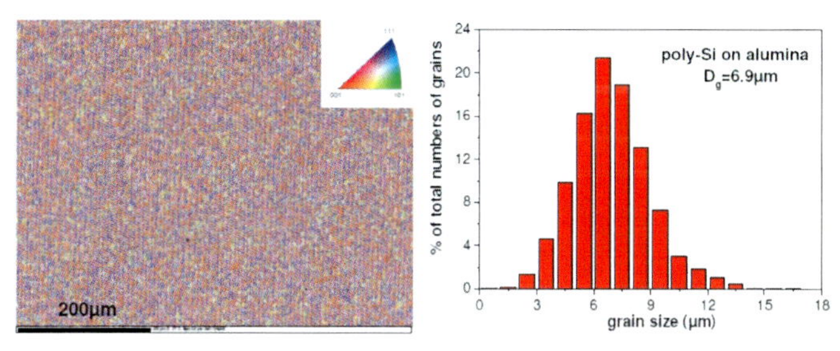

Figure 5.32 (a) Orientation-mapping analysis and (b) grain size distribution of epitaxial poly-Si layers on AIC seed layers grown on FOx-coated alumina at a temperature of ~1000°C. The surface was prepared by chemical mechanical polishing for EBSD measurements. Reprinted with permission from Ref. [95], Copyright 2010, Elsevier.

The distribution of the deviation angle (θ), which is accumulated by segments of 5° compared to the <100> orientation, is plotted in Fig. 5.33. The minimum deviation (5°) means a fully <100>-oriented layer, while the maximum deviation of 54.7° corresponds to the (111) plane. The majority of grains (64.6%) in these poly-Si films have an orientation deviation between 5° and 25°, which is indicative of <100> (5°–25°)-oriented grains. Besides, the distribution of <110> (30°–45°) and <111> (45°–55°) orientation for films is 17.1% and 18.3%, respectively. This result indicates that the preferred orientation of the poly-Si epilayer is <100>.

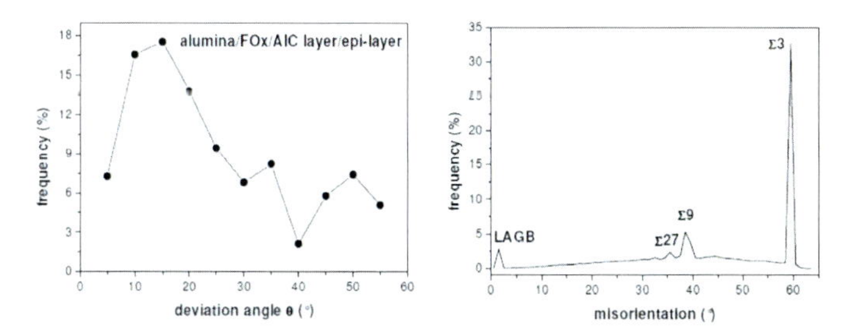

Figure 5.33 (a) Deviation angle from the <100> fiber texture and (b) distributions of CSL boundaries for the poly-Si layer deposited by CVD at 1000°C on FOx-coated alumina using the AIC seed layer approach. Reprinted with permission from Ref. [95], Copyright 2010, Elsevier.

The main crystallographic defects present in the continuous poly-Si layers are the LAGB (angle < 2°), and the CSL boundaries are mainly twin boundaries of the first order (Σ3) and of less proportion in the second order (Σ9) and third-order twins (Σ27), as shown in Fig. 5.33b. The high fraction of high-angle boundaries around 60° in poly-Si is due to the existence of Σ3 boundaries that have around 33% distribution. Such defects are present irrespective of the deposition temperature and the substrate.

The results on epitaxial layers reveal that the defects are inherent because of the microstructural defects in AIC seed layers. The coherent first-order (Σ3) twins are electrically inactive. Higher-order twins and faceted twin boundaries, for example, in most cases, coherent first-order twins with incoherent segments are only weakly electrically active [102, 103]. A secondary dislocation network frequently leads to electronic levels in the bandgap and hence electrical activity in terms of minority carrier recombination and potential barriers. Random grain boundaries are generally electrically active [104]. Thus these highly disordered grain boundaries, if not neutralized, could be detrimental for electronic transport in silicon solar cells devices. Hydrogen passivation of such defected silicon layers might be needed

5.6.2 Solar Cell Results

With p-type cells made on alumina substrates using AIC and thermal CVD followed by plasmatexturing and optimization of the cell structure, a maximum cell efficiency of 8.0% was achieved at IMEC, Belgium [96], as shown in Fig. 5.34. The corresponding structure of the poly-Si thin-film solar cell is in an alumina substrate/spin-on oxide/p+-type AIC seed layer/p+-type Si back-surface field (BSF) (epitaxial)/p-type Si absorber layer (epitaxial)/i/n+-type a-Si:H emitter (PECVD)/indium tin oxide (ITO). Plasmatexturing of these thin cells improved the current and voltage and hence the efficiency of 1.3%.

Both p-type (p^+pn^+) and n-type (n^+np^+) polysilicon heterojunction–based solar cells were fabricated on an alumina substrate at CNRS-InESS, Strasbourg, France, and their photovoltaic performances compared. The results of *I*–*V* measurements for p- and n-type solar cells are shown in Fig. 5.35. The n-type cells showed J_{sc} = 16.8 mAcm^{-2}, V_{oc} = 462 mV, fill factor (*FF*) = 0.645, and η = 5.0%. The n-type cell exhibits a much higher short-circuit current J_{sc}, which is almost two times more than that of the p-type cell. These cell parameters are obtained without control over the doping profile, emitter, and hydrogen passivation and without light trapping. As a result of the current increase, using an n-type cell led to an increase in efficiency from 2.9% to 5.0% compared to the p-type cell. Importantly, the spectral response of the n-type cell is also much widened than that of the p-type cell over a large part of the spectrum. This led to an enhancement in L_{eff} for the n-type cell (~2.6 μm) compared to that of p-type (0.9 μm). The efficiency of 5.0% was realized for the n-type cell, which is the best efficiency so far for n-type poly-Si solar cells based on the AIC approach [95].

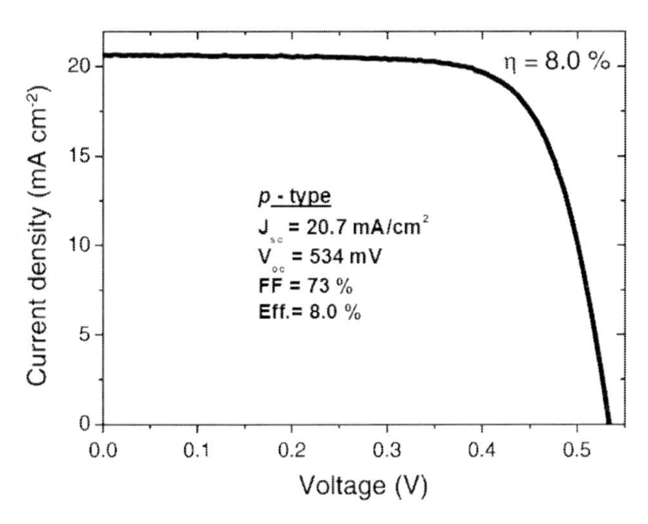

Figure 5.34 Illuminated current–voltage curve of a p-type pc-Si solar cell employing an AIC seed layer alumina substrate. Reprinted with permission from Ref. [96], Copyright 2008, Elsevier.

Studies concerning the efficiency of poly-Si thin-film solar cells based on AIC seed layers reveal that efficiency is not limited by grain size but by intragrain defects. Cells with epitaxial layers deposited on an AIC seed layer using high-rate electron beam evaporation at 600°C showed an efficiency of 3.2%. The current density was limited to 11.9 mA/cm^2, together with an open circuit voltage, V_{oc}, of 407 mV and an *FF* of 67% [105]. This proves the prevalent influence of the intragrain defects on the ultimate photovoltaic conversion efficiency of the poly-Si solar cells.

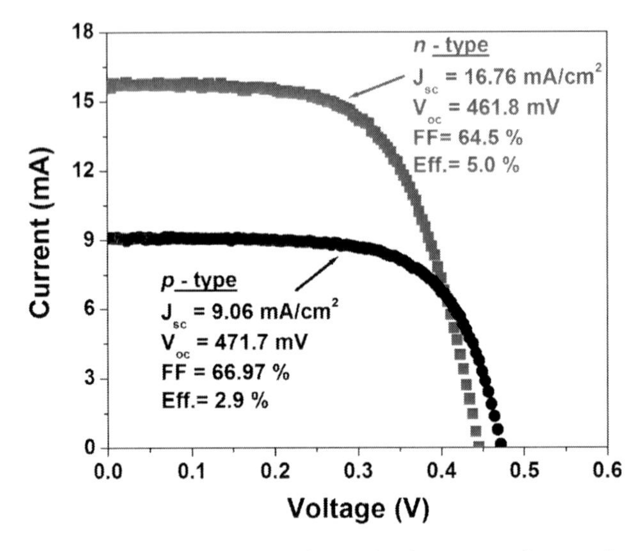

Figure 5.35 Illuminated current–voltage (*I*–*V*) curves of p- and n-type poly-Si solar cells on an alumina substrate. Reprinted with permission from Ref. [88], Copyright 2011, Elsevier.

5.7 Summary and Conclusions

In this chapter, crystallization of precursor a-Si films by AIC is introduced elaborately, together with the physics of the AIC process. The AIC process and the microstructural quality of the resultant poly-Si film are highly influenced by the annealing temperature, interface of Al/a-Si, nature of a-Si, etc. The as-grown films are found to have high density of microstructural defects, which are electrically active. These defects are brought forward to the epitaxial layer—hence a degradation in the ultimate electronic quality of the absorber layer. Relatively, the defect density in high-temperature-deposited epitaxial films is lower compared to low-temperature-deposited films. The solar cell employing the AIC seed layer showed a limited photovoltaic conversion efficiency of 8%, which is due to the poor microstructural quality as well as impurities diffused from the substrates during high-temperature processing. Though the studies to fabricate thin-film solar cells using poly-Si seed layers formed by

AIC are well progressed, fundamental physical characteristics of AIC-grown poly-Si films still need to be improved as a solar cell material. Therefore, future developments in this process are still expected for highly productive solar cells.

Acknowledgments

The authors would like to acknowledge I. Gordon, IMEC, Belgium, for fruitful collaboration in the fabrication of solar cell devices. Also, the research on AIC process development has been funded from ANR through the projects POLYSIVERRE, CRISILAL, and SILASOL. We appreciated the contributions of our colleagues at InESS-CNRS-UdS, Strasbourg, France.

References

1. http://www.solarbuzz.com/facts-and-figures/retail-price-environment/module-prices.
2. Bergmann, R. B., Rinke, T. J. (2000) Perspectives of crystalline Si thin film solar cells: a new era of thin monocrystalline Si films. *Progress in Photovoltaics: Research and Applications,* **8**, 451–464.
3. Zhao, J., Wang, A., Green, M. A. (1999) 24.5% efficiency silicon PERT cells on MCZ substrates and 24.7% efficiency PERL cells on FZ substrates. *Progress in Photovoltaics: Research and Applications*, **7**, 471–474.
4. Green, M. A. (2000) The future of crystalline silicon solar cells. *Progress in Photovoltaics: Research and Applications,* **8**, 127–139.
5. Brendel, R. (2003) *Thin-Film Crystalline Silicon Solar Cells* (Wiley VCH Verlag GmbH).
6. Tiedje, T., Yablonovitch, E., Cody, G. D., Brooks, B. G. (1984) Limiting efficiency of silicon solar cells. *IEEE Transactions on Electron Devices*, **ED-31**, 711–716.
7. Matsuyama, T., Tanaka, M., Tsuda, S., Nakano, S., Kuwano, Y. (1993) Improvement of n-type poly-Si film properties by solid phase crystallization method. *Japanese Journal of Applied Physics*, **32**, 3720–3728.
8. Morimoto, R., Izumi, A., Masuda, A., Matsumura, H. (2002) Low-resistivity phosphorus-doped polycrystalline silicon thin films formed by catalytic chemical vapor deposition and successive rapid thermal annealing. *Japanese Journal of Applied Physics*, **41**, 501–506.

9. Werner, J. H., Dassow, R., Rinke, T. J., Kohler, J. R., Bergmann, R. B. (2001) From polycrystalline to single crystalline silicon on glass. *Thin Solid Films*, **383**, 95–100.
10. Bergmann, R. B. (1999) Crystalline Si films on foreign substrates for electronic applications. *Recent Research Developments in Crystal Growth Research*, **1**, 241.
11. Bergmann, R. B. (1999) Crystalline Si thin-film solar cells: a review. *Applied Physics A: Materials Science & Processing*, **69**, 187–194.
12. Prathap, P., Slaoui, A., Ducros, C., Baclet, N., Reydet, P. L. (2009) Development of polycrystalline silicon films on flexible metallic substrates by aluminium induced crystallization. *Applied Physics A: Materials Science & Processing*, **97**, 45–54.
13. Reber, S., Hurrle, A., Eyer, A., Willeke, G. (2004) Crystalline silicon thin-film solar cells-recent results at Fraunhofer ISE, *Solar Energy*, **77**, 865–875.
14. Takami, A., Arimoto, S., Morikawa, H., Hamamoto, S., Ishihara, T., Kumabe, H., Murotani, T. (1994) High efficiency (16.45%) thin film silicon solar cells prepared by zone-melting recrystallization. *Proceedings of the 12th European Photovoltaic Solar Energy Conference*, **59**.
15. Hebling, C., Glunz, S. W., Schumacher, J. O., Knobloch, J. (1997) High-efficiency (19.2%) silicon thin-film solar cells with interdigitated emitter and base front-contacts. *Proceedings of the 14th European Photovoltaic Solar Energy Conference*, 2318–2321.
16. Green, M. A. (2009), Polycrystalline silicon on glass for thin-film solar cells. *Applied Physics A: Materials Science & Processing*, **96**, 153–159.
17. Basore, P. A. (2006) CSG-2: expanding the production of a new polycrystalline silicon PV technology. *Proceedings of the 21st European Photovoltaic Solar Energy Conference*, 544.
18. Bergmann, R. B., Kohler J., Dassow, R., Zaczek, C., Werner, J. H. (1998) Nucleation and growth of crystalline silicon films on glass for solar cells. *Physica Status Solidi A*, **166**, 587–602.
19. Nast, O., Puzzer, T., Koschier, L. M., Sproul, A. B., Wenham, S. R. (1998) Aluminum-induced crystallization of amorphous silicon on glass substrates above and below the eutectic temperature. *Applied Physics Letters*, **73**, 3214–3216.
20. Ohdaira, K., Shiba, K., Takemoto, H., Fujiwara, T., Endo, Y., Nishizaki, S., Jang, Y. R., Matsumura, H. (2009) Precursor cat-CVD a-Si films for the formation of high-quality poly-Si films on glass substrates by flash lamp annealing. *Thin Solid Films*, **517**, 3472–3475.

21. Tuzun, O. Slaoui, A. Maurice, C. Vallon, S. (2010). Growth kinetics and polysilicon formation by aluminium-induced crystallization on glass-ceramic substrates, *Applied Physics A*, **99**, 53–61.

22. Olson, G. L., Roth, J. A. (1994) *Thin Films and Epitaxy*, Vol. 3 of *Solide Phase Epitaxy Handbook of Crystal Growth* (ed. Hurle, D. T. J.) (Elsevier Science).

23. Radnoczi, G., Robertson, A., Hentzell, H. T. G., Gong, S. F., Hasan, M. A. (1991) Al-induced crystallization of a-Si. *Journal of Applied Physics*, **69**, 6394–6399.

24. Cheng, C. F., Poon, V. M. C., Kok, C. W., Chan, M. (2003). Modelling of grain growth mechanism by nickel silicide reactive grain boundary effect in metal-induced-lateral-crystallisation. *IEEE Transactions on Electron Devices*, **50**, 1467–1474.

25. Hayzelden, C., Bastone, J. L. (1993) Silicide formation and silicide-mediated crystallization of nickel-implanted amorphous silicon thin films. *Journal of Applied Physics*, **73**, 8279–8290.

26. Joo, S K. (2005) Understanding of metal-induced lateral crystallization mechanism: a low temperature crystallization phenomenon. *Electronic Materials Letters,* **1**, 7–10.

27. Joshi, A. R., Krishnamohan, T., Saraswat, K. C. (2003) A model for crystal growth during metal induced lateral crystallization of amorphous silicon. *Journal of Applied Physics*, **93**, 175–181.

28. Lee, S. W., Joo, S. K. (1996) Low temperature poly-Si thin-film transistor fabrication by metal-induced lateral crystallization. *IEEE Electron Device Letters*, **17**, 160–162.

29. Yoon, S. Y., Kim, K. H., Kim, C. O., Oh, J. Y., Jang, J. (1997) Low temperature metal induced crystallization of amorphous silicon using a Ni solution. *Journal of Applied Physics*, **82**, 5865–5867.

30. Slaoui, A., Siffert, P. (2004) Polycrystalline silicon films for electronic devices, in *Silicon-Evolution and Future of a Technology* (ed. Krimmel, S. E.) (Springer Verlag).

31. Ottaviani, G., Sigurd, D., Marrello, V., Mayer, J. W., McCaldin, J. O. (1974) Crystallization of Ge and Si in metal films. *Journal of Applied Physics*, **45**, 1730–1740.

32. Maa, J.-S., Lin, S.-J. (1979) Low temperature crystallization of amorphous silicon films. *Thin Solid Films*, **64**, 63–64.

33. Bian, B., Yie, J., Li, B., Wu, Z. (1993) Fractal formation in a-Si:H/Ag/a-Si:H films after annealing. *Journal of Applied Physics*, **73**, 7402–7406.

34. Hultman, L., Robertsson, A., Hentzell, H. T. G., Engström, I., Psaras, P. A. (1987) Crystallization of amorphous silicon during thin-film gold reaction. *Journal of Applied Physics*, **62**, 3647–3655.

35. Gong, S. F., Hentzell, H. T. G., Robertsson, A. E., Hultman, L., Hörnström, S.-E., Radnoczi, G. (1987) Al-doped and Sb-doped polycrystalline silicon obtained by means of metal-induced crystallization. *Journal of Applied Physics*, **62**, 3726–3732.

36. Russell, S. W., Li, J., Mayer, J. W. (1991) In situ observation of fractal growth during a-Si crystallization in a Cu_3Si matrix. *Journal of Applied Physics*, **70**, 5153–5155.

37. Lee, S. W., Jeon, Y. C., Joo, S. K. (1995) Pd induced lateral crystallization of amorphous Si thin films. *Applied Physics Letters*, **66**, 1671–1673.

38. Liu, G., Fonash, S. J. (1989) Selective area crystallization of amorphous silicon thin films by low-temperature rapid thermal annealing. *Applied Physics Letters*, **55**, 660–662.

39. Shim, J. Y., Park, S. W., Baik, H. K. (1997) Silicide formation in cobalt/amorphous silicon, amorphous Co-Si and bias-induced Co-Si films. *Thin Solid Films*, **292**, 31–39.

40. Kim, J., Piwowar, A. M., Nowak, R., Gradella, J. A., Jr., Anderson, W. A. (2007) Cobalt-induced polycrystalline silicon film growth. *Applied Surface Science*, **253**, 3053–3056.

41. (1957) Alloys of platinum metals with boron, phosphorus and silicon. *Platinum Metals Review*, **1**(4), 136–137.

42. Woodward, J. T., IV, Zasadzinski, J. A. (1996) Thermodynamic limitations on the resolution obtainable with metal replicas. *Journal of Microscopy*, **184**, 157–162.

43. Hayzelden, C., Batstone, J. L. (1993) Silicide formation and silicide-mediated crystallization of nickel implanted amorphous silicon thin films. *Journal of Applied Physics*, **73**, 8279–8289.

44. Jin, Z., Bhat, G. A., Yeung, M., Kwok, H. S., Wong, M. (1998) Nickel induced crystallization of amorphous silicon thin films. *Journal of Applied Physics*, **84**, 194–200.

45. Tu, K. N., Mayer, J. W., Feldman, L. C. (1992) *Electronic, Thin Film Science for Electrical Engineers and Materials Scientists* (Macmillan Publishing, New York, USA), 274.

46. Sze, S. M. (1981) *Physics of Semiconductor Devices*, 2nd ed. (Wiley, New York, USA), 21.

47. Murray, J. L., McAlister, A. J. (1984) The Al-Si (aluminum-silicon) system. *Journal of Phase Equilibria and Diffusion*, **5**, 74–84.

48. Nast, O., Hartmann, A. J. (2000) Influence of interface and Al structure on layer exchange during aluminum-induced crystallization of amorphous silicon, *Journal of Applied Physics*, **88**, 716–724.

49. Davis, J. R. (193) *ASM Specialty Handbook: Aluminium and Aluminium Alloys* (ASM International, OH, USA).

50. McCaldin, J. O., Sankur, H. (1971) Diffusivity and solubility of Si in the Al metallization of integrated circuits. *Applied Physics Letters*, **19**, 524–527.

51. Seto, S., Sakaguchi, T., Nakabayashi, Y., Matsumoto, S., Murota, J., Wada, K., Abe, T. (2004) Determination of silicon self-interstitial diffusivity using isotopically pure ^{30}silicon multi-layer. *Materials Science and Engineering: B*, **114–115**, 334–338.

52. Zhao, Y. H., Wang, J. Y., Mittemeijer, E. (2004) Microstructural changes in amorphous Si/crystalline Al thin bilayer films upon annealing. *Applied Physics A: Materials Science & Processing,* **79**, 681–690.

53. He, D., Wang, J. Y., Mittemeijer, E. J. (2005) Reaction between amorphous Si and crystalline Al in Al/Si and Si/Al bilayers: microstructural and thermodynamic analysis of layer exchange. *Applied Physics A: Materials Science & Processing*, **80**, 501–509.

54. Khalifa, F. A., Naseem, H. A., Shultz, J. L., Brown, W. D. (1999) Large-grained poly-silicon thin films by aluminium-induced crystallisation of microcrystalline silicon. *Thin Solid Films*, **355–356**, 343–358.

55. Benedictus, R., Böttger, A., Mittemeijer, E. J. (1996) Thermodynamic model for solid-state amorphization in binary systems at interfaces and grain boundaries. *Physical Review B*, **54**, 9109–9125.

56. Jeurgens, L. P. H., Sloof, W. G., Tichelaar, F. D., Mittemeijer, E. J. (2000) Thermodynamic stability of amorphous oxide films on metals: application to aluminum oxide films on aluminum substrates. *Physical Review B*, **62**, 4707–4719.

57. Wang, J. Y., Wang, Z. M., Mittemeijer, E. J. (2007) Mechanism of aluminum-induced layer exchange upon low-temperature annealing of amorphous Si/polycrystalline Al bilayers. *Journal of Applied Physics*, **102**, 113523-1–113523-9.

58. Murray, J. L., McAlister, A. J. (1984). The aluminium-silicon system. *Bulletin of Alloy Phase Diagrams*, **5**, 74–84.

59. Wang, J. Y., He, D., Zhao, Y. H., Mittemeijer, E. J. (2006) Wetting and crystallization at grain boundaries: origin of aluminum-induced crystallization of amorphous silicon. *Applied Physics Letters*, **88**, 061910–061912.

60. Hiraki, A. (1984) Low temperature reactions at Si/metal interfaces: what is going on at the interfaces? *Surface Science Reports*, **3**, 357–412.
61. He, D., Wang, J. Y., Mittemeijer, E. J. (2006) Thermodynamic and kinetic criteria for layer exchange in amorphous silicon/crystalline aluminium bilayers during annealing. *Scripta Materialia*, **54**, 559–561.
62. Klein, J., Schneider, J., Muske, M., Gall, S., Fuhs, W. (2004) Aluminium-induced crystallisation of amorphous silicon: influence of the aluminium layer on the process. *Thin Solid Films*, **451–452**, 481–484.
63. Schneider, J., Klein, J., Muske, M., Gall, S., Fuhs, W. (2004) Aluminum-induced crystallization of amorphous silicon: preparation effect on growth kinetics. *Journal of Non-Crystalline Solids*, **338–340**, 127–130.
64. Sarikov, A., Schneider, J., Muske, M., Gall, S., Fuhs, W. (2006) Theoretical study of the kinetics of grain nucleation in the aluminium-induced layer-exchange process. *Journal of Non-Crystalline Solids*, **352**, 980–983.
65. del Canizo, C., Luque, A. (2000) A comprehensive model for the gettering of lifetime-killing impurities in silicon. *Journal of the Electrochemical Society*, **147**, 2685–2692.
66. Schneider, J., Klein, J., Muske, M., Gall, S., Fuhs, W. (2005) Depletion regions in the aluminum-induced layer exchange process crystallizing amorphous Si. *Applied Physics Letters*, **87**, 031905-1–031905-3.
67. Nast, O., Wenham, S. R. (2000) Elucidation of the layer exchange mechanism in the formation of polycrystalline silicon by aluminum-induced crystallization. *Journal of Applied Physics*, **88**, 124–132.
68. Zener, C. (1949) Theory of growth of spherical precipitates from solid solution. *Journal of Applied Physics*, **20**, 950–953.
69. Pihan, E., Slaoui, A., Roca i Cabarrocas, P., Focsa, A. (2004) Polycrystalline silicon films by aluminium-induced crystallisation: growth process vs. silicon deposition method. *Thin Solid Films*, **451–452**, 328–333.
70. Sarikov, A., Schneider, J., Klein, J., Muske, M., Gall, S. (2006) Theoretical study of the initial stage of the aluminium-induced layer-exchange process. *Journal of Crystal Growth*, **287**, 442–445.
71. Ekanayake, G., Quinn, T., Reehal, H. S. (2006) Large-grained poly-silicon thin films by aluminium-induced crystallisation of microcrystalline silicon. *Journal of Crystal Growth*, **293**, 351–358.
72. Sze, S. M. (1981) *Physics of Semiconductor Devices*, 2nd ed. (Wiley, New York, USA), 21.
73. Pihan, E., Slaoui, A., Maurice, C. (2007) Growth kinetics and crystallographic properties of polysilicon thin films formed by

aluminium-induced crystallization. *Journal of Crystal Growth*, **305**, 88–98.

74. Kim, H., Lee, G., Kim, D., Lee, S. H. (2002) A study of polycrystalline silicon thin films as a seed layer in liquid phase epitaxy using aluminum-induced crystallization. *Current Applied Physics*, **2**, 129–133.

75. Gall, S., Muske, M., Sieber, I., Nast, O., Fuhs, W. (2002) Aluminum-induced crystallization of amorphous silicon. *Journal of Non-Crystalline Solids*, **299–302**, 741–745.

76. Crespo, D., Pradell, T., Clavaguera, N., Clavaguera-Mora, M. T. (1997) Kinetic theory of microstructural evolution in nucleation and growth processes. *Materials Science and Engineering: A*, **238**, 160–165.

77. Ohring, M. (2002) *Material Science of Thin Films*, 2nd ed. (Academic Press, San Diego, USA), 685.

78. Schneider, J. (2005) *Nucleation and Growth during the Formation of Polycrystalline Silicon Thin Films.* PhD thesis (Universität Berlin, Berlin).

79. Gall, S., Schneider, J., Muske, M., Sieber, I., Nast, O., Fuhs, W. (2002) Poly-Si seeding layers by aluminium-induced crystallisation. *Proceedings of the PV in Europe*, 87–90.

80. Jeurgens, L. P. H., Sloof, W. G., Tichelaar, F. D., Mittemeijer, E. J. (2002) Structure and morphology of aluminium-oxide films formed by thermal oxidation of aluminium. *Thin Solid Films*, **418**, 89–101.

81. Tuzun, O. (2009) *Polycrystalline Silicon Films by Aluminium Induced Crystallization and Epitaxy: Synthesis, Characterizations and Solar Cells.* PhD thesis (Université de Strasbourg, France).

82. Nast, O. (2000) *The Aluminium-Induced Layer Exchange Forming Polycrystalline Silicon on Glass Fort Hin-Film Solar Cells.* PhD thesis (Philipps-Universität, Marlburg).

83. Graper, E. B. (1971) Deposition of aluminum from an electron beam source. *Journal of Vacuum Science & Technology*, **9**, 33–36.

84. Tuzun, O., Auger, J. M., Gordon, I., Focsa, A., Montgomery, P. C., Maurice, C., Slaoui, A., Beaucarne, G., Poortmans, J. (2008) EBSD analysis of polysilicon films formed by aluminium induced crystallization of amorphous silicon. *Thin Solid Films*, **516**, 6882–6887.

85. Dillon, S. J., Rohrer, G. S. (2009) Mechanism for the development of anisotropic grain boundary character distributions during normal grain growth. *Acta Materialia*, **57**, 1–7.

86. Kawamoto, N., Masuda, A., Matsuo, N., Seri, Y., Nishimori, T., Kitamon, Y., Matsumura, H., Hamada, H., Miyoshi, T. (2006) Grain enlargement of

polycrystalline silicon by multipulse excimer laser annealing: role of hydrogen. *Japanese Journal of Applied Physics*, **45**, 2726–2730.

87. Prathap, P., Tuzun, O., Roques, S., Schmitt, S., Maurice, C., Slaoui, A. (2011) Microstructural tuning of polycrystalline silicon films from hydrogen diluted amorphous silicon films by AIC. *Physica Status Solidi C*, **8**, 859–862.
88. Prathap, P., Tuzun, O., Madi, D., Slaoui, A. (2011) Thin film silicon solar cells by AIC on foreign substrates. *Solar Energy Materials and Solar Cells*, **95**, S44–S52.
89. Pihan, E., Focsa, A., Slaoui, A., Maurice, C. (2006) Crystallographic analysis of polysilicon films formed on foreign substrates by aluminium induced crystallisation and epitaxy. *Thin Solid Films*, **511–512**, 15–20.
90. Schneider, J., Sarikov, A., Klein, J., Muske, M., Sieber, I., Quinn, T., Reehal, H. S., Gall, S., Fuhs, W. (2006) A simple model explaining the preferential (1 0 0) orientation of silicon thin films made by aluminum-induced layer exchange. *Journal of Crystal Growth*, **287**, 423–427.
91. Sarikov, A., Schneider, J., Muske, M., Sieber, I., Gall, S. (2007) A model of preferential (100) crystal orientation of Si grains grown by aluminium-induced layer-exchange process. *Thin Solid Films*, **515**, 7465–7468.
92. Nast, O., Brehme, S., Neuhaus, D. H., Wenham, S. R. (1999) Polycrystalline silicon thin films on glass by aluminum-induced crystallization. *IEEE Transactions on Electron Devices*, **46**, 2062–2068.
93. Tsaur, B. Y., Turner, G. W., Fan, J. C. C. (1981) Efficient Si solar cells by low-temperature solid-phase epitaxy. *Applied Physics Letters*, **39**, 749–751.
94. Tuzun, O., Slaoui, A., Gordon, I., Focsa, A., Ballutaud, D., Beaucarne, G., Poortmans, J. (2008) N-type polycrystalline silicon films formed on alumina by aluminium induced crystallization and overdoping. *Thin Solid Films*, **516**, 6892–6895.
95. Tuzun, O., Qiu, Y., Slaoui, A., Gordon, I., Maurice, C., Venkatachalam, S., Chatterjee, S., Beaucarne, G., Poortmans, J. (2010) Properties of n-type polycrystalline silicon solar cells formed by aluminium induced crystallization and CVD thickening. *Solar Energy Materials and Solar Cells*, **94**, 1869–1874.
96. Gordon, I., Carnel, L., Gestel, D. V., Beaucarne, G., Poortmans, J. (2008) Fabrication and characterization of highly efficient thin-film polycrystalline-silicon solar cells based on aluminium-induced crystallization. *Thin Solid Films*, **516**, 6984–6988.

97. Rau, B., Sieber, I., Selle, B., Brehme, S., Knipper, U., Gall, S., Fuhs, W. (2004) Homo-epitaxial Si absorber layers grown by low-temperature ECRCVD, *Thin Solid Films*, **451–452**, 644–648.

98. Rau, B., Sieber, I., Schneider, J., Muske, M., Stoger-Pollach, M., Schattschneider, P., Gall, S., Fuhs, W. (2004) Low-temperature Si epitaxy on large-grained polycrystalline seed layers by electron–cyclotron resonance chemical vapor deposition. *Journal of Crystal Growth*, **270**, 396–401.

99. Gestel, D. V., Dogan, P., Gordon, I., Bender, H., Lee, K. Y., Beaucarne, G., Gall, S., Poortmans, J. (2009) Investigation of intragrain defects in pc-Si layers obtained by aluminum-induced crystallization: comparison of layers made by low and high temperature epitaxy. *Materials Science and Engineering: B*, **159–160**, 134–137.

100. Gestel, D. V., Romero, M. J., Gordon, I., Carnel, L., D'Haen, J., Beaucarne, G., Al-Jassim, M., Poortmans, J. (2007) Electrical activity of intragrain defects in polycrystalline silicon layers obtained by aluminum-induced crystallization and epitaxy. *Applied Physics Letters*, **90**, 092103-1–092103-3.

101. Gestel, D. V., Gordon, I., Bender, H., Saurel, D., Vanacken, J., Beaucarne, G., Poortmans, J. (2009) Intragrain defects in polycrystalline silicon layers grown by aluminum-induced crystallization and epitaxy for thin-film solar cells. *Journal of Applied Physics*, **105**, 114507-1–114507-11.

102. Cavalcoli, D., Cavallini, D., Capperdoni, C., Palmeri, D., Martinelli,G. (1995) On the electrical activity of first- and second-order twin boundaries in silicon. *Semiconductor Science and Technology*, **10**, 660–665.

103. Girginoudi, S., Girginoudi, D., Thanailakis, A., Georgoulas, N., Papaioannou, V. (1998) Electrical and structural properties of poly-Si films grown by furnace and rapid thermal annealing of amorphous Si. *Journal of Applied Physics*, **84**, 1968–1972.

104. Christiansen, S., Lengsfeld, P., Krinke, J., Nerding, M., Nickel, N. H., Strunk, H. P. (2001) Nature of grain boundaries in laser crystallized polycrystalline silicon, thin films. *Journal of Applied Physics*, **89**, 5348–5354.

105. Gall, S., Becker, C., Conrad, E., Dogan, P., Fenske, F., Gorka, B., Lee, K. Y., Rau, B., Ruske, F., Rech B. (2009) Polycrystalline silicon thin-film solar cells on glass. *Solar Energy Materials and Solar Cells*, **93**, 1004–1008.

Chapter 6

Applications of Metal-Induced Crystallization Polycrystalline Silicon for Advanced Flat-Panel Displays

Man Wong, Hoi Sing Kwok, Shuyun Zhao, Zhiguo Meng, Pengfei Sun, Wei Zhou, and Tsz Kin Ho
Department of Electronic and Computer Engineering, The Hong Kong University of Science and Technology, Clear Water Bay, Kowloon, Hong Kong.
eemwong@ece.ust.hk

Metal-induced crystallization (MIC) polycrystalline silicon (poly-Si) thin-film transistors (TFTs) have attracted more and more interest in their use in flat-panel displays, not only due to their superior device performances, but also due to the mass production potential. In this chapter, the main applications of MIC poly-Si TFTs in flat-panel displays will be presented.

6.1 Introduction

It is well known that the interaction between certain metals, for example, aluminum (Al) or gold (Au) and amorphous silicon (a-Si),

Metal-Induced Crystallization: Fundamentals and Applications
Edited by Zumin Wang, Lars P. H. Jeurgens, and Eric J. Mittemeijer

ISBN 978-981-4463-40-9 (Hardcover), 978-981-4463-41-6 (eBook)
www.panstanford.com

will induce the crystallization of a-Si at a temperature below that of a-Si crystallization itself (~600°C). Most of the experimental evidence shows that the addition of a small amount of metallic impurities can often dramatically enhance the crystallization of a-Si films at low temperatures. Nickel (Ni) is widely used for silicide-mediated crystallization of a-Si by forming a silicide crystallite of nickel silicide ($NiSi_2$)—not only because the lattice constant of $NiSi_2$ is 5.406 Å, which is 0.4% less than that of Si (5.430 Å), but also because the quality of polycrystalline silicon (poly-Si) material from Ni metal-induced lateral crystallization (MILC) is the best [1]. The lateral crystallization behavior was first reported in the metal-induced crystallization (MIC) of a-Si after Ni ion implantation. The Ni atoms aggregate to form $NiSi_2$ precipitate, which causes the lateral crystallization of the a-Si [2].

Normally, Ni-induced crystallization of a-Si film is realized via a three-step process: (1) silicide crystallite formation, (2) growth of $NiSi_2$ crystallites until the critical size and breakup of the silicide layer into small nodules, and (3) migration of silicide nodules in a-Si, leaving behind Si needles [3]. Kawazu, et al. reported the silicide formation temperature of Ni_2Si, NiSi, and $NiSi_2$ on a-Si. It increases in the order of Ni_2Si, NiSi, and $NiSi_2$, which can be formed at a temperature lower than 400°C. [4]. $NiSi_2$ is the final phase in the formation of nickel silicide and has a cubic lattice with a lattice constant of 5.406 Å. So, the $NiSi_2$ crystallites function as nuclei for low-temperature crystallization of a-Si. At the beginning of annealing, Ni reacts with the a-Si and converts it to $NiSi_2$ [5]. Further growth of the $NiSi_2$ crystalline grains eventually leads to the breakup of the $NiSi_2$ crystallization nuclei. Subsequent to the breakup, small nodules of $NiSi_2$ move away from the crystallized top region of the a-Si film. The nodules 50–70 nm wide and ~1000 nm long were observed in the a-Si network by transmission electron microscopy (TEM) [5]. The thinner nodules generally move faster. In the case of MIC, the fastest-moving nodules maintain the crystallization front, which is the interface between the crystallized Si and the a-Si. The slower-moving nodules are trapped inside the crystallized region. Once the crystallization front reaches the bottom buried oxide, all remaining moving nodules are stopped. Therefore, there is a small peak of the Ni signal in the X-ray photoelectron spectroscopy (XPS) results, shown in Fig. 6.1.

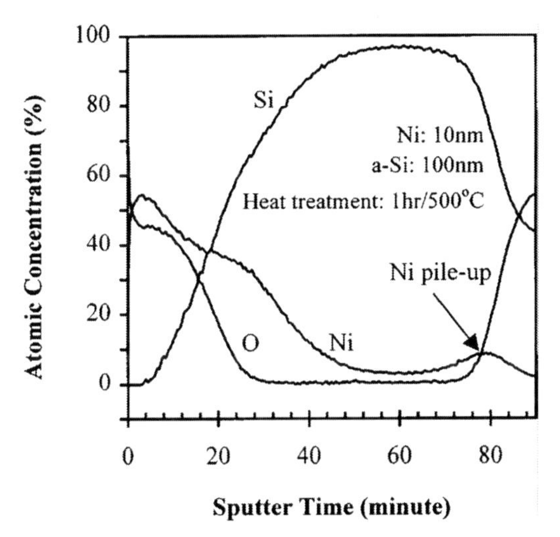

Figure 6.1 XPS depth profiles of Si, Ni, and O concentrations in Ni-covered silicon after 1 hour of heat treatment at 500°C. Reprinted with permission from Ref. [7], Copyright 1998, American Institute of Physics.

On the basis of the proposed model for MIC, metal-induced lateral crystallization (MILC) can be easily explained. At the edges of a Ni-covered region, a certain number of the breakaway $NiSi_2$ nodules will move laterally into the a-Si region not originally covered by Ni. As the nodules move laterally, any a-Si along the path of the moving nodules will be crystallized. Because the film is thin, crystallization occurs rapidly and any slow-moving nodules will be quickly trapped within a short distance from the edges of the Ni-covered region, leaving only the fast-moving nodules at the MILC front. Figure 6.2 shows the Ni distribution across the MIC, MILC, and a-Si regions [6]. The Ni concentration near the front is approximately 0.4 at.% [6]. Furthermore, a rough estimation of the Ni concentration in the MILC region is 0.02 at.%. This finite concentration of Ni in the MILC region is consistent with the proposition that slower-moving Ni-containing nodules would be trapped in the grain boundaries (GBs) in the MILC region [7].

The Ni diffusion coefficient D_{Ni} of $\sim 10^{-10}$ cm^2/s was extrapolated using the reported D_{Ni} measured between 270°C and 435°C [8]. For 1-hour diffusion, the diffusion length is ~6 μm, which is significantly shorter than the normal MILC rate of ~28 μm at 600°C. To further investigate the relationship between the MILC and Ni diffusion, a serial of experiments was carried out. The different placements of crystallization-inducing windows, a-Si islands with different length L_{Is}, and phosphorous-doped areas are shown in Fig. 6.3. The 30 μm wide regions were implanted with phosphorus at a dose of 10^{16} cm^{-2}. The 10 nm–thick Ni was deposited in the inducing windows. Then the samples were annealed in nitrogen ambience at 600°C for 2.5 hours. The resulting MILC length L_{MILC} was measured using an optical microscope. Shown in Fig. 6.4 is the dependence of the MILC length L_{MILC} on crystallization-inducing windows on one or both ends of the islands. It can be seen that L_{MILC} decreases with increasing island length L_{Is} no matter whether there is one or two inducing windows. This trend eventually levels off at $L_{\mathrm{Is}} \approx 200$ μm or 400 μm, respectively, for islands with one or two windows. Before the leveling off, L_{MILC} measured on an island containing two windows is longer than that on an island with the same L_{Is} but containing only one window. Furthermore, L_{MILC} on an island with length L_{Is} and two windows is roughly the same as that on an island with length $L_{\mathrm{Is}}/2$ and one window.

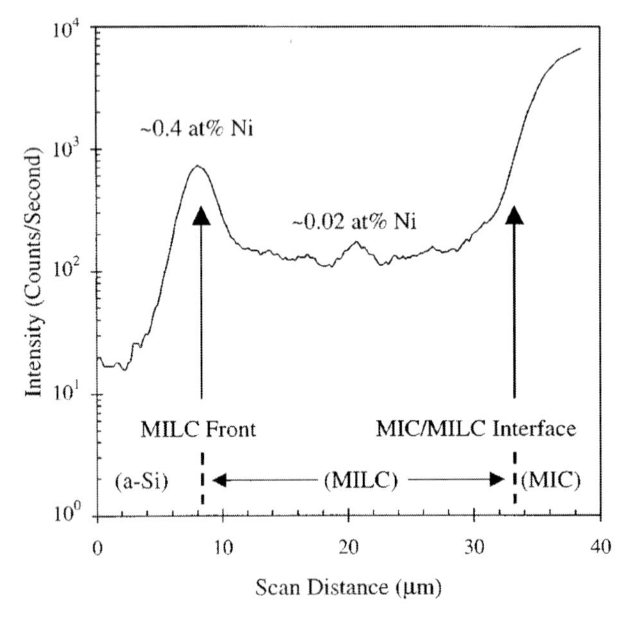

Figure 6.2 Distribution of Ni across the MIC, the MILC, and the a-Si regions obtained by SIMS. Abbreviation: SIMS, secondary ion mass spectrometry. Reprinted with permission from Ref. [6], Copyright 2000, IEEE.

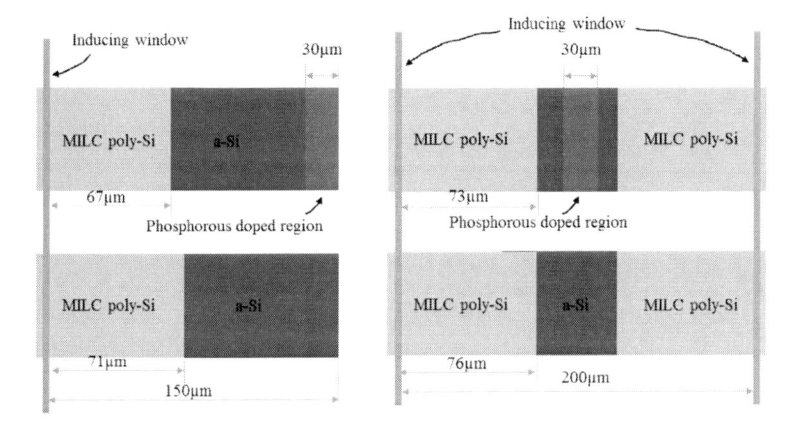

Figure 6.3 The schematic layout of inducing windows, phosphorous doping, and a-Si islands.

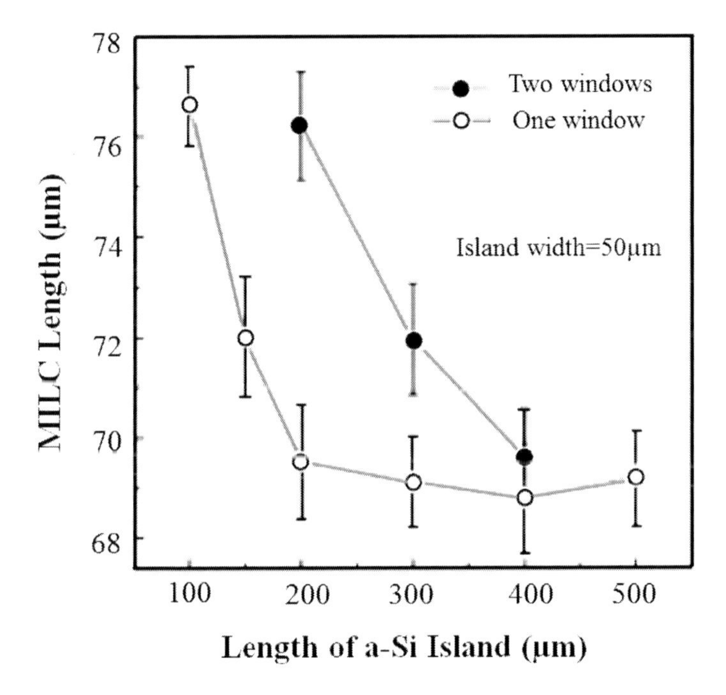

Figure 6.4 Dependence of the MILC length on the a-Si island length for 50 μm wide islands with a crystallization-inducing window at one or both ends of the Reprinted with permission from Ref. [9], Copyright 2005, John Wiley and Sons.

As shown in Fig. 6.3 30 μm wide phosphorus-implanted regions were selectively inserted, either in the middle of an island containing two windows or at the opposite end of an island containing only one window. It should be noted that the MILC fronts on these islands have not advanced into the phosphorus-implanted regions. Compared to L_{MILC} measured on islands containing phosphorus-implanted regions, those islands without such implanted regions are clearly longer. The difference between the two lengths is calculated, and its statistical distribution over many samples is summarized in Fig. 6.5 for both types of islands containing one or two windows [9].

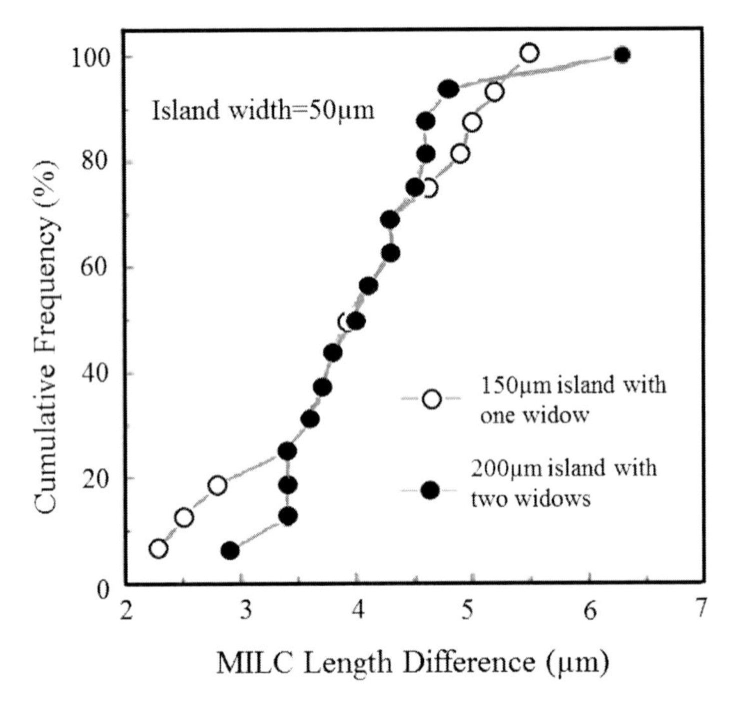

Figure 6.5 Statistical distributions of the MILC length difference on 150 μm and 200 μm long islands with one and two crystallization-inducing windows, respectively. Reprinted with permission from Ref. [9], Copyright 2005, John Wiley and Sons.

It is clear that the MILC rate is affected by the island length, the presence of an extra crystallization-inducing window, or the presence of a phosphorus-implanted region. However, what is the underlying mechanism? According to the diffusion model, a shorter island length results in an increase in the concentration of Ni in the a-Si host matrix. Such increase near the MILC front may enhance the reaction rate of Ni and Si at the silicide/a-Si interface. This would be consistent with the slight but consistently increase in the MILC rates with decreasing L_{Is}. Since phosphorus is a known getter of Ni [10], it lowers the concentration of Ni by removing it from the a-Si host matrix. The overall effect is similar to the dilution effect resulting from an increase in the length of an a-Si island. The diffusion model implies the existence of a Ni diffusion "front" moving ahead of the MILC front. For an island containing only one crystallization-inducing window, the MILC rate reduction eventually levels off when the island is sufficiently long. At 600°C for 2.5 hours, this length is ~200 μm. For the opposite end of an island to become "visible" to the MILC front, the length of this island must be comparable to the diffusion length $L_{Ni} = (D_{\mathrm{Ni}}t)^{1/2}$ of Ni, where t is the MILC heat treatment time. Setting $L_{\mathrm{Ni}} = 200$ μm and $t = 2.5$ hours at 600°C in the present case, one estimates a value of $\sim 4 \times 10^{-8}$ cm^2/s for D_{Ni}. This is about 400 times larger than the value of $\sim 10^{-10}$ cm^2/s estimated earlier [8]. From the similarity between L_{MILC} measured on an island with length L_{Is} and two windows and that on an island with length $L_{\mathrm{Is}}/2$ and one window, it can be concluded that the "free" end of a low-temperature oxide (LTO)-covered a-Si island gives rise to a largely "reflecting" boundary condition for the diffusing Ni flux in a-Si [9].

Figure 6.6 shows the sample configurations of different methods, including MIC, self-aligned metal-induced lateral crystallization (SA-MILC), and metal-induced unilateral crystallization (MIUC). Figure 6.6b shows that the SA-MILC proceeds from the two metal/a-Si contacts using the gate as a mask. The a-Si between the contacts (source/drain) can be laterally crystallized completely after annealing [11]. The thin-film transistor (TFT) showed the maximum field effect mobility as high as 90 cm^2/Vs, which was almost three times higher than that of the conventional SPC poly-Si TFT fabricated at approximately 600°C. The advantage of this process is that the crystallization of the gate, active channel, as well as the gate/source/drain dopants' activation, could be achieved simultaneously in one-step annealing. However, the GB forms at the middle of the TFT channel as well as the MILC/MIC grain boundary (MMGB) located at the interface between the TFT channel and the source/drain region.

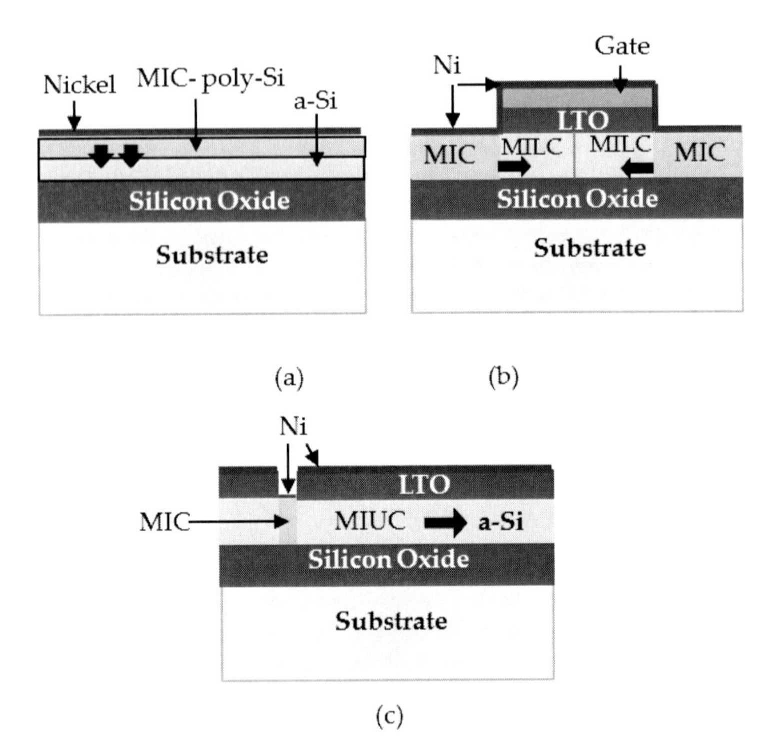

Figure 6.6 Schematic of MIC (a), SA-MILC (b), and MIUC (c) techniques.

In 2000, Meng et al. proposed a new technology, MIUC, which removed all major GBs transverse to the drain current flow in both edges and within the channel region [11], as shown in Fig. 6.6c. The TFT results show that the MIUC TFT presents higher field effect mobility, smaller threshold voltage, smaller subthreshold swing, and lower off-state leakage current than the SA-MILC TFT for both n- and p-channel devices. They also reported that the MIUC TFT had better uniformity and higher drain breakdown voltage than the SA-MILC TFT by removing all the MMGB and MILC/MILC grain boundaries (LLGBs) from the channel region.

However, an MIUC TFT has problems of subsequent mask misalignment induced by glass substrate shrinkage during the annealing process. Additionally, residual Ni in the poly-Si channel can affect the long-term stability of the TFT. The electrical performance of an MIUC TFT may shift and still suffer from higher off-state leakage current and early drain breakdown.

There have been several attempts to reduce the Ni content in MIC-based TFTs. Normally, metal films are deposited by the physical vapor deposition (PVD) method, which is time consuming and expensive [13]. To further reduce nickel residual as well as process complexity, metal-salt solutions were utilized to replace PVD in Ref. [14]. Since there is no vacuum process involved, the solution method is much simpler, faster, and cheaper than the traditional PVD method. They used a diluted acid such as hydrochloric acid (HCl) and nitric acid (HNO_3) or a photoresist as the solvent. The coating of a metal solution on Si films is simple, and metal-salt solutions can well control the amount of metals deposited on a-Si films through the metal concentration in the solution [14]. With the electroless plating Ni method used by Chao et al. [15], a poly-Si TFT with a mobility of 41.9 cm^2/Vs was obtained. The Ni concentration in the MIC poly-Si using an Ni solution is about 1.2×10^{18} cm^{-3}, which is much lower than that ($>10^{19}$ cm^{-3}) in the Ni-MIC poly-Si using a thin (~0.4 nm) Ni layer on a-Si:H.

Large-grain [or disk-like/giant grain silicon (GGS)/MIC with a nitride cap layer (MICC)] poly-Si silicon has been obtained by Ni-mediated crystallization of a-Si with a siliconnitride (SiN_x) cap layer [16]. In this process, Ni was sputtered onto the SiN_x/a-Si layer and then annealed at around 600°C. The cap SiN_x controlled the Ni content inside the MIC layer to within tolerable levels. However, this process is quite complicated. It also has a high cost because of the vacuum process involved and the need to remove the cap layer afterward. Moreover, the random distribution of crystalline nuclei leads to a longer annealing time, which is not desirable for a large-area glass substrate.

Samsung Electronics successfully fabricated a 21.3-inch UXGA (1600 × RGB × 1200) active-matrix liquid-crystal display (AMLCD) with integrated gate drivers and multiplexing circuits based on an improved GGS process called nanocap-assisted crystallization (NAC) [17].

TFTs are mostly used for driving display pixels, including AMLCDs and active-matrix organic light-emitting diodes (AMOLEDs). As we know, a-Si TFT has a low manufacturing cost and low equipment investment due to few process steps. So, it has continuously attracted interest within the industry [18]. However, a-Si TFTs still have a long-term reliability issue for driving organic light-emitting diodes (OLEDs), even though many compensation circuits and new driving methods have been proposed recently [19–22]. Low-temperature polycrystalline Si (LTPS) TFTs have attracted more and more attention due to their superb current-driving capability and electrical reliability. It is well known that LTPS has the best potential for peripheral circuit integration to realize a system on panel (SOP) due to its higher mobility than the a-Si TFTs. To reduce the manufacturing cost and mask number, a lot of p-channel metal-oxide semiconductor (PMOS) circuits for display have been proposed, including shift registers, level shifters, multiplexers, and DC–DC converters [23]. Regardless of liquid-crystal display (LCD) or OLED, poly-Si TFT is a good candidate to realize a low-power-consumption and high-resolution display panel.

In this chapter, some new applications for MIC TFTs in flat-panel displays will be presented. First, the common use of pixel circuits for AMLCDs and AMOLEDs will be briefly introduced. An MIC TFT shows a much higher field effect mobility, which enables a higher driving frequency for LCDs. So it can be applied for field-sequential color (FSC) LCDs without color filter (CF). Then, an SOP system realized by a CMOS MIC TFT process will be introduced. To further reduce the process cost and improve the yield, PMOS SOP is also a hot topic. Besides application in peripheral circuit integration, the MIC TFT process also has the capability of integration of p-i-n photodiode sensor, which can be applied to in-cell touch panels for future display.

6.2 Pixel Circuits for AMLCDs and AMOLEDs

LCDs have been dominating the market since the year 2000. In that year, LCDs occupied more than 80% of the market [24]. High-resolution and full-color video LCDs, such as modern LCD computer monitors, LCD TVs, and mobile displays, typically use the active-matrix (AM) addressing structure. In AM displays, each pixel has its own active devices, as shown in Fig. 6.7, which normally are TFTs and capacitance to control the optical state of each pixel [25, 26]. The schematic of the AMLCD structure is shown in Fig. 6.8. When the TFTs in one row are turned on, the data signal is written into the LCD electrode as well as the storage capacitor (Cs) of the pixels of this row. Pixels in other rows will not be affected. When the TFTs are turned off, the charges stored in the Cs will be held to maintain the liquid crystal (LC) at a certain state until the next refresh cycle. AMLCDs provide improved image quality over a passive matrix. By using the AM addressing method, the cross-talk effect is eliminated. Thus the quantity of scan lines can be increased to 1000 for higher-resolution displays [27].

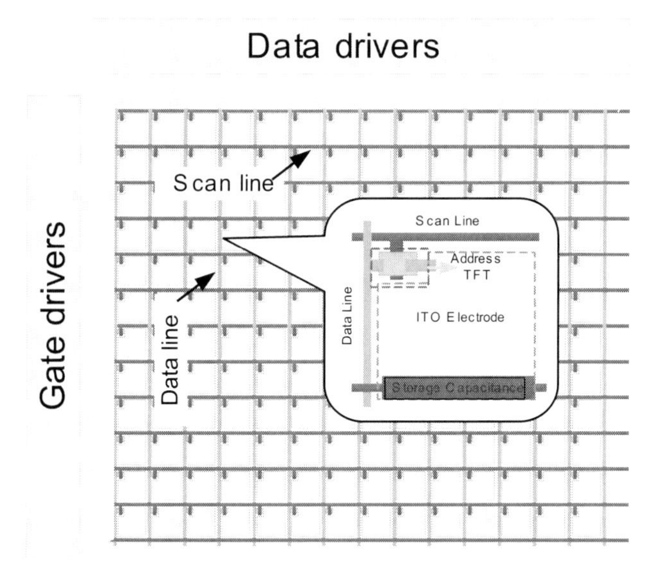

Figure 6.7 AM of a TFT for an AMLCD.

It is claimed that OLED technology is competing with current LCD technology and can even replace it in the future [28]. Generally, OLEDs offer the following advantages compared to LCDs: They are thin and lightweight, and they have low power consumption, high contrast, a fast response time, a wide viewing angle, and a broad color gamut [29]. An AM TFT backplane is also desirable for high-resolution displays [30].

Unlike the optical state of a pixel in an LCD that can be voltage-switched using a single TFT, the current-controlled emission of an OLED pixel requires at least two TFTs [31], one as a switch and the other as a current supplier (Fig.6.8b). Four or more TFTs are used to compensate the dispersed threshold voltages and drain currents of the driving TFTs to improve the display uniformity. The switch TFT's function is similar to that in the AMLCD to control the transfer of the data signals. When the switch TFT is turned on, the data signals are transferred to the gate of the driving TFT and a corresponding driving current will be supplied to the OLED. The Cs will be charged up at the same time. The data signals maintained by the Cs allow the driving TFT to continue supplying current to the OLED when the switch TFT is turned off.

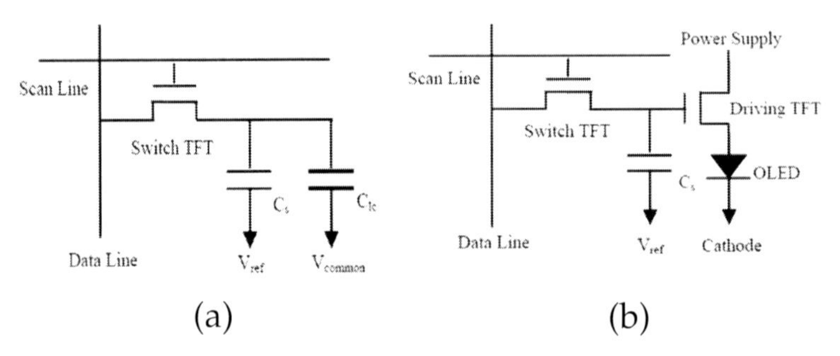

Figure 6.8 Schematic pixel circuit of a pixel in an (a) AMLCD and an (b) AMOLED.

6.3 Driver-Integrated LCD Panels

The a-Si:H TFT technology is well developed and widely used in commercial electronic displays, from large-size TVs to small-size applications, such as cell phones, PDAs, and digital cameras. Even though a-Si TFTs are currently so popular, there is always demand for poly-Si TFTs. It is well known that the fact that a-Si:H TFTs have a relatively low field effect mobility (μ_{FE}) of 0.4–1.5 cm^2/Vs for electrons and a much lower field effect mobility of 2×10^{-4}–7×10^{-4} cm^2/Vs for holes limits the application of a-Si:H TFTs to pixel switches. Due to low μ_{FE}, TFTs that perform the address driver process must have a large width. The typical value of W/L of an a-Si:H TFT is 100 μm/10 μm. The large dimension reduces the aperture ratio (AR) of the pixel. The brightness of the display is reduced, especially in higher-resolution displays, and the power consumption of the backlighting is increased. However, an optimized MIC process can provide device-quality poly-Si films that can be used for high-quality poly-Si TFTs with enhanced TFT uniformity and reduced surface morphology at low temperatures due to the reduced metal contents in the crystallized poly-Si films [32]. So, with an integrated driver, a poly-Si TFT LCD has the potential to reduce the output connections and further increase panel connection reliability.

In the 2006 conference of the Society for Information Display (SID), Samsung Electronics successfully demonstrated a 21.3-inch UXGA (1600 × RGB × 1200) LTPS AMLCD manufactured by a CMOS MIC-TFT process. Gate driver circuits composed of shift registers and buffer circuits were integrated on both sides of a panel for dual-driving methods, and the multiplexing circuit was implemented on its data circuits [17].

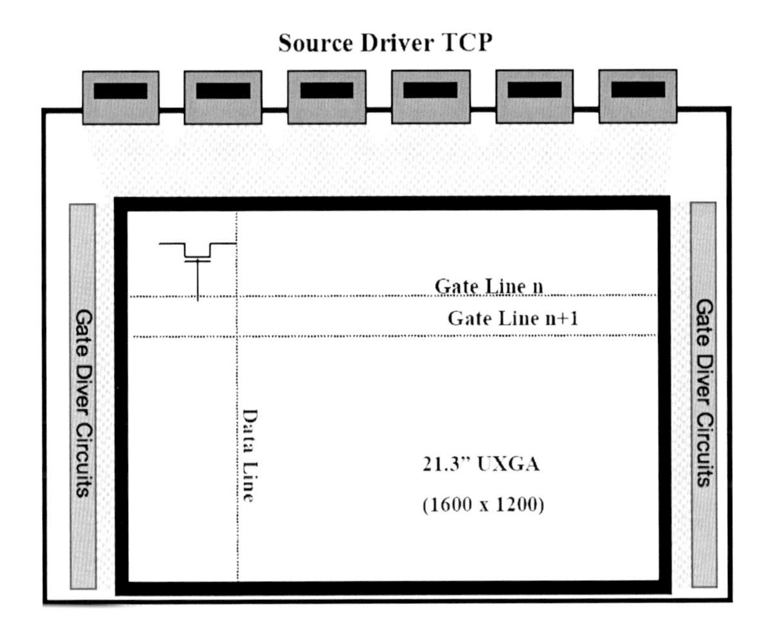

Figure 6.9 Schematic diagram and circuit construction of the 21.3-inch UXGA MIC AMLCD. Reprinted with permission from Ref. [17], Copyright 2006, John Wiley and Sons.

Figure 6.9 shows a schematic diagram of the 21.3-inch MIC UXGA AMLCD. Dual CMOS gate-driving circuits consisting of shift registers and output buffers were implemented on both sides of the pixel arrays to reduce the resistance of scan lines for the uniform display without cross talk and flicker effects. This driving method has been known as an effective way to decrease the time constant of scan lines, which are proportional to the diagonal panel size. In addition to dual-driving gate circuits, the multiplexing scheme was implemented on data signal arrays as a transmission gate. By implementing the multiplexing scheme (2:1), the number of source driver integrated circuits (ICs) with tape carrier packages (TCPs) can be reduced by half since each transmission gate is distributed to two signal lines per one channel of source IC.

The resolution is 1600 × RGB × 1200, the AR is over 60%, luminance is 500 cd/m^2, and the contrast ratio is over 500:1. Figure 6.10 shows the display result of the 21.3-inch MIC AMLCD.

Figure 6.10 Image of the 21.3-inch UXGA MIC AMLCD. Reprinted with permission from Ref. [17], Copyright 2006, John Wiley and Sons.

6.4 MIC TFTs Applied to Field-Sequential Color LCDs for Small-Size Panels

6.4.1 Introduction to Field-Sequential Color LCDs

TFTs are essential for high-resolution flat-panel displays [33, 34]. LCDs are dominant among all the flat-panel technologies. TFT LCD technology has a wide range of applications, ranging from small consumer products such as cell phones and digital cameras to larger displays such as desktop computer monitors and televisions [35]. TFT LCD technology has a huge, mature manufacturing base [36]. High optical efficiency and low cost are becoming the critical factors for LCD displays, especially for portable applications [37, 38]. FSC technology is a great candidate for new high-definition displays with high contrast, high resolution, high color purity, and clear moving images. Such displays, based on digital light processing (DLP) or liquid crystal on silicon (LCoS) technology, have been successfully introduced to the market [39].

For color filter (CF) based LCDs, red, green, and blue (RGB) subpixels are needed, as shown in Fig. 6.11a. In this arrangement, high optical efficiency and high color purity cannot coexist [40, 41]. High color purity often needs a thicker CF and hence poorer optical transparency [42, 43]. Even if color purity is compromised, CF-based displays still suffer from poor optical efficiency. Even in the best case, less than 10% of the backlight can be utilized [44]. Also, with this RGB subpixel configuration, it is hard to improve the resolution under a certain process level.

An FSC LCD reproduces RGB colors in a single pixel in a time-sequential manner using synchronously pulsed colored LED backlights [45, 46]. The pixel structure is shown in Fig. 6.11b. This method can produce a bright display with good optical efficiency since there is no CF that absorbs the backlight. Moreover, the number of pixels of an FSC display is only one-third of that of a CF display. As a result, an FSC LCD is expected to have a higher aperture ratio (AR) than a CF display with the same resolution, thus making the optical efficiency even higher. Alternatively, the same AR can be used with a three times higher resolution using the same technology [47].

Thus, FSC LCD is an important green technology. An important issue of FSC LCD technology is the color breakup artifact [48]. To avoid color breakup, a higher frame rate such as 90 Hz instead of 60 Hz is needed. Thus the subframe rate is 270 Hz with a frame period of only 3.7 ms. To realize the FSC, fast LCD mode together with fast electronic addressing of the TFT panel is indispensable. It is believed that a poly-Si TFT with much higher mobility is the best candidate for fast data loading on the TFT panel. The detailed calculation results will be shown in Section 6.4.2. In this part, the considerations for designing an FSC LCD will be discussed in detail on the basis of mathematical calculations such as the pixel TFT and storage capacitance, layout, and process parameters. It will be shown that fast loading is able to realize a high-quality FSC LCD. A fast-addressing TFT using continuous zonal domain (CZD) poly-Si technology was discussed in Ref. [49]. The TFT backplane based on CZD poly-Si TFT technology will be introduced, and the display results of a 3-inch QVGA prototype will be shown.

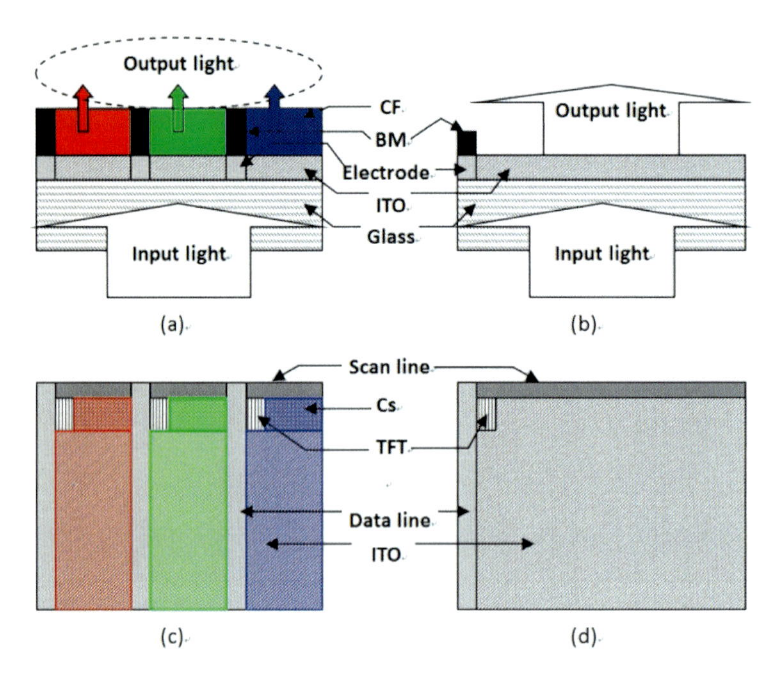

Figure 6.11 Schematic cross section for (a) traditional CF LCD, (b) schematic cross section of FSC LCD, (c) pixel layout of traditional CF LCD, and (d) pixel layout of FSC LCD. Reprinted with permission from Ref. [49], Copyright 2010, IEEE.

6.4.2 Design of the Active-Matrix Array

6.4.2.1 Working principle of the FSC LCD

The working principle of an FSC LCD is to display the RGB subframes of a display time sequentially. If the frame frequency is fast enough, human eyes will integrate the fields and the observer will see the original colors of the image. The RGB subframes can be achieved by using an independently controlled RGB LED backlight. By controlling the gray level of the subframes, a full-color display can be obtained with a color saturation that is usually better than CF-type displays.

The subframe rate of the FSC LCD should be three times as fast as a conventional CF LCD, assuming the same frame rate. However, to reduce the color breakup artifacts, a higher subframe rate should be employed. The principle of the driving mechanism for an FSC LCD is shown in Fig. 6.12b. For a QVGA display with a 90 Hz frame rate, one frame should have a duration period of approximately 11.1 ms. The duration of each subframe is therefore only 3.7 ms. Assuming a fast response time for an LCD of 1.7 ms and a minimum LED illumination duty cycle of 30%, only approximately 1 ms is left for the data loading, as shown in Fig. 6.12b. For a CF LCD with the same resolution, the same frame rate and data loading, LCD response, and LED illumination can occur at the same time, as shown in Fig. 6.12a. That means the loading time for the FSC LCD is one-tenth of that for the CF LCD, assuming the same resolution. Thus, fast data loading is the key issue for FSC LCDs, which means a-Si TFTs with a field effect mobility of ~1 cm^2/Vs will not meet the requirement. Poly-Si TFTs with a much higher field effect mobility therefore become vital for the FSC LCD.

6.4.2.2 Design of the pixel TFT

What is required of the pixel TFT in an AM FSC LCD is analyzed in this section. With V_D representing the voltage on the data line, the voltage level on the pixel as a function of time (t) is approximately described by the following equations:

$$V_{write} = V_D(1-e^{-t/\tau_{on}});\ \tau_{on} = R_{on}C_S \text{ (when selected)} \quad (6.1)$$

$$V_{hold} = V_D(1-e^{-t/\tau_{off}});\ \tau_{off} = R_{off}C_S \text{ (when unselected)} \quad (6.2)$$

R_{on} and R_{off} are the resistances of the pixel TFT at the "on" and "off" states, respectively. V_{write} and V_{hold} are the voltages on the pixel electrode during the charging process and the holding process, respectively. V_D is the voltage applied on the data line, and C_s is the capacitance of the LC.

When writing images to the LC, it is required that

$$V_{write} > 0.99 \times V_D = \gg T_{writing} > 4.6 \times \tau_{on} \quad (6.3)$$

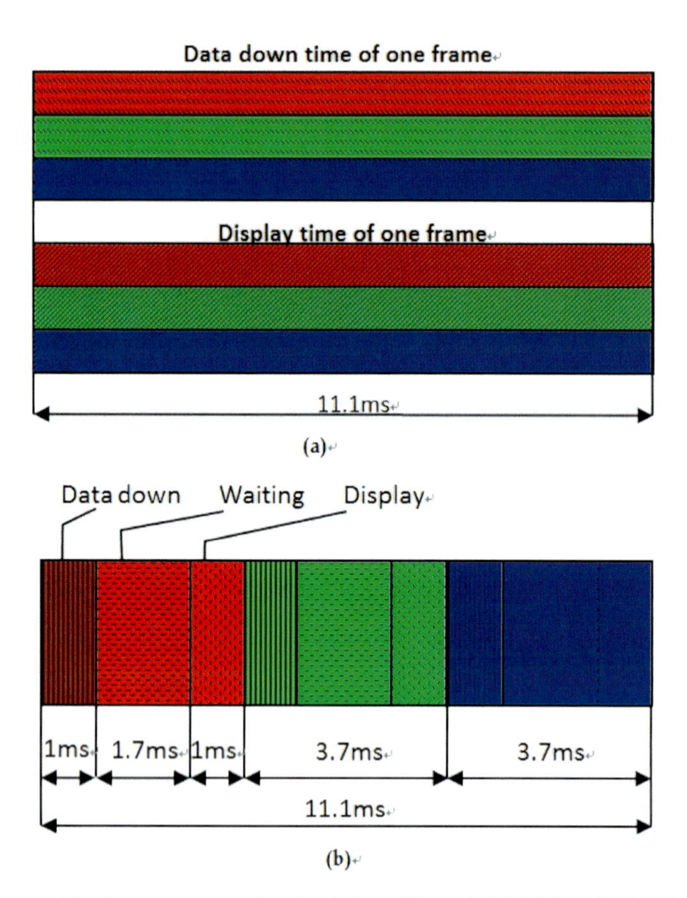

Figure 6.12 Driving principle of (a) CF LCD and (b) FSC LCD. Reprinted with permission from Ref. [49], Copyright 2010, IEEE.

In the images holding process, it is required that

$$V_{\text{hold}} > 0.95 \times V_{\text{D}} = \gg T_{\text{holding}} < \tau_{\text{ff}} / 19.5 \tag{6.4}$$

So

$$\tau_{\text{on}} < {T_{\text{writing}}}\big/{4.6} = \gg R_{\text{on}} < T_{\text{writing}} / \left(4.6 \times C_{\text{s}}\right) \tag{6.5}$$

$$\tau_{\text{off}} > 19.5 \times T_{\text{holding}} = \gg R_{\text{off}} > \left(19.5 \times T_{\text{holding}}\right) / C_{\text{s}} \tag{6.6}$$

T_{writing} is the time of writing the image, and T_{holding} is the time of holding the image. This is a typical frame period in the display standard. In the AM FSC LCD display, C_{s} is ~1.0 pF in the present design. As mentioned before, the time limitation for each subframe is only 3.7 ms. So the holing time is 3.7 ms, and R_{off} should be $R_{off} > (19.5 \times T_{holding})/C_S = 7.2 \times 10^{10}\ \Omega$.

An FSC LCD should be driven three times as fast as a conventional LCD, assuming the same frame rate. The data loading time for one frame is only about 1 ms, so it requires a much smaller R_{on} than a conventional LCD. The address TFT works in the linear region, and the drain current ($\boldsymbol{I}_d$) is given by

$$I_{ds} = \mu_{FE} \times (\varepsilon_0 \varepsilon / t_{ox}) \times (W / \mathrm{L}) \times (V_{gs} - V_T) \times V_D \tag{6.7}$$

So

$$R_{on} = \frac{V_D}{I_{ds}} = \frac{(t_{ox} L)}{(\mu_{FE} \varepsilon_0 \varepsilon W (V_{gs} - V_T))} \tag{6.8}$$

where μ_{FE} is the field effect mobility, $\varepsilon_0\varepsilon$ is the dielectric constant of the gate oxide, W and L are the effective gate width and gate length of the TFT, V_{gs} is the applied gate to source voltage, and V_T is the threshold voltage. Combined with Eq. 6.5, the ratio of gate width to gate length (W/L) of the TFT can be written as

$$W / L > (4.6 \times t_{ox} C_s) / (\mu_{FE} \varepsilon_0 \varepsilon (V_{gs} - V_T)) \tag{6.9}$$

Here, we assume that at the "on" state V_{gs} of the TFT is 15 V and C_s = 1.0 pF. With the main parameters of the a-Si:H TFTs and the poly-Si TFTs listed in Table 6.1, the required ratio of channel width (W) to length (L) of the TFTs can be derived with Eq. 6.9, as listed in Table 6.1. For a QVGA display with a 90 Hz frame rate, the writing time is 4 μs for each pixel. The required W/L ratio of the a-Si:H TFT is at least 14. At the same time, the required W/L ratio of the poly-Si TFT is only 0.06. So for the fabrication process with a 5 μm feature size, the width of the TFT channel is 70 μm for the a-Si:H TFT and 0.3 μm for the poly-Si TFT theoretically. For a higher-resolution display, the writing time for each pixel will decrease. So a TFT with a larger W/L is required. Taking an SXGA display, for example, the limited writing time is only 0.9 μs. The channel width of the a-Si:H TFT requires at least 300 μm on the basis of a 5 μm feature size, which is too large for a high-definition display. However, it is only ~1.5 μm if poly-Si TFT technology is used. From the calculation, it is clear that a higher AR can be reached using the poly-Si TFT due to the smaller TFT size and higher field effect mobility as compared to the a-Si:H TFT.

Table 6.1 Typical parameters of the poly-Si TFT and the a-Si:H TFT

	μ_{Max} (cm^2/Vs)	$\mu_{Work\ point}$ (cm^2/Vs)	Gate insulator	T_{ox} (nm)	ε
a-Si:H TFT	0.8	0.4	SiN_x	300	7.0
Poly-Si TFT	60	30	LTO	120	4.0

Source: Reprinted with permission from Ref. [49], Copyright 2010, IEEE.

Here, to make the calculation even simpler, we assume that the design rules are as follows: the pixel size is fixed at 200μm × 200 μm, the width of both the scan line and the data line is 12 μm, the minimum separation at the same layer is 5 μm, and the minimum separation for different layers is 2 μm. Normally, the gate length of a TFT is 5 μm, or 10 μm in the LCD display industry. Figure 6.13 shows the schematic diagram of the layout design rules for a 5μm feature size process. On the basis of these process parameters, the typical layout for a pixel that has only one TFT and the W/L of the TFT for different resolutions employing the a-Si:H TFT and the poly-Si TFT are listed in Table 6.2. It is known that the photolithography process is unable to provide a line width or length smaller than the technology feature ($2\lambda = 5$ μm), which means the W and L of the TFT cannot be smaller than 2λ. For example, if $L = 5$ μm, the width of the TFT has to be equal to or larger than 5 μm, which means W/L must be equal or larger than 1. If $L = 10$ μm, the W of the TFT has to be larger than 5 μm, which means the W/L ratio must be equal to or larger than 0.5. So, in Table 6.2, for the poly-Si TFT, all of the W/L ratios are set to 0.5 and 1 in the AR calculation for L equal to 5 μm and 10 μm, respectively.

Table 6.2 The ratio of width to length of the addressing TFT

Resolution	Rows (lines)	$T_{writing}$ (μs)	W/L (a-Si:H TFT)	W/L (poly-Si TFT)
QVGA	240	4.0	14	0.06 (0.5 or 1)
VGA	480	2.0	28	0.12 (0.5 or 1)
XGA	768	1.3	47	0.24 (0.5 or 1)
SXGA	1024	0.9	60	0.3 (0.5 or 1)

Source: Reprinted with permission from Ref. [49], Copyright 2010, IEEE.

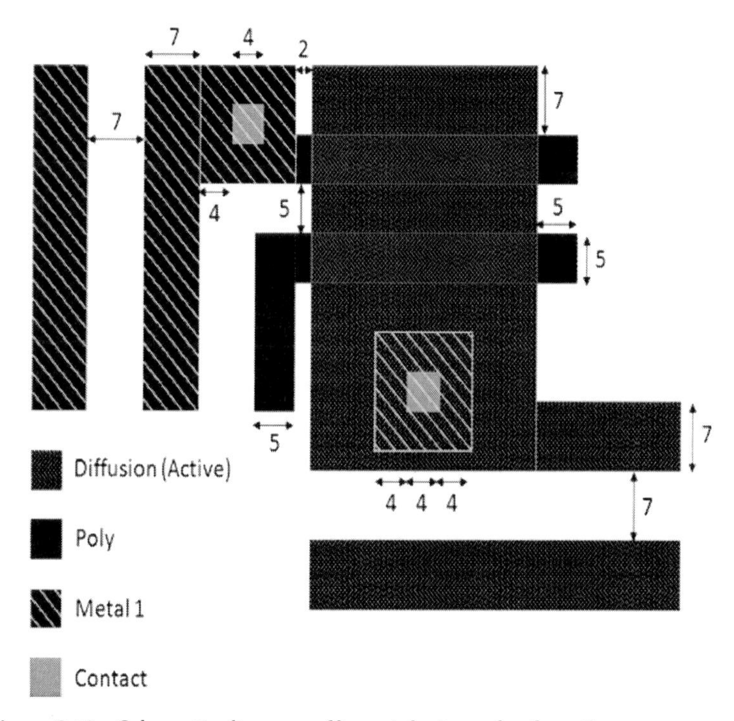

Figure 6.13 Schematic diagram of layout design rules for a 5 μm process.

Additionally, for an FSC LCD, the storage capacitance can be the gate capacitance of the TFT. The storage capacitance can be eliminated on the basis of our calculations, which will be introduced later in this section. So, only the addressing TFT size is considered in AR derivation. Figure 6.14 shows the AR as a function of the number of scan lines using the a-Si:H TFT and the poly-Si TFT as the addressing TFT based on the 5 μm and 10 μm gate length, respectively. For the a-Si:H TFT, the AR of the pixel decreases with an increase of the scan lines. For the poly-Si TFT, the AR of the pixel remains at 77%–78% due to the constant value of the W/L ratio. So, the poly-Si TFT is necessary for achieving a high-resolution and high-AR display.

6.4.2.3 Pixel capacitance

Capacitance is given by $C = \varepsilon A/d$. Here, A and d are the area and thickness of the dielectric material of the capacitance. For an LC, there is one more complication that makes it useful. The dielectric constant is anisotropic, that is, it depends on the direction of the electric field. Figure 6.15 shows the capacitance of an LC as a function of the applied voltage. For a positive LC, $C_{\parallel} > C_{\perp}$, the LC would like to align parallel to the electric field at a high voltage. Here, we assume $C_{\perp} = 0.31$ pF, $C_{\parallel} = 1.0$ pF, V_{on} = 10V, and R_{on} = 0.1 MΩ.

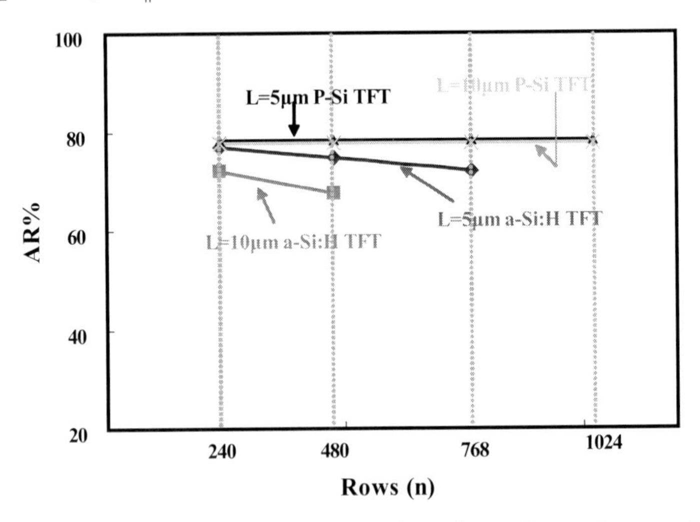

Figure 6.14 AR as a function of the number of scan lines using an a-Si:H TFT and a poly-Si TFT based on a 5 μm and 10 μm TFT gate length, respectively. Reprinted with permission from Ref. [49], Copyright 2010, IEEE.

Normally, C is considered as a constant, assuming $C_C = (C_{\parallel} + C_{\perp})/2$.

The charging process can be described using the following formula:

$$(V_D - V)/R_{on} = (C_{\parallel} + C_{\perp})/2 \times dV/dt \tag{6.10}$$

Here V_D represents the voltage applied on the data line. V is the voltage applied on the LC. The red line in Fig. 6.16 shows the time response of the voltage on the LC. The writing time for achieving $V_{signal} > 0.99V_D$ is about 0.4 μs.

To describe this process more precisely, the variation of C at different electric fields should be taken into account.

As shown in Fig. 6.16, assume that

$$V \in (0, V_{on}), \text{ and } C = C_{\perp} + ((C_{\parallel} + C_{\perp})/V_{on}) \times V \tag{6.11}$$

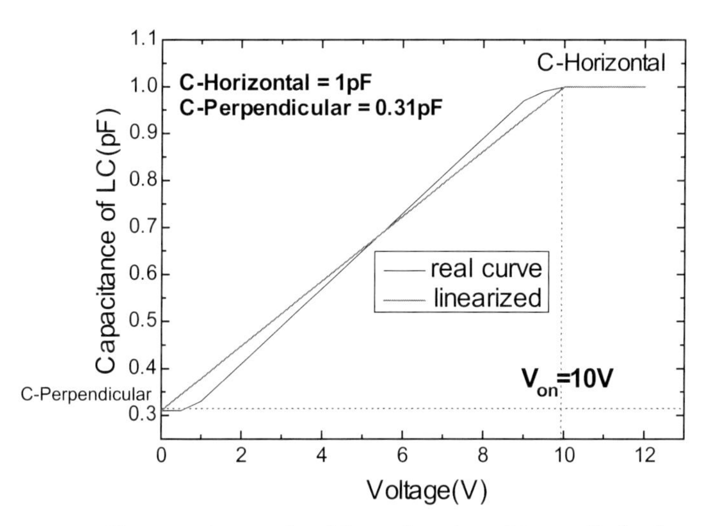

Figure 6.15 The capacitance of an LC as a function of the applied voltage.

So, the charging process can be more precisely described as

$$\frac{V_{\mathrm{D}}-V}{R_{\mathrm{on}}}=\frac{dQ}{dt}=\frac{d\left(C_{\perp}+\dfrac{C_{\parallel}-C_{\perp}}{V_{\mathrm{on}}}\times V\right)\times V}{dt},$$

which also can be written as

$$\frac{V_{\mathrm{D}}-V}{R_{\mathrm{on}}}=C_{\perp}\times\frac{dV}{dt}+2\frac{C_{\parallel}-C_{\perp}}{V_{\mathrm{on}}}\times V\times\frac{dV}{dt} \tag{6.12}$$

The blue line in Fig. 6.16 shows the time response of the voltage on the LC. The writing time for achieving $V_{\mathrm{signal}} > 0.99V_{\mathrm{D}}$ is about 0.7 μs.

For the holding process, the same simulation has also been done. If C is treated as a constant, the holding process can be described as

$$V/R_{\mathrm{off}} = -(C_{\parallel} + C_{\perp})/2 \times dV/dt \tag{6.13}$$

If C is treated as a variable, the holding process can be described as

$$\frac{V}{R_{\mathrm{off}}}=-C_{\perp}\times\frac{dV}{dt}-2\frac{C_{\parallel}-C_{\perp}}{V_{\mathrm{on}}}\times V\times\frac{dV}{dt} \tag{6.14}$$

Employing the above Eqs. 6.13 and 6.14, the time responses of the image voltage are shown in Fig. 6.17. The requirement of the holding process is $V_{\mathrm{signal}} > 0.95V_{\mathrm{D}}$. So the maximum holding time for the above situation is 4 ms and 6 ms, respectively. That means the capacitance of the LC is large enough for the QVGA FSC LCD with a 90 Hz frame rate. So, in our design, additional storage capacitor in the pixel circuit for the FSC LCD is not needed any more.

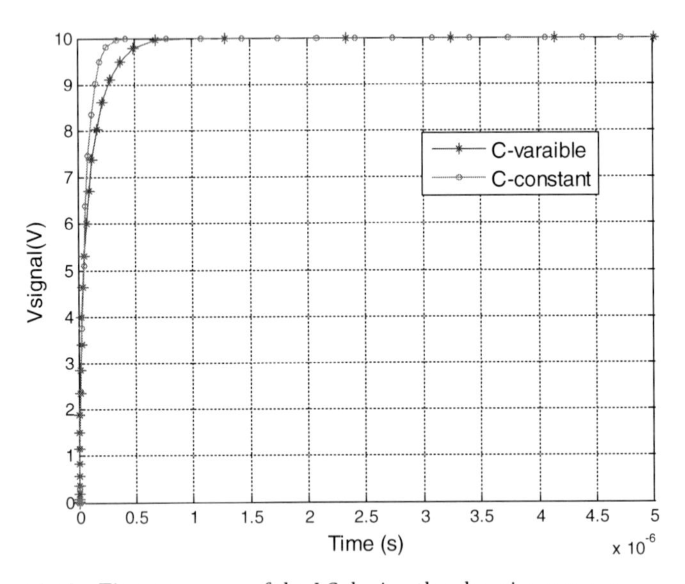

Figure 6.16 Time response of the LC during the charging process.

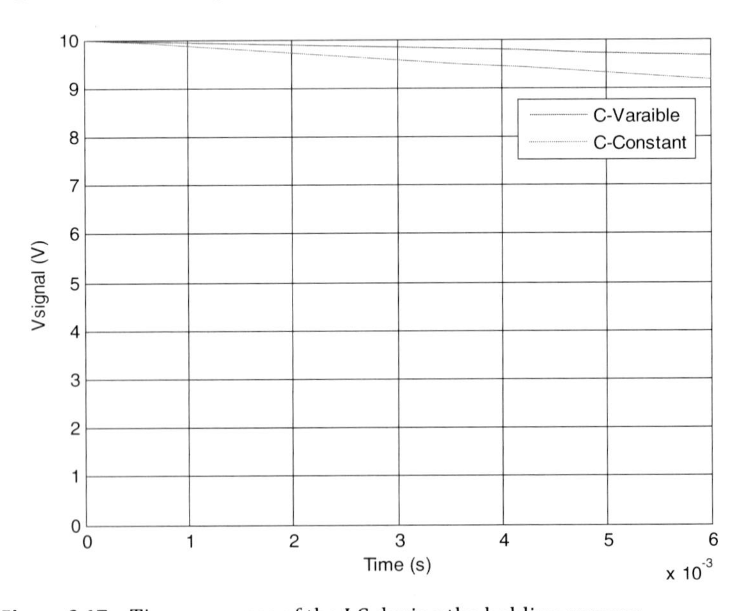

Figure 6.17 Time response of the LC during the holding process.

The layout of the pixel circuit became only one TFT, as shown in Fig. 6.18. The W/L of our TFT is 24 μm/5 μm × 2, which is different from the calculation results of 5 μm/5 μm. Two separated gates are used to reduce the leakage current of the CZD poly-Si TFT. A larger W/L was designed to increase the "on" state current and reduce the writing time for each pixel on the premise of not decreasing the AR of the display panel by 1%, depending on the pixel structure mentioned before.

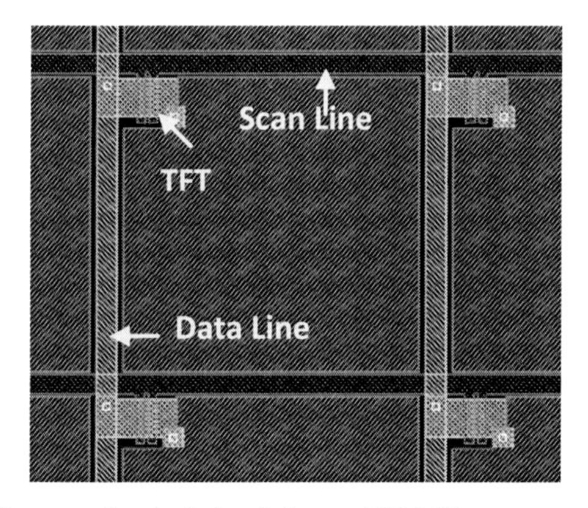

Figure 6.18 Layout of a pixel circuit for an FSC LCD.

6.4.2.4 Pixel structure

On the basis of the above calculations, a 3-inch QVGA prototype FSC LCD was designed and fabricated. The advantage of the FSC LCD is that the additional storage capacitor is not needed any more. So the pixel circuit for one pixel only includes one addressing TFT with W/L of 24 μm/2 × 5 μm. Due to the material and equipment limitation in our lab, 200 nm thick Al, which had an in-line measured sheet resistance of 0.176 Ω/sqr, was used as the scan line. A 50 nm LTO was deposited as a gate insulation layer, and a 600 nm plasma-enhanced chemical vapor deposition (PECVD) oxide was deposited as insulation between the scan line and the data line. It should be noted that the oxide thickness for calculation is 120 nm, which is normally used in the referenced papers. For our real process, to reduce the threshold voltage of the TFT, a 50 nm LTO was deposited. As mentioned before, the W/L ratio of our design is 2.4, which is much larger than the calculated value of 0.06. So the thinner oxide thickness will not affect the TFT writing speed. For the image-holding process, the thinner oxide could provide large capacitance, which would contribute to the voltage-holding process. The width of the scan line and the data line is fixed at 12 μm. Figure 6.19 shows the layout of the 3-inch QVGA TFT AM backplane.

Figure 6.20a shows optical microscopic photos of the 3-inch QVGA AM panel, and Fig. 6.20b shows a magnified photograph of a pixel region.

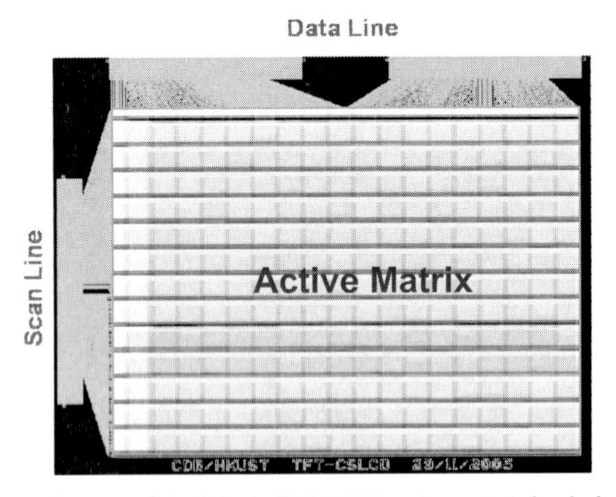

Figure 6.19 Layout of the 3-inch QVGA TFT active matrix backplane.

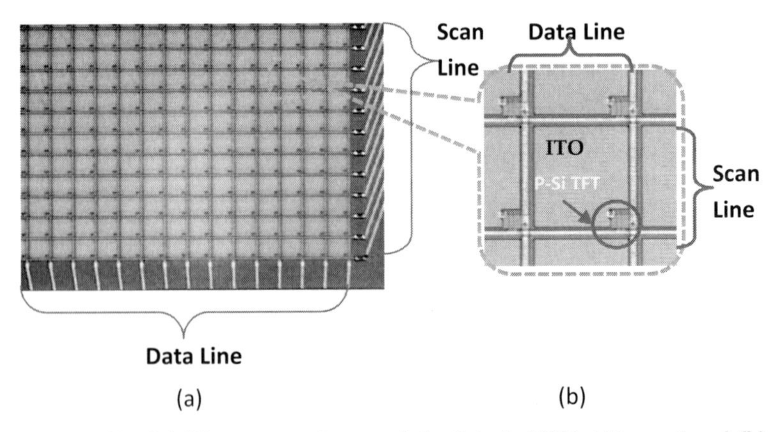

Figure 6.20 (a) Microscope picture of the 3-inch QVGA AM panel and (b) magnified photograph of a pixel region.

6.4.2.5 LCD integration

The finished AM backplane was used to fabricate an LCD. As mentioned before, the LCD mode has to be very fast for FSC applications. A transient mode based on optical rebounce was employed in Ref. [50]. This mode has previously been used for making a passively driven FSC display. It relies on optical bounce, which is a transient effect [51]. Thus, it overcomes the difficult requirement of having to use a very fast LCD mode. Details of the operation and principle of this LCD have been discussed before [51].

Essentially, it is noted that with a subframe time of only 3.7 ms (subframe frequency is 90 × 3 Hz), it is very difficult to have a transition from one stable LC alignment to another LC alignment. However, the transition can occur very fast. All of the gray level can be obtained within 1.8 ms when the LC cell gap is 5 μm. The driving voltage is also very low, <4 V. In the FSC LCD, the LED turns on for a very short time (1 ms), so it is not necessary for the LC alignment to be a stable state in deed. A transient state works perfectly. Figure 6.21 shows the representative image from this 3-inch QVGA AM FSC LCD.

Figure 6.21 Representative images from the 3-inch QVGA AM FSC LCD prototype. Reprinted with permission from Ref. [49], Copyright 2010, IEEE.

6.5 CMOS-Based SOP Systems for AMOLEDs

Among existing display technologies, AMOLEDs are regarded as the best candidates for an "ultimate display" due to their superior characteristics, such as self-emission, fast motion picture response, vivid colors, wide view angle, high contrast, and superslimness. However, the question of which backplane is most suitable for the AMOLED display has no answer as yet. It is known that an OLED is a current driving device, and the driving TFT needs to supply a constant current to the OLED whenever the pixel is on. The a-Si TFTs still have a long-term reliability issue for driving the OLED, even though many compensation circuits and new driving methods have been proposed recently. Low-temperature polycrystalline silicon (LTPS) TFTs have attracted more and more attention due to their superb current-driving capability and electrical reliability. It is well known that an LTPS TFT has the best potential for peripheral circuit integration to realize an SOP due to its higher mobility compared to a-Si TFTs.

In SOP technology, the driver circuits, the AM, and other function parts are integrated on the same substrate, which has a lot of advantages, for example, the wire connection becomes easy and the parasitic capacitance from the wires is decreased. Therefore, the working frequency is increased and the cost of the system is decreased. Thus, SOP technology becomes the main current technology for super-higher-density and superthinner displays. At the early stage, most of the basic units of the digital circuits are composed of NAND gates, multiplexers, and D-type flip-flops, which are realized by a complementary metal-oxide semiconductor (CMOS) process.

6.5.1 System Design

Driver circuits are formed with units that include an edge trigger D-type flip-flop, control circuits, and output buffer circuits. These are shown in Fig. 6.22.

Figure 6.23 shows the schematic and layout of one stage of the data driver that is composed of an edge trigger D-type flip-flop (D_{ff}), a middle control circuit, and an output buffer. One D_{ff} is performed with four transmit gates and four inverters. The simulation results show that a maximum delay time of one stage of the shift circuit is smaller than 25 ns; thus the shift circuit has the potential of a

40 MHz working frequency. The enable signal *OE* and the *Q* signal are formed in D_{ff} and connected to NAND gate input pins. The output of the NAND gate is connected to the first inverter input pin. Meanwhile, the output of the first inverter is connected to the input pin of the second inverter. The outputs of the two inverters are connected to the output *C* and *NC* pins of this circuit to control the next-stage circuit, that is, the output buffer. The functions of output buffers for data and scan circuits are the same, which are the CMOS multiplexers. Wider-channel TFTs are used in the output buffer circuits for higher output current. The width of the channel in n- and p-channel TFTs is 500 μm. Using this output buffer push of 100 pF capacity, the rise time is 170 ns and the fall time is 150 ns. The output buffers have excellent pushing capacity with high and low voltage states.

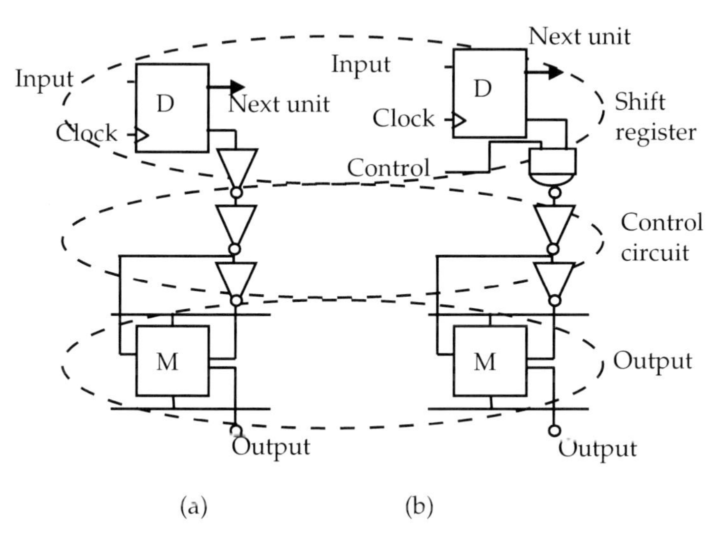

Figure 6.22 Schematic of a scan line circuit unit (a) and a data line circuit unit (b). Reprinted with permission from Ref. [52], Copyright 2001, John Wiley and Sons.

There are eight electrodes in the data driver: V_{dd}, V_{gnd}, gray-scale direct voltage (GDV), clock signal for data line (CLK_d), complementary clock signal for data line ($CLKB_d$), input data signal (*D*), output enable (*OE*), and the output signal (D_o). The data circuit works in the following way. First, the data signals are shifted, and 160 data signals are registered in Q_1–Q_{160} at 10 μs. At this time, *OE* is 0, and the output signals (D_{o1}–D_{o160}) are 0. After this operation step, the CLK is stopped and the *OE* is changed to 1. D_{o1}–D_{o160} are the same as the signals of Q_1–Q_{160} and load to the data lines. This time is 6.5 μs.

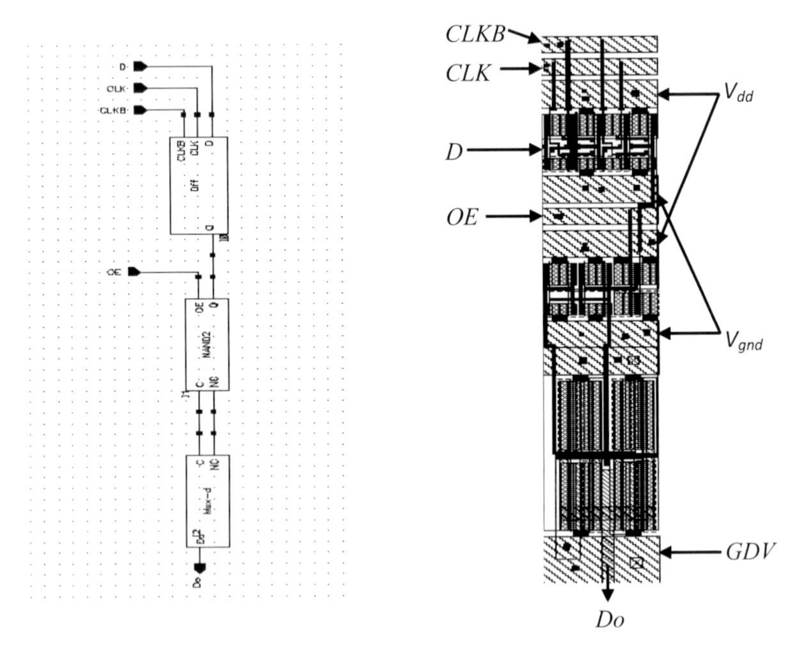

Figure 6.23 The schematic and layout of a one-stage data circuit.

The unit of an integrated scan driver is composed of one edge trigger D_{ff}, a middle driver circuit, and an output buffer. The schematic and layout are shown in Fig. 6.24. There are seven electrodes in this circuit: V_{dd1}, V_{dd2}, V_{gnd}, CLK_s, $CLKB_s$, enable signal (G), and output signal (G_o). The widths of the power line and clock lines are 100 μm and 60 μm, respectively. The length of the electrical wires in the display system is about 4 cm. The scan circuit works in the following way. When signal *G* comes, a pulse signal is shifted and it pushes the output buffer to drive the scan lines. It is the middle driver circuit of the scan circuit where a circuit combines with three series inverters. The channel widths of the second- and third-stage inverters are two times that of the first-stage inverter. The purpose is to improve the push capacity.

Lastly, the SOP panel for an AMOLED is shown in Fig. 6.25. The area of this panel is 65 × 70 mm^2. It was designed and fabricated at HKUST. At the vertical left side is the gate driver, and at the horizontal bottom is the data driver [12]. There are 120 output pins in the scan circuit output shift pulse for 120-stage gate drivers. There are 160 output pins in the data circuit for writing the data signals to the data lines. The data signals input the eight parallel pins. The purpose of the design is to decrease the working frequency of the data circuit. There are 39 pins in the system to connect the outside signals. The common cathode and power line are connected to two large pins. The relation between the number of pins and the outside signals is shown in Table 6.3.

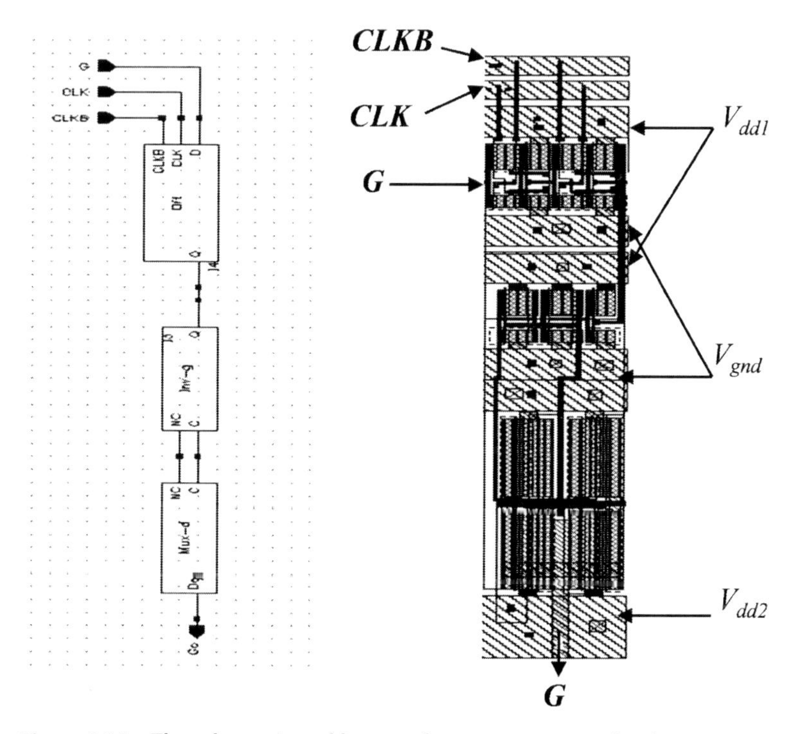

Figure 6.24 The schematic and layout of a one-stage scan circuit.

The typical one-poly-Si and two-metal-layer electrode technology and recrystallized metal-induced unilateral crystallization (RC-MIUC) TFT technology are used in the SOP fabrication process. Both n- and p-channel RC-MIUC TFTs are realized [52].

Figure 6.25 Photo of the SOP RC-MIUC TFT AMOLED display system.

Table 6.3 The relation between the number of pins and outside signals

Scan circuit	Pin no.	1	2	3	4	5	6	7	
	Signal	V_{dd2}	V_{gnd}	V_{dd1}	V_{gnd}	V_{dd1}	CLK_s	$CLKB_s$	
	Pin no.	8	9	10	11	12	13	14	15
	Signal	$CLKB_s$	CLK_s	V_{dd1}	G	V_{gnd}	V_{dd1}	V_{gnd}	V_{dd2}
Data circuit	Pin no.	16	17	18	19	20	21	22	23
	Signal	DGV	V_{gnd}	V_{dd}	OE	V_{gnd}	V_{dd}	$CLKB_d$	CLK_d
	Pin no.	24	25	26	27	28	29	30	31
	Signal	D_{I1}	D_{I2}	D_{I3}	D_{I4}	D_{I5}	D_{I6}	D_{I7}	D_{I8}
	Pin no.	32	33	34	35	36	37	38	39
	Signal	CLK_d	$CLKB_d$	V_{dd}	V_{gnd}	OE	V_{dd}	V_{gnd}	DGV

6.5.2 Characteristics of Driving Circuits

Microscopic photos of fabricated circuits are shown in Fig. 6.26. The major circuit blocks include a shift register and control and output buffer circuits.

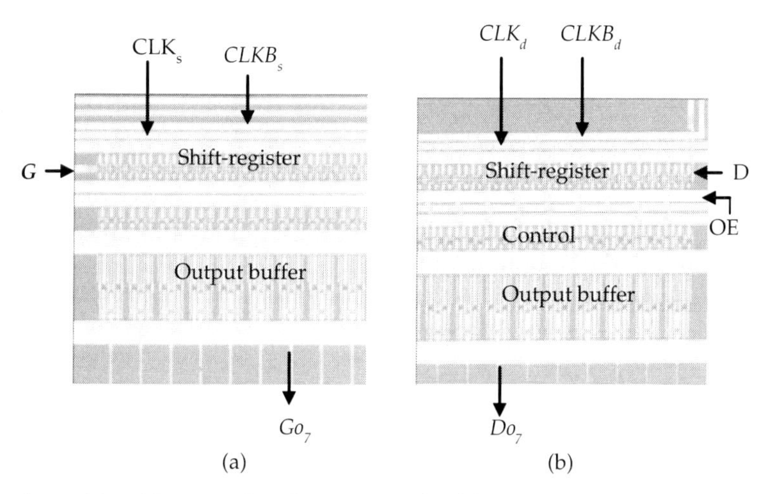

Figure 6.26 Photographs of a portion of a (a) scan circuit and a (b) data circuit. Reprinted with permission from Ref. [52], Copyright 2001, John Wiley and Sons.

For testing driving circuits, a testing signal's source circuit was designed and fabricated. The input signal is the 5 V square wave, and the frequency can change from 1 Hz to 30 MHz. The four output signals all are 5 V square waves, and two output signals are complementary with the same frequency of the input signal for the CLK and CLKB of the driving circuits. The third output signal is the synchronic signal of CLK with 1/8 of the frequency of CLK for the data or enable signal of the driving circuits. The fourth output signal is the synchronic signal of CLKB with 1/16 of the frequency of CLKB for the *OE* signal of the data circuit.

The testing waves of G_{o7} are shown in Fig. 6.27. These are the experimental results of the scan circuit with one output end loading of the 100 pF capacitor. The scan unit is driven by 3.0 MHz CLK, 0.37 MHz *G* signals. The signal has correct logic of operation. The rise time is 145 ns, and the fall time is 135 ns. The output buffers have excellent pushing capacity with high and low voltage states. The testing result is very close to the simulation result in Fig. 6.27.

The testing waves of D_{o7} are shown in Fig. 6.28. These are the experimental results of the data circuit with one output end loading of the 5 pF capacitor. It is driven by 30.0 MHz CLK, 3.7 MHz *D* signals and a 1.8 MHz *OE* signal. The signal has correct logic of operation. The testing result shows that the circuit fabricated with RC-MIUC

TFTs has high performance. These data circuits are suitable for use in the frame sequence AMOLED SOP. Much improvement is anticipated with further scaling of L [53, 54]. The circuits are suitable for driving the QVGA AM by the frame sequence driving mode.

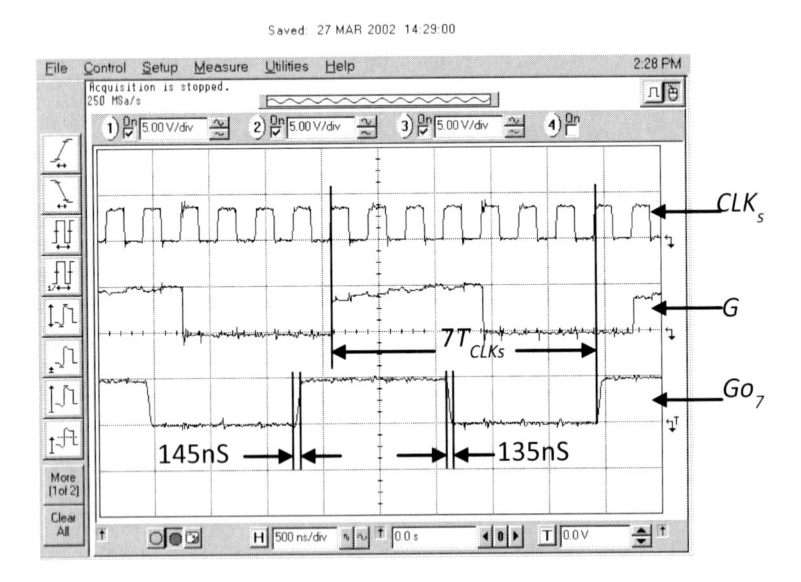

Figure 6.27 Testing waves of the CLK_s G and G_{o7} of a scan circuit.

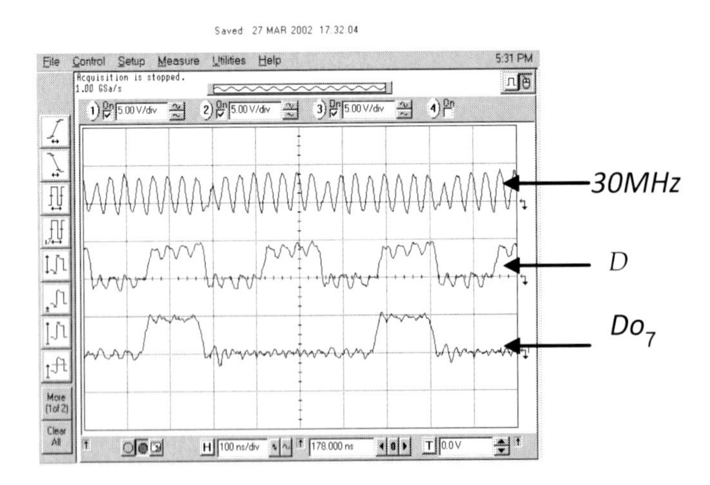

Figure 6.28 Testing waves of the CLK_d D and D_{o7} of a data circuit.

6.5.3 Integrated Top Emitting OLED on a System Panel

A top emitting structure was employed in this application. When an RC-MIUC AM and a circuit system were performed, the first step was the deposition of a 500 nm PECVD oxide and the opening of holes to Al-1%Si. Then a lift-off photoresist mask was defined prior to the room-temperature sputtering of the 20 nm titanium (Ti) and 30 nm platinum (Pt) anode [55]. Following the liftoff process, a 2 mm thick polymer insulation layer was coated to enhance the interlevel insulation and reduce the parasitic capacitance.

At an initial pressure of 20 Torr, sequential evaporation through appropriately designed shadow masks of a 1 nm PrO_x quantum layer, a 20 nm copper phthalocyanine (CuPc) anode buffer layer, a 50 nm 5% rubrene-doped *N,N'*-diphenyl-*N,N'*-bis-(3-methylphenyl) 1,1'-bipheny-4,4' diamine (TDP:Ru) hole transport layer, and a 60 nm tris-8-hydroxyquinoline aluminum (Alq_3) electron transport and emission layer was performed. The various organic films were deposited at rates of ~0.2 nm/s. A 10 nm top Al layer was evaporated, and the top indium tin oxide (ITO) electrode was sputtered by low power density at room temperature. A cross-sectional schematic of the OLED pixel is shown in Fig. 6.29. Light is emitted from the top electrode. The top emitting display system has the advantage of the full scale of the emitting area [56].

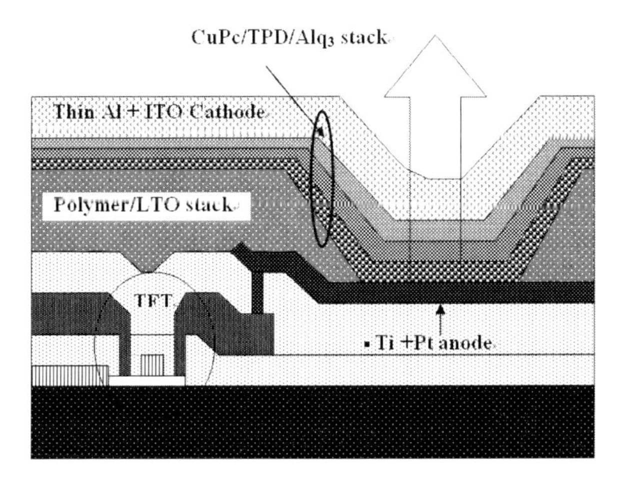

Figure 6.29 Schematic cross section of an integrated AM panel and OLED.

Figure 6.30 shows the display results of this 160 × 129 SOP AMOLED panel.

Figure 6.30 Typical image of a 160 × 129 SOP AMOLED display. Reprinted with permission from Ref. [36], Copyright 2001, John Wiley and Sons.

6.6 PMOS MIC TFT Applied to a Driver-Integrated AMOLED

To reduce the manufacturing cost and mask number, a lot of PMOS circuits for display have been proposed, including shift registers, level shifters, multiplexers, and DC–DC converters. Thus low-cost, high-performance, and reliable LTPS processing technologies are greatly required. The whole PMOS TFT SOP AMOLED system will be introduced, including PMOS gate drivers and data drivers.

6.6.1 SOP AMOLED Panel Design

The function block diagram in Fig. 6.31 shows the architecture of the slow-release metal-induced lateral crystallization (SR-MILC) TFT SOP panel. It consists of a PMOS shift register with two inverted clocks, a 1:6 multiplexer (MUX), an on panel electrostatic discharge (ESD) protection circuit, and a 2T1C pixel circuit. An HX5116 timing controller is designed to be mounted on the TFT substrate by chip-on-glass (COG) technology. The resolution of this panel is 320 × 180 ×

3 with a pixel size of 183 × 61 μm. Table 6.4 shows the specifications of this AMOLED panel. The system design is in accordance with the equipment feasibility in NFF. In our work, the 5 μm process design rule is used with a gate length of 5 μm, a contact hole of 4 × 4 μm^2, a minimum metal line width of 10 μm, and a minimum metal line separation of 5 μm. Inducing lines must be placed with special consideration. Figure 6.32 shows the layout of pixel circuits, and Fig. 6.33 shows the layout of a PMOS shift register.

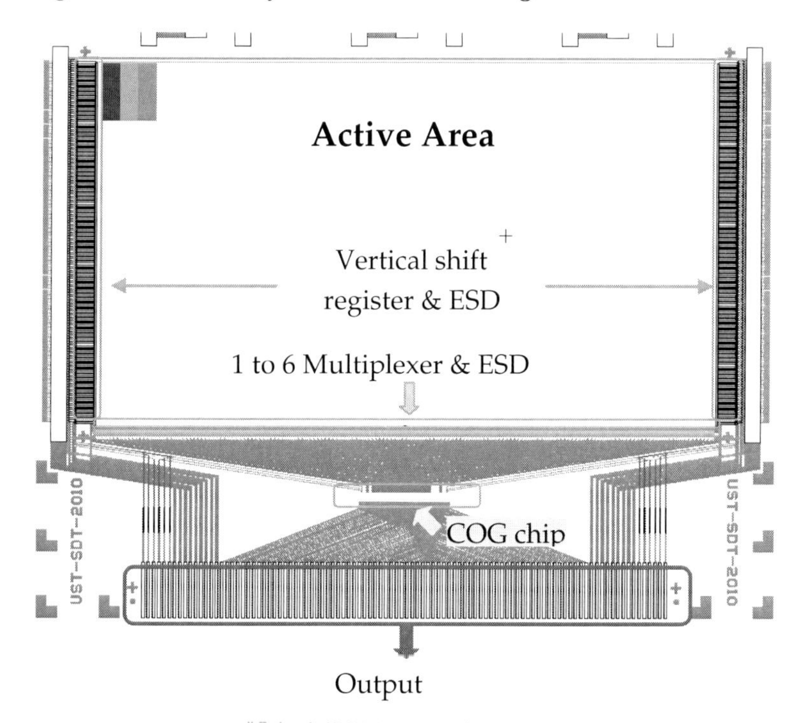

Figure 6.31 Layout of 3-inch SOP TFT panel.

Table 6.4 Specifications of the developed 3-inch AMOLED panel

Display size	**3-inch diagonal**
Number of pixels	360 × 160 × 3
Pixel pitch	183 × 61 μm
Pixel circuit	Voltage programming type 2T1C
Emission type	Bottom emission type
Process	LTPS PMOS

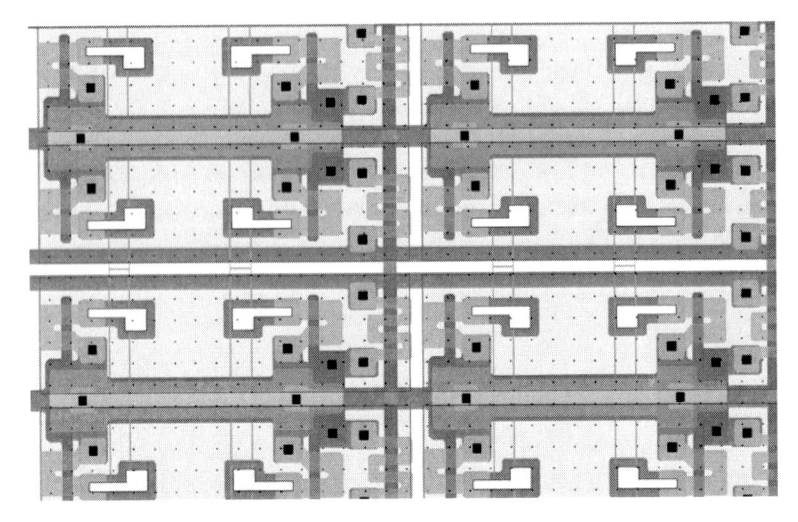

Figure 6.32 Layout of a pixel circuit.

The fabrication process began with a fully crystallized SR-MILC poly-Si film on a 4-inch Corning® EAGLE^{2000TM} glass substrate covered with 300 nm low-pressure chemical vapor deposition (LPCVD) silicon oxide. Total seven masks were included: active channel (AC), gate and scan line (GT), contact hole (CH), data line and internal connection (M1), passivation (PA), pixel electrode ITO (ITO), and planarization (PLN) layer. The schematic diagram of the process flow is shown in Fig. 6.34.

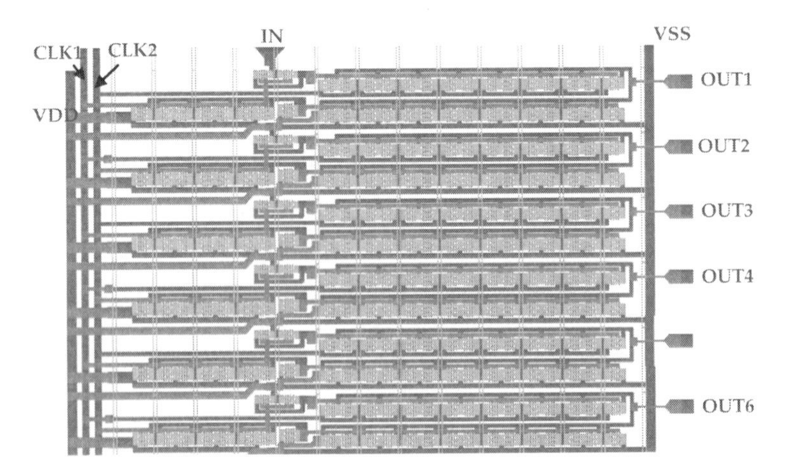

Figure 6.33 Layout of a PMOS two-clock shift register.

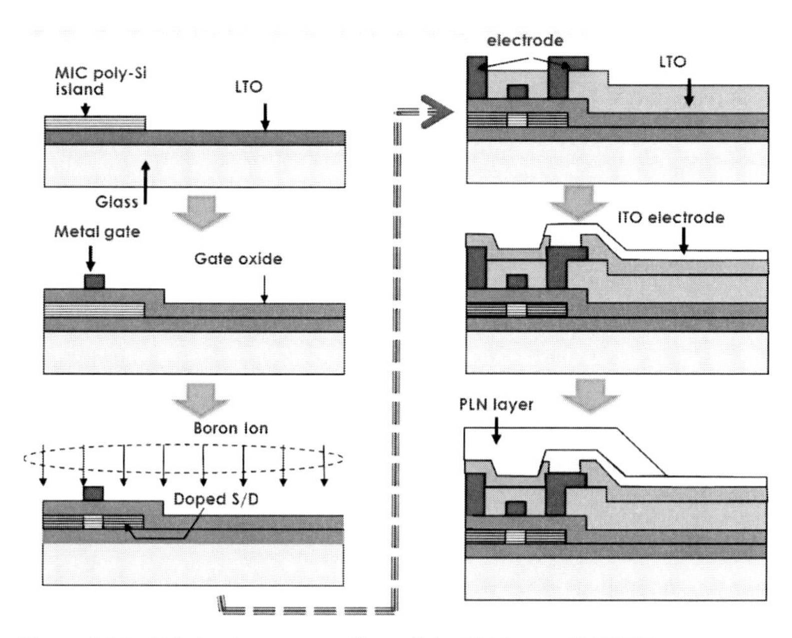

Figure 6.34 Fabrication process flow of the AM for an AMOLED.

6.6.1.1 PMOS poly-Si TFT circuit design considerations

With high yield and reliability as the main design priorities, a circuit topology containing only five PMOS TFT devices in each stage, which has the minimum number of transistors among conventional shift register circuits, is widely reported and used [57, 58].

A one-stage schematic is shown in Fig. 6.35. The five PMOS TFTs can be divided into pull-up and pull-down networks to implement the output scan pulse. In a conventional CMOS circuit, a pull-up network is implemented by PMOS transistors, while a pull-down network is implemented by an n-type metal-oxide semiconductor (NMOS). However, in this PMOS TFT shift register, both the pull-up and pull-down networks are composed of only PMOS devices.

Pn1 acts as a switch transistor, and Pn2 acts as a driver transistor in the pull-down network. By this transmission gate, a voltage of IN is being transmitted to the gate of Pn2 periodically when Pn1 is turned on by CLK1. When an IN pulse of low voltage is transmitted, TFT Pn2 is turned on. Then an OUT pulse of low voltage is generated and synchronized by CLK2.

A cross-couple design composed of Pn3, Pn4, and Pn5 implements the function of the pull-up network with a function similar to a CMOS six-transistor cross-couple topology of a static random-access memory (SRAM) cell. The differential logic gate of Pn3 and Pn4 pulls up the OUT node to high voltage. Pn5 acts as a transmission gate that can be turned on periodically by CLK1 so that the differential logic gate would hardly change state, due to poly-Si TFT leakage, unless the OUT node was pulled down to low voltage by the pull-down network of Pn1 and Pn2.

Transmission logic can minimize the number of transistors required that could suppress the variation introduced by poly-Si TFT devices. However, it also brings in the problem of clock feed-through, causing voltage transient change and noise at the OUT node. In this design, clock feed-through is just due to CLK1 and CLK2 through the parasitic capacitance of the transistor in the transmission logic, which is shown in Fig. 6.35. One feed-through path from CLK1 to the OUT node is composed of C_{1gd} and C_{2gs}, while the other path is composed of C_{3gd}, C_{4gd}, and C_{5gs}. The feed-through path from CLK2 to the OUT node is composed of C_{2gd} and C_{2gs}. To minimize the effect caused by clock feed-through, CLK1 and CLK2 are precisely complementary so that the voltage transient change brought in by the CLK1-to-OUT path can be compensated by the CLK2-to-OUT path. From the measurement, the clock feed-through effect on the OUT node can be suppressed a lot and the high voltage of the OUT pulse has little ripple [59, 60].

The OUT_n signal acts as IN_{n+1} signal so that the output pulse can be transmitted stage by stage in a series. Figure 6.36 shows the timing of this three-stage circuit, and Fig. 6.37 shows the schematic of a three-stage shift register. It can be extended to more stages in a large-scale system.

6.6.1.2 Measurement and analysis

The layout of a PMOS shift register is shown in Fig. 6.33. The output waveforms of the 180-stage shift register probed by a four-channel oscilloscope are shown in Fig. 6.38. The chip was bonded to a PCB for measurement. The whole circuit has robust output-driving ability and performance, showing that variation caused by the process can be compensated effectively. From the result, it can be seen that the output pulse signal can be transmitted in a series at the power supply voltage of 11 V and the circuit can implement the function, as designed. Considering the falling edge and rising edge time, the circuit can work well under different frame rates from 60 Hz to 220 Hz (oscilloscope load: C_{load} = 40 pF, R_{load} = 1 MΩ; parallel-connected to ground).

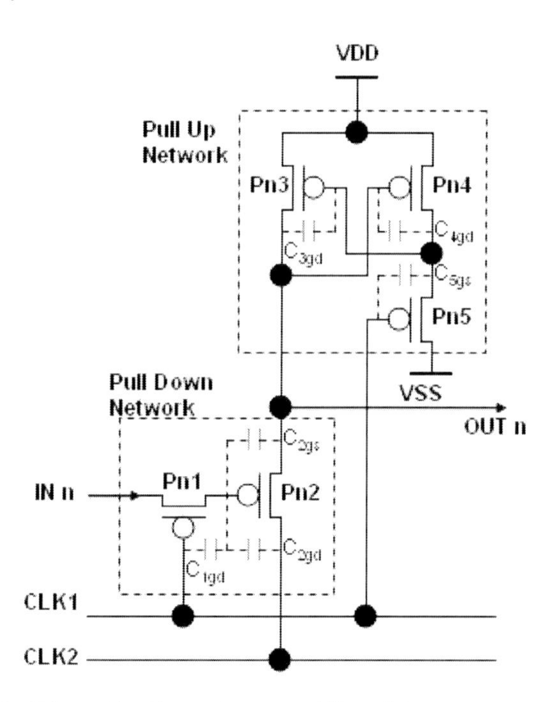

Figure 6.35 Schematic of a single-stage shift register.

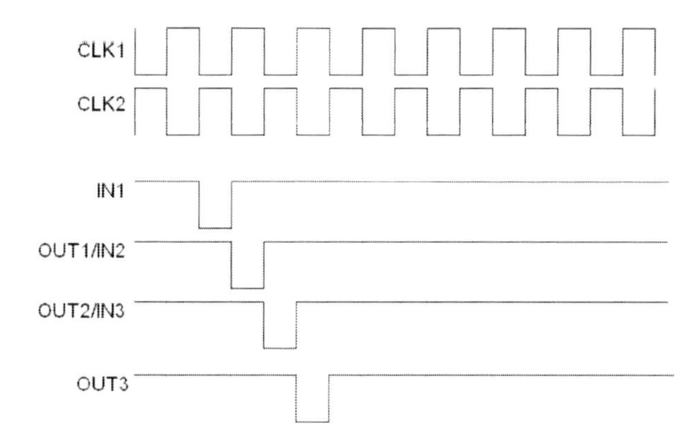

Figure 6.36 Timing of a three-stage shift register.

Reliability and distortion of the transient performance are important for the TFT shift register circuit [61]. Due to the TFT device characteristic and variation compared with single-crystalline silicon, it presents some deduction and distortion in the shift register outputs. Also, logic network design affects the transient performance to some extent, which can be characterized by the rising time and falling time of the output pulse.

There is some asymmetry between the pull-up network and pull-down network in a PMOS-only circuit, which is different from the CMOS logic family. So it is always difficult for designers to trade off the transient performance of falling time and rising time.

In this work, all the outputs are designed to be synchronized by complementary CLK signals, so there is no delay time between stages. As a result the logic fault caused by transient performance variation can be minimized. The measurement on uniformity of the edge rates in Fig. 6.39 indicates that the falling time is more sensitive to device variation than the rising time. For falling edge time, it is about 2–3 μs, with a variation coefficient σ/μ of 23.53%. For rising edge time, it is within less than 1 μs, with a variation coefficient of 4.26%.

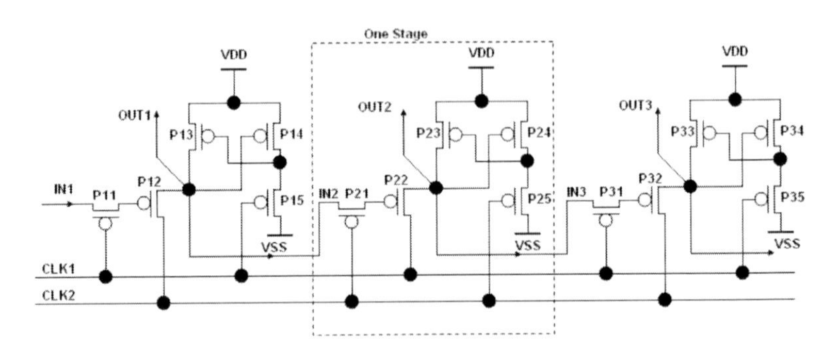

Figure 6.37 Schematic of a three-stage shift register.

The noise margin, which shows the self-repair ability of the circuit when deduction happens, is investigated. While the cross-couple design composed of Pn3, Pn4, and Pn5 holds the high voltage level of the output, the low voltage level of the output generated by the pull-down network more easily suffers from the deduction, causing

ΔV. As the OUT_n signal acts as the IN_{n+1} signal in the series-stage circuit, the attenuated V_{IL} of IN_{n+1} cannot implement the function well because the on-state channel resistance of Pn2 depends on the gate bias, which is directly related to the voltage of V_{IL}.

Figure 6.40b,c shows that when the deduction ΔV of stage n is within the range of the noise margin, the operation is fault tolerant and can be self-repaired back to normal. However, if ΔV of stage n is out of the noise margin, the fault operation would happen as shown in Fig. 6.40d. In this work, the noise margin can be about 2.7–3 V when the circuit works under a power supply of 11 V.

If the threshold voltage shift or device performance varies more, the probability of fault operation would increase. Since the OUT signal is bootstrapped by CLK, the negative effect from devices could also be suppressed by optimizing the ratio of gate oxide capacitance and parasitic capacitance of Pn2.

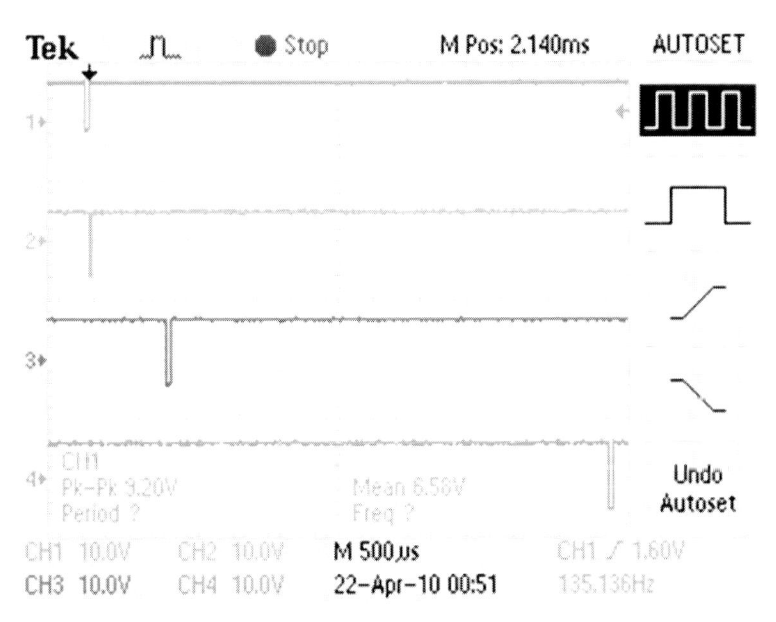

Figure 6.38 Output results of a shift register, including input, output1, 28, and 180.

Figure 6.41 shows the 8 × 8 checkbox display results of a green-color SOP AMOLED panel.

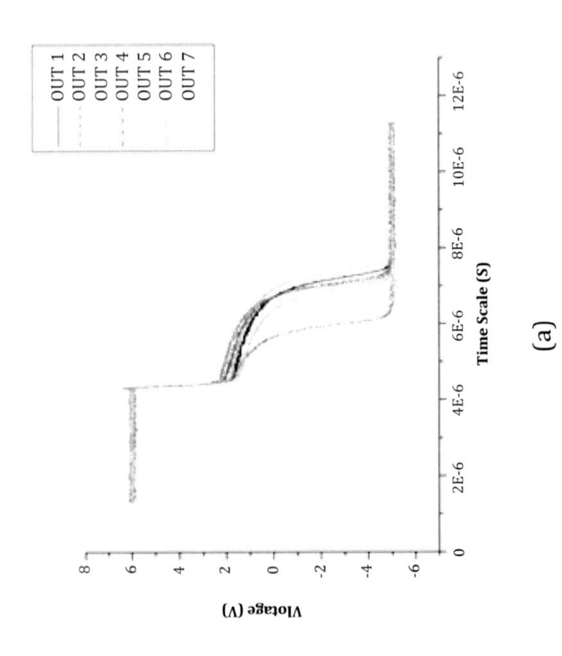

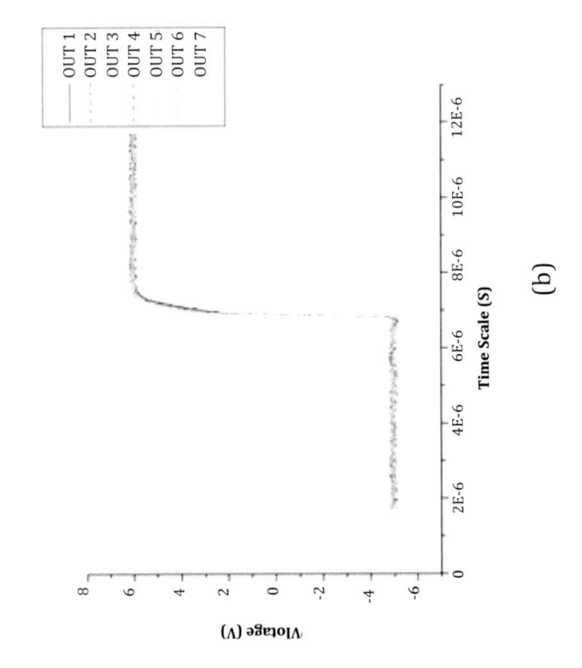

Figure 6.39 Falling edge variation (a) and rising edge variation (b).

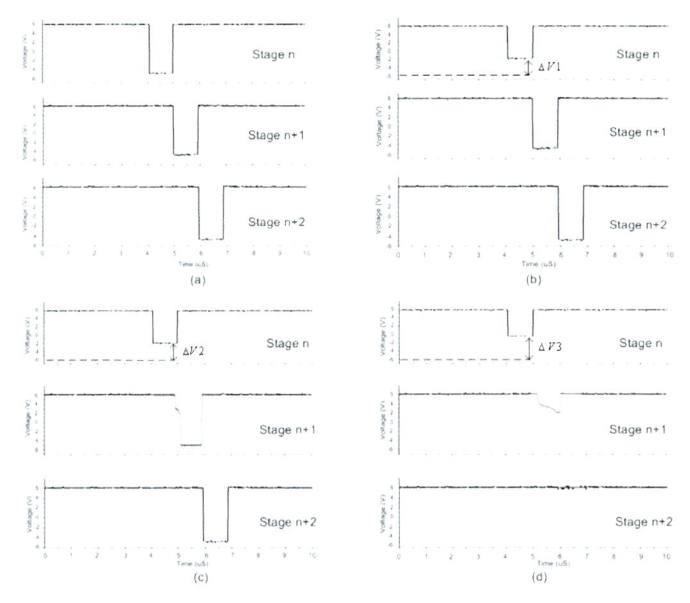

Figure 6.40 (a) Normal operation, (b) deduction-tolerant operation, (c) self-repair operation ($\Delta V1 < \Delta V2 < \Delta V3$), and (d) fault operation.

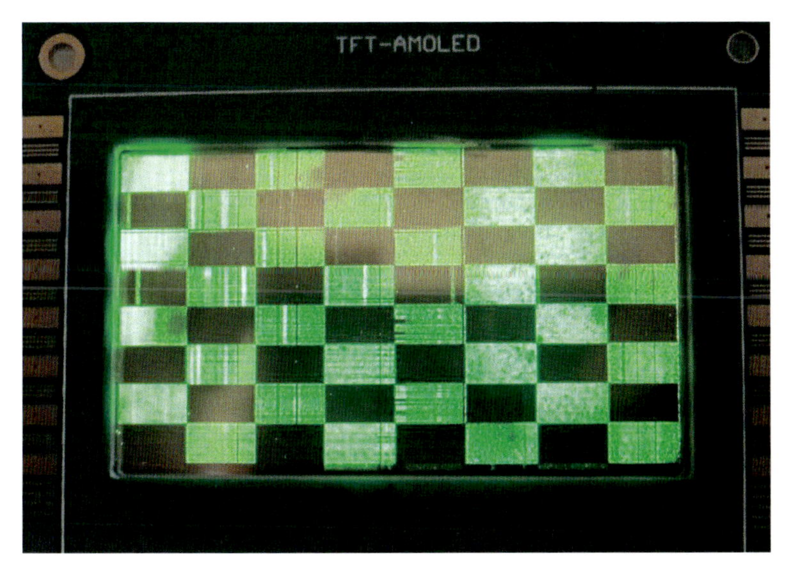

Figure 6.41 Display image of a 3-inch PMOS SOP AMOLED.

6.7 MIC TFT with In-Cell p-i-n Photodiode Sensors for AMOLEDs

As mentioned before, besides application in peripheral circuit integration, the MIC TFT process also has the capability of integration of a p-i-n photodiode sensor, which can be applied to an in-cell touch panel for future display.

Figure 6.42 shows the schematic of a unit pixel of an LTPS AMOLED with an embedded p-i-n photodiode sensor. The pixels are divided into an AMOLED and a lateral p-i-n photodiode, in which a p-i-n sensor is located at the center of four pixels to make a touch function in AMOLED display and also guarantee a higher AR for display [62].

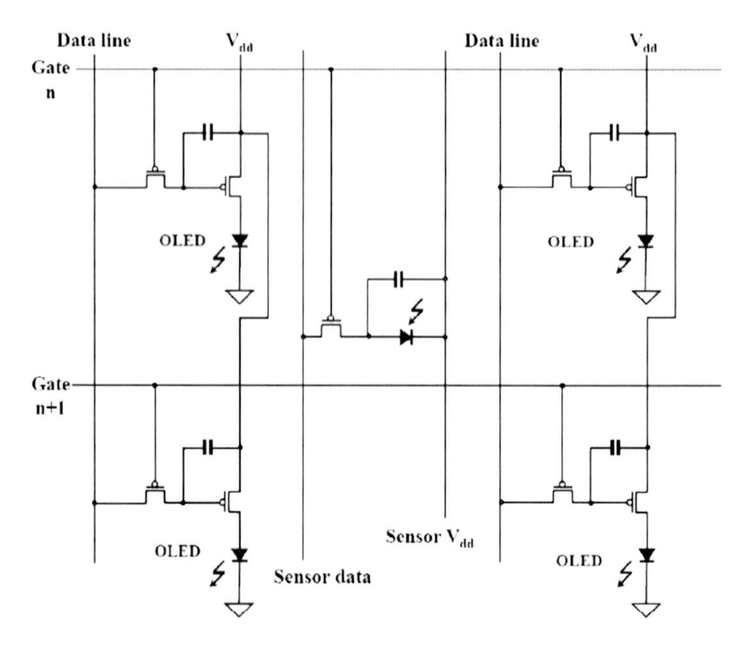

Figure 6.42 Schematic of an AMOLED with an embedded p-i-n photodiode. Reprinted with permission from Ref. [62], Copyright 2008, John Wiley and Sons.

The pixel numbers are 160 × 128, and the AR is ~40%. The storage capacitors for the AMOLED (COst) and the sensor (CSst) are 0.4 pF and 2.7 pF, respectively. The capacitance of the storage capacitor is relatively large, which is because the photosensitivity of the poly-Si p-i-n diode is very low. This is due to the low absorption coefficient of the poly-Si because its thickness is ~50 nm. Figure 6.43 shows the design configuration of an AMOLED with an embedded p-i-n photodiode sensor.

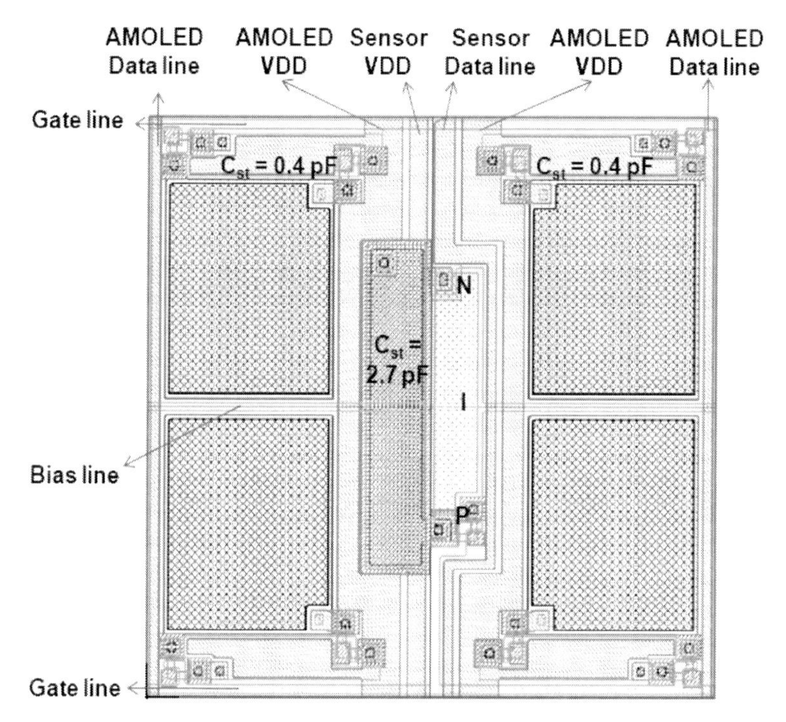

Figure 6.43 Layout of a unit touch pixel of an AMOLED. Reprinted with permission from Ref. [62], Copyright 2008, John Wiley and Sons.

Figure 6.44 shows the electrical properties of a p-channel TFT and a p-i-n diode. Figure 6.44a shows the transfer characteristics of a p-channel poly-Si TFT, exhibiting a field effect mobility of 49 cm^2/Vs, a threshold voltage of –0.8 V, and a subthreshold voltage swing of 0.31 V/decade. Figure 6.44b shows the photocurrents of the LTPS p-i-n photodiode under various illuminations. The photocurrent is at least 10 times higher than that of dark current under 100 lux. Therefore, the photocurrents under low-level illumination can be detected using a readout IC.

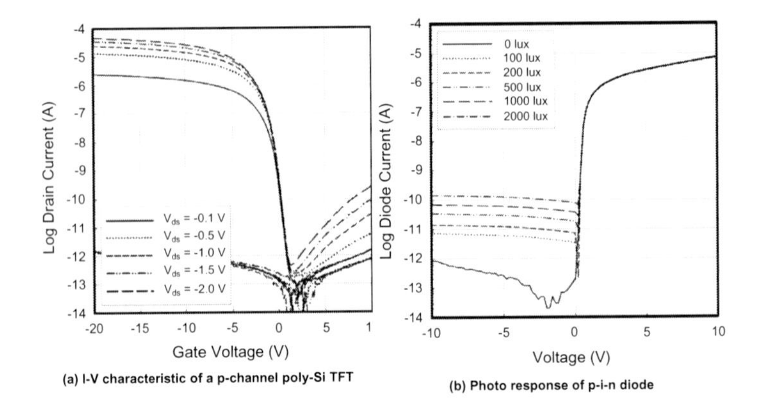

Figure 6.44 Transfer characteristics of (a) poly-Si TFT and current–voltage characteristics of (b) p-i-n diode. Reprinted with permission from Ref. [62], Copyright 2008, John Wiley and Sons.

6.8 Summary

The most common applications for MIC TFTs in flat-panel displays have been introduced. These are pixel circuits for AMLCDs and AMOLEDs, FSC LCDs without a CF, SOP systems realized by CMOS MIC TFTs, and the PMOS SOP system. Besides application in peripheral circuit integration, the MIC TFT process also has the capability for integration of a p-i-n photodiode sensor, which can be applied to in-cell touch panels for future display.

References

1. Yoon, S. Y., Oh, J. Y., Kim, C. O., Jang, J. (1998) Low temperature solid phase crystallization of amorphous silicon at 380 °C. *Journal of Applied Physics*, **84**(11), 6463–6465.
2. Van Bockstael, C., Detavernier Christophe, Vanmeirhaeghe, R. L., Jordan-Sweet, J. L., Lavoie, C. (2009) In situ study of the formation of silicide phases in amorphous Ni-Si mixed layers. *Journal of Applied Physics*, **106**(6), 064515–064518.
3. Hayzelden, C., Batstone, J. L., Cammarata, R. C. (1992) In situ transmission electron microscopy studies of silicide-mediated crystallization of amorphous silicon. *Applied Physics Letters*, **60**(2), 225–227.

4. Kawazu, Y., Kudo, H., Onari, S., Arai, T. (1990) Low-temperature crystallization of hydrogenated amorphous silicon induced by nickel silicide formation. *Japanese Journal of Applied Physics*, **29**, 2698.

5. Kuo, Y. (2004) *Thin Film Transistors, Materials and Processes. Polycrystalline Silicon Thin Film Transistors*, Vol. 2 (Kluwer Academic), 237.

6. Wong, M., Jin, Z., Bhat, G.A., Wong, P.C., Kwok, H.S. (2000) Characterization of the MIC/MILC interface and its effects on the performance of MILC thin-film transistors. *IEEE Transactions on Electron Devices*, **47**(5), 1061–1067.

7. Jin, Z., Bhat, G.A., Yeung, M., Kwok, H.S., Wong (1998) Nickel induced crystallization of amorphous silicon thin films. *Journal of Applied Physics*, **84**(1), 194–200.

8. Kuznetsov, A. Y., Svensson, B.G. (1995) Nickel atomic diffusion in amorphous silicon. *Applied Physics Letters*, **66**(17), 2229–2231.

9. Zhang, D. and Wong, M. (2005) Effects of trace nickel on the growth kinetics and the electrical characteristics of metal-induced laterally crystallized polycrystalline silicon and devices. *Journal of the Society for Information Display*, **13**(10), 815–822.

10. Kang, J. S., Schroder, D.K. (1989) Gettering in silicon. *Journal of Applied Physics*, **65**(8), 2974–2985.

11. Lee, S.-W., Ihn, T.-H., Joo, S.-K. (1996) Fabrication of high-mobility p-channel poly-Si thin film transistors by self-aligned metal-induced lateral crystallization. *IEEE Electron Device Letters*, **17**(8), 407–409.

12. Meng, Z. G., Wang, M.X., Wong, M. (2000) High performance low temperature metal-induced unilaterally crystallized polycrystalline silicon thin film transistors for system-on-panel applications. *IEEE Transactions on Electron Devices*, **47**(2), 404–409.

13. Yoon, S. Y., Kim, K.H., Kim, C.O., Oh, J.Y., Jang, J. (1997) Low temperature metal induced crystallization of amorphous silicon using a Ni solution. *Journal of Applied Physics*, **82**(11), 5865–5867.

14. Ahn, J. H., Ahn, B.T. (2001) Crystallization of amorphous silicon thin films using a viscous nickel solution. *Journal of the Electrochemical Society*, **148**(9), H115–H119.

15. Chao, C. W., Wu, Y.C.S., Hu, G.R., Feng, M.S. (2002) Device characteristics of polysilicon thin-film transistors fabricated by electroless plating Ni-induced crystallization of amorphous Si. *Japanese Journal of Applied Physics*, **42**, 1556–1559.

16. Kim, J. C., Choi, J.H., Kim, S.S., Kim, K.M., Jang, J. (2003) Single-grain thin-film transistor using Ni-mediated crystallization of amorphous

silicon with a silicon nitride cap layer. *Applied Physics Letters*, **83**(24), 5068–5070.

17. Chang, Y.-J., Kim, Y.-I., Shim, S.-H., Park, S., Ahn, K.-W., Song, S.-C., Choi, J.-B., Min, H.-K., Kim, C.-W. (2006) 28.3: World's largest (21.3 in.) UXGA non-laser LTPS AMLCD. *SID Symposium Digest of Technical Papers*, **37**(1), 1276–1279.
18. Lih, J. J., Sung, C.F., Li, C.-H., Hsiao, T.-H., Lee, H.-H.(2004) 57.1: Invited paper: comparison of a-Si and poly-Si for AMOLEDs. *SID Symposium Digest of Technical Papers*, **35**(1), 1504–1507.
19. Sanford, J. L., Libsch, F.R. (2003) 4.2: TFT AMOLED pixel circuits and driving methods. *SID Symposium Digest of Technical Papers*, **34**(1), 10–13.
20. Lee, J.-H., You, B.-H., Han, C.-W., Shin, K.-S., Han, M.-K. (2005) P-2: A new a-Si:H TFT pixel circuit suppressing OLED current error caused by the hysteresis and threshold voltage shift for active matrix organic light emitting diode. *SID Symposium Digest of Technical Papers*, **36**(1), 228–231.
21. Taro, H., Shinji, T., Kanoh, K., Yoshinao, K. (2006) 46.2: New OLED pixel circuit and driving method to suppress threshold voltage shift of a-Si:H TFT. *SID Symposium Digest of Technical Papers*, **37**(1), 1547–1550.
22. Hekmatshoar, B., Kattamis, A.Z., Cherenack, K., Wagner, S., Sturm, J.C. (2008) A novel TFT-OLED integration for OLED-independent pixel programming in amorphous-Si AMOLED pixels. *Journal of the Society for Information Display*, **16**(1), 183–188.
23. Yeh, S.-H., Sun, W.-T., Yu, J.-S., Chen, C.-C., Lee, J., Yang, C.-S. (2005) P-173L: late-news poster: a 2.2-inch QVGA system-on-glass LCD using p-type low temperature poly-silicon thin film transistors. *SID Symposium Digest of Technical Papers*, **36**(1), 352–355.
24. *Further Strengthening Global Domination* (2007) Available from http://www.ednasia.com/article-18676-furtherstrengtheningglobaldomination-Asia.html.
25. Gofuku, E., Takahashi, S., Numano, Y. (2000) 30.4L: late-news paper: 15-in. quad-VGA TFT-LCD high resolution for multimedia applications. *SID Symposium Digest of Technical Papers*, **31**(1), 464–467.
26. Higuchi, T., Hanari, J., Nakamura, N., Mametsuka, K., Watanabe, M., Murai, T., Watanabe, R., Seiki, M., Azuma, R., Hori, Y., Nakamura, K., Aoki, Y., Sakurai, H., Nakazono, T., Harada, N. (2000) 48.4: Development of a 15-inch UXGA low-temperature poly-Si TFT-LCD. *SID Symposium Digest of Technical Papers*, **31**(1), 1121–1123.

27. Martin, R., Turner, W. (1997) Design of high -resolution AMLCDs with greater than 2000 gate lines. *SID 97 DIGEST*, 7–10.
28. Kim, H. D., Jeong, J.K., Chung, H.-J., Mo, Y.-G. (2008) 22.1: Invited paper: technological challenges for large-size AMOLED display. *SID Symposium Digest of Technical Papers*, **39**(1), 291–294.
29. Reza, C. G., Arokia, N. (2006) Low-cost stable a-Si:H AMOLED display for portable applications. *2006 IEEE North-East Workshop on Circuits and Systems.*
30. Sambandan, S., Striakhilev, D., Nathan, A. (2006) Device and circuit level optimization for high performance a-Si:H TFT-based AMOLED displays. *Journal of Display Technology*, **2**(1), 52–59.
31. Hack, M., Brown, J.J., Mahon, J.K., Kwong, R.C., Hewitt, R. (2001) Performance of high-efficiency AMOLED displays. *Journal of the Society for Information Display*, **9**(3), 191–195.
32. Cammarata, R. C., Thomson, C.V., Hayzelden, C., Tu, K.N. (1990) *Journal of Material Research*, (5), 2133–2135.
33. Kinoshita, H., Scheleupen, K., Colgan, E.G., Nunes, R., Kitahara, H., Kodate, M., Tagsugi, S. (1999) High resolution AMLCD made with a-Si:H TFTs and with an Al-gate and IZO last structure. *SID Symposium Digest of Technical Papers*, **30**(1), 736–739.
34. Chang, W. (2000) 7.2: Invited paper: fourth-generation TFT-LCD production line. *SID Symposium Digest of Technical Papers*, **31**(1), 64–68.
35. Lee, Y., Park, H., Moon, S.-H., Kim, T., Lee, K.-C., Berkeley, B.H., Kim, S.-S. (2006) 16.2: Advanced TFT-LCD Data Line Reduction Method. *SID Symposium Digest of Technical Papers*, **37**(1), 1083–1086.
36. Meng, Z. G.,Chen, H.Y., Qiu, C.F., Kwok, H.S., Wong, M. (2001) 24.3: Active-matrix organic light-emitting diode display implemented using metal-induced unilaterally crystallized polycrystalline silicon thin-film transistors. *SID Symposium Digest of Technical Papers*, **32**(1), 380–383.
37. Okumura, H. (2007) Low power double scanning MFD for multi-media LCDs. *Proceedings of Asia Display 2007*, **1**, 335–338.
38. Lee, J.-B., Park, H.-S., Choi, J.-W., Won, T. (2007) Gamma curve control for image data processing. *Proceedings of Asia Display 2007*, **1**, 63–66.
39. Liu, K.-H., Liao, J.-S., Chu, Y.-N., Tseng, W.-T., Yu, H.-T. (2007) P-68: Color-sequential TFT-LCD with SAM (shared array mask) structure and color breakup reduction schema. *SID Symposium Digest of Technical Papers*, **38**(1), 445–448.

40. Han, S.-S., Lim, K.-M., Yoo, J.S., Jeong, Y.-S., Lee, K.-E., Park, J., Nam, D.H., Lee, S.-W., Yoon, J.-M., Jung, Y.-H., Seo, H.S., Kim, C.-D. (2003) P-2: 3.5 inch QVGA low-temperature poly-Si TFT LCD with integrated driver circuits. *SID Symposium Digest of Technical Papers*, **34**(1), 208–211.

41. Taniguchi, Y., Inoue, H., Sawasaki, M., Tanaka, Y., Hasegawa, T., Sasaki, T., Koike, Y., Okamoto, K. (2000) 25.3: An ultra-high-quality MVA-LCD using a new multi-layer CF resin spacer and black-matrix. *SID Symposium Digest of Technical Papers*, **31**(1), 378–381.

42. Meng, Z. G., Peng, H.J., Wu, C.Y., Qiu, C.F., Li, K.K., Wong, M., Kwok, H.S. (2004) Room-temperature deposition of thin-film indium tin oxide on micro-fabricated color filters and its application to flat-panel displays. *Journal of the Society for Information Display*, **12**(1), 113–118.

43. Takatori, K.-i., Sekine, H., Saitoh, G., Svetlana, K., Masumura, K., Sumiyoshi, K., Imai, M., Sato, T., Sato, Y., Okumura, F. (2005) 29.3: A 1450-ppi field-sequential system-on-glass LCD capable of operating over a wide temperature range. *SID Symposium Digest of Technical Papers*, **36**(1), 1182–1185.

44. Hong, W.-S., Jung, K.-W., Choi, J.-H., Hwang, B.-K., Chung, K. (2004) High transmittance TFT-LCD panels using low-k CVD films. *IEEE Electron Device Letters*, **25**(6), 381–383.

45. Koma, N., Miyashita, T., Uchida, T., Mitani, N. (2000) P-28: Color field sequential LCD using an OCB-TFT-LCD. *SID Symposium Digest of Technical Papers*, **31**(1), 632–635.

46. Baron, P. C., Walter, C. (2004) P-29: Variability in susceptibility to the color flash effect in field sequential color displays. *SID Symposium Digest of Technical Papers*, **35**(1), 338–341.

47. Yamada, F., Nakamura, H., Sakaguchi, Y., Taira, Y. (2000) 52.2: Invited paper: color sequential LCD based on OCB with an LED backlight. *SID Symposium Digest of Technical Papers*, **31**(1), 1180–1183.

48. Chen, C.-H., Lin, F.-C., Hsu, Y.-T., Huang, Y.-P., Shieh, H.P.D. (2009) A field sequential color LCD based on color fields arrangement for color breakup and flicker reduction. *Journal of Display Technology*, **5**(1), 34–39.

49. Zhao, S. Y., Meng, Z.G., Wong, M., Kwok, H.S. (2010) Metal-induced continuous zonal domain (CZD) polycrystalline silicon thin-film transistors and its application on field sequential color liquid crystal display. *Journal of Display Technology*, **6**(4), 135–141.

50. Li, Y. W., Tan, L., Kwok, H.S. (2008) Passive-matrix-driven field-sequential-color displays. *Journal of the Society for Information Display*, **16**(3), 429–434.

51. Li, Y. W., Tan, L., Kwok, H.S. (2008) 5.2: Field-sequential-color LCDs based on transient modes. *SID Symposium Digest of Technical Papers*, **39**(1), 32–35.

52. Meng, Z. G., Lee, K.K., Kwok, H.S., Wong, M. (2001) Application of Re-crystalled metal-induced unilaterally crystallized polycrystalline silicon thin-film transistor technology to reflective liquid-crystal display. *Journal of the Society for Information Display*, **9**(4), 319–323.

53. Chan, V. W. C.,Chan, P.C.H., Chan, M. (2000) Three dimensional CMOS integrated circuits on large grain polysilicon films, in *IEDM* (San Francisco, CA, USA).

54. Sakamoto, H., Makita, N., Hijikigawa, M. (2000) 2.6 inch HDTV Panel Using CG Silicon. *SID International Symposium Digest of Technical Paper*, XXXI, 1190–1193.

55. Bulovic, V., Tian, P., Burrows, P.E., Gokhale, M.R., Forrest, S.R., Thompson, M.E. (1997) A surface-emitting vacuum-deposited organic light emitting device. *Applied Physics Letters*, **70**(22), 2945–2956.

56. Sasaoka, T., Sekiya, M., Yumoto, A., Yamada, J., Hirano, T., Iwase, Y., Yamada, T., Ishibashi, T., Mori, T., Asano, M., Tamura, S., Urabe, T. (2001) A 13.0-inch AM-OLED display with top emitting structure and adaptive current mode programmed pixel circuit (TAC). *Society for Information Display 2001 International Symposium Digest of Technical Papers*, XXXII, 384–387.

57. Yang, C.-S. (2007) High integrated 10.2-inch WVGA LTPS LCD manufactured by PMOS process. *SID 07 DIGEST*, 245–248.

58. Kim, B. (2011) A depletion-mode a-IGZO TFT shift register with a single low-voltage-level power signal. *IEEE Electron Device Letters*, **32**(8), 1092–1094.

59. *SmartSpice User Manual* (2008) (Silvaco International, Santa Clara, CA, USA).

60. Ohkawa, S., Aoki, M., Masuda, H. (2003) Analysis and characterization of device variations in an LSI chip using an integrated device matrix array. *International Conference on Microelectronic Test Structures (ICMTS)*, 2003, 70–75.

61. Allee, D. (2009) Circuit-level impact of a-Si:H thin-film-transistor degradation effects. *IEEE Transactions on Electron Devices*, **56**(6), 1166–1175.

62. Kim, S. H., Lee, S.H., Park, W.H., Son, N.K., Kim, A.R., Hur, J.H., Kwon, J.H., Jang, J. (2008) 48.3: A 2 Inch LTPS AMOLED with an embedded lateral p-i-n photodiode sensors. *SID 08 DIGEST*, 724–727.

Chapter 7

Laser-Assisted Metal-Induced Crystallization and Its Applications in Data Storage

Yung-Chiun Her
Department of Materials Science and Engineering, National Chung Hsing University, 250 Kuo-Kuang Road, Taichung, 40254, Taiwan, Republic of China
ycho@dragon.nchu.edu.tw

7.1 Introduction

Due to the rapid development of multimedia technology, there is an increasing demand for data storage media with larger recording capacities and a faster data transfer rate. Optical disks have become the most popular storage media for data storage in computers and content distribution in audio/video applications, due to their advantages of having low cost and large capacities and of having portability and interchangeability. In the past two decades, various formats of compact disk (CD) and digital versatile disk (DVD) were developed for different applications. However, requirements of high-density optical media for high-definition television (HDTV) and network download applications had stimulated the development of

Metal-Induced Crystallization: Fundamentals and Applications
Edited by Zumin Wang, Lars P. H. Jeurgens, and Eric J. Mittemeijer

ISBN 978-981-4463-40-9 (Hardcover), 978-981-4463-41-6 (eBook)
www.panstanford.com

the Blu-ray disk (BD) system by nine leading consumer electronic companies [1–5]. The BD system employs a blue laser pickup with a wavelength (λ) of 405 nm and an objective lens with a numerical aperture (NA) of 0.85 and can achieve a recording capacity of 22.5 GB on a 120 mm disk. Among those three different formats, write-once Blu-ray disks (BD-Rs) are predominantly suitable for the storage of permanent or nonalterable information, such as clinical charts, official documents, and financial transitions. On the basis of the great success of the recordable compact disk (CD-R) and the recordable digital versatile disk (DVD-R), the commercial demand for BD-Rs will be expected to increase inevitably. For write-once CD-Rs and DVD-Rs, organic dyes such as cyanine, phthalocyanine, or azo dyes are commonly employed as recording materials. The organic dye is deposited on a polycarbonate (PC) substrate by a spin-coating method and then covered by a metal reflective layer. When the recording laser power is applied, the organic dye will decompose due to laser heating, leading to a change of optical properties of the disk. However, organic dyes are thought to be difficult to apply to a BD-R due to some practical issues encountered in the blue ray conditions, such as low optical absorption, poor light resistance, poor thermal stability, poor filling properties and bad adhesion in the narrow track pitch with the spin-coating process, and toxicity. Besides, an enormous amount of recorded organic write-once disks will be thrown away in the future. From both functional and ecological points of view, it is necessary to develop environmentally friendly inorganic recording materials with high recording sensitivity for BD-Rs.

In 2001, Ohkubo et al. first proposed an inorganic BD-R with only an amorphous Si (a-Si) layer on the substrate [6]. The reflectivity of 38% and the absorptivity of 44% were obtained at a wavelength of 400 nm, and a carrier-to-noise ratio (CNR) of about 43 dB was achieved for a 0.25 μm length mark with a recording power of 6 mW. However, they also pointed out that the recording sensitivity still needs to be improved. Besides, the crystallization of a-Si has been observed to take place at about 700°C [7], indicating that high blue laser power is required during recording that will lead to an increase in the cost of the blue laser pickup. To increase the recording speed and lower the recording power, improvement of recording sensitivity and reduction of crystallization temperature of the a-Si recording film are required. It is well known that metal-induced crystallization (MIC) using metals such as Al, Ag, Au, Cu, Co, Ni, Pd, and Ti can dramatically reduce the crystallization temperature of a-Si and increase the crystal growth rate [8]. Therefore, the metal/a-Si bilayers may be considered as promising candidates for use in the high-data-transfer-rate BD-R.

The crystallization of a-Si can be induced by metals such as Cu, Pd, and Ni forming silicides with Si [9–11] or by metals such as Ag, Au, and Al forming a single eutectic system with Si [12–14]. This enhancement of crystallization occurs due to an interaction of the free electrons from the metal with the silicon covalent bonds near the growing front, originating of a metal silicide, or a eutectic alloy, the migration of the metal silicide, or some combination of these that reduces the crystallization temperature [15]. Up to date, several metal/a-Si bilayers have been proposed for high-speed write-once Blu-ray recording and the recording mechanisms have also been investigated. Inoue et al. developed an inorganic BD-R adopting a 5 nm Cu/5 nm a-Si bilayer as the recording film and obtained a jitter value of less than 8% with optimal recording power of 5 mW at data transfer rates ranging from 36 Mbps to 144 Mbps [16]. Her and Wu also confirmed the feasibility of the Cu/a-Si bilayer for high-data-transfer-rate write-once Blu-ray recording and estimated that a data transfer rate as high as 223 Mbps can be achieved at a recording power of 10 mW [17]. Later, Schleipen et al. proved that the recording material CuSi is a good candidate for a 12X BD, without the need for complex write strategies [18]. Besides the Cu/a-Si bilayer system, Chen et al. proposed a write-once blue laser disk with AlSi alloy and obtained a jitter of less than 5% at a clock frequency of 66 MHz, a linear velocity of 8.25 m/s, and a track pitch of 0.45 μm [19]. Her et al. demonstrated that the maximum value of CNR for 3T marks can reach 43 dB at a linear velocity of 6.6 m/s for the BD-R with a Ni (1 nm)/Si (20 nm) bilayer recording film [20]. Hu et al. also developed a new inorganic BD-R with a recording layer composed of a multilayered structure of an alternating Mg-based alloy layer and a Si layer, and jitter values below 7.5% can be achieved at both 1X and 2X speeds [21].

Compared to Si, germanium (Ge) has the same crystal structure and similar physical and chemical properties. However, amorphous germanium (a-Ge) exhibits a much lower crystallization temperature (~450°C) than that of a-Si [22]. Therefore, lower recording power and higher recording speed will be expected for the a-Ge/metal bilayer recording film used in the BD-R. The feasibilities of a-Ge/Au, a-Ge/Ni, and a-Ge/Cu bilayer recording films for write-once blue laser recording have been evaluated, where the partial response signal-to-noise ratio (PRSNR) values of higher than 15 dB at 2X recording speed was measured in a-Ge/Au [23], a maximum data transfer rate of 112 Mbits/s could be achieved in a-Ge/Ni at a recording power of 6 mW [24], and a maximum data transfer rate of 44 Mbits/s could be achieved in a-Ge/Cu at a recording power of 3 mW [25].

In the past two decades, MICs of a-Si and a-Ge have been extensively investigated. However, most of the works were conducted under thermal annealing. It should be noted that the heating rates applied to the a-Si/metal or a-Ge/metal bilayers by a high-power pulsed blue laser for Blu-ray recording is much higher than those under thermal annealing. Consequently, the crystallization characteristics of a-Si and a-Ge thin films induced by metals with the assistance of laser irradiation might be different. Although the potential of using a-Si/metal and a-Ge/metal bilayer recording films in the BD-R has been demonstrated, laser-assisted MICs of a-Si and a-Ge, which are closely related to recording sensitivity and mechanism, are rarely investigated. In this chapter, we will discuss and compare the crystallization characteristics of a-Si and a-Ge induced by various metals with the assistance of laser irradiation. The recording performance of various a-Si/metal and a-Ge/metal bilayers for use in BD-Rs will also be presented.

7.2 Laser-Assisted Metal-Induced Crystallization of Amorphous Silicon

Metal/Si systems can be divided into two categories [26], compound-forming systems and simple eutectic systems. For compound-forming systems such as Cu/Si, Pd/Si, and Ni/Si, the low-temperature formation of metal silicide crystallites, which act as heterogeneous nucleation sites, can induce the subsequent crystallization of a-Si. On the other hand, simple eutectic systems such as Al/Si, Au/Si, and Ag/Si, do not form stable silicides. The crystallization of a-Si starting within the metallic layer [27, 28] or initiated by diffusion of metallic elements into a-silicon [29] has been suggested. The formation of a metastable metal silicide thin layer at the metal/Si interface before crystallization has also been proposed [30, 31]. Nevertheless, MIC of a-Si at low temperature is believed to be a solid-state process [32], which can be a nucleation-controlled process or a diffusion-limited process.

Explosive crystallization (EC) is a different mechanism observed in various amorphous films, including Sb [33], Ge [34], and Si [35], when they are stimulated by mechanical impact with a sharp stylus, pulsed laser heating, or touching of the films with a heated glass tip. The propagation speed of the crystallization wave front from the initiation point can be as high as 1 m/s [36]. For the propagation to be self-sustaining, the heat released from the crystallization of one region has to be sufficient to stimulate the crystallization of an adjacent region. It is the volume heat of crystallization that differentiates the EC phenomenon from MIC. Nucleation- or diffusion-controlled phase transformation takes place when the energy difference between phases is small [37]. However, EC requires a large energy release after phase transformation so that a certain thickness is necessary for the occurrence of EC in an a-Ge or a-Si film at room temperature [38]. For optical data storage applications, a pulsed laser is applied on the recording film to form recording marks with different optical properties against the unrecorded region through phase transformation. As the metal/a-Si bilayers are adopted as the recording film for a BD-R, low recording power (high sensitivity) and high data transfer rate can be achieved due to the combined effects of MIC and EC. In the following sections, we will discuss the crystallization characteristics of various a-Si/metal systems with the assistance of pulsed laser irradiation.

7.2.1 The a-Si/Cu System

The a-Si/Cu bilayer was first proposed as a recording thin film for BD-Rs by Inoue et al. from TDK in 2003 [16] and is still being considered as the most promising candidate for high-speed (12X) BDs [18]. The phase formation and crystallization behaviors of a-Si/Cu bilayers with different stacks under thermal annealing and/or pulsed laser irradiation have been investigated by different research groups to reveal the operative mechanism.

Inoue et al. [16] proposed a possible recording mechanism for a BD-R consisting of a reflective layer (Ag alloy), a protective layer ($ZnS–SiO_2$), a recording stack (Cu alloy layer and Si layer), and a protective layer ($ZnS–SiO_2$). At recording, when a laser pulse is irradiated on to the recording stack, the irradiated region will be heated and the laser-induced mixing of the two different films of Cu alloy and Si will form a recorded mark with a reflection coefficient that differs significantly from the nonirradiated region.

Her and Wu investigated the crystallization kinetics of an a-Si/Cu bilayer under thermal and pulsed laser annealing [9]. In their experiment, a recording stack with 20 nm a-Si/5 nm Cu was deposited on pregrooved PC substrates. For comparison, a recording stack with only 20 nm a-Si was also prepared. Pulsed laser annealing was carried out by a two-laser static tester from Tueoptics. The reflectivity variation of the a-Si/Cu bilayer recording film with time during the heating and cooling periods of the recording process was monitored. Laser 1 with a wavelength of 399 nm was used in pulsed mode for recording, while laser 2 with a wavelength of 422 nm was used in continuous-wave (cw) mode for monitoring reflectivity variations. The recording powers of laser 1 were chosen to be 6 mW, 8 mW, and 10 mW, and the pulse duration varied from 20 ns to 100 ns. Figure 7.1a shows the transmission electron microscopy (TEM) images of an a-Si/Cu bilayer and an a-Si single layer after irradiation by a 405 nm blue pulsed laser with a power of 6 mW for 50 ns, 100 ns, 150 ns, 200 ns, and 250 ns. It is seen that recording marks were formed in the a-Si/Cu bilayer recording film, whereas no structural change was observed in the a-Si single layer. Obviously, the recording power of the a-Si/Cu bilayer used for the BD was much lower than that of the a-Si single layer because of the lower crystallization temperature and activation energy with the aid of Cu-induced crystallization. As they closely examined the microstructure of the recording mark formed in the a-Si/Cu bilayer recording film, as shown in Fig. 7.1b, Cu_3Si precipitates with sizes of tens of nanometers were found to be uniformly dispersed in the polycrystalline Si matrix, which was the same as the microstructure found in the a-Si/Cu bilayer recording film after thermal annealing at 500°C for three minutes. On the basis of the observed results, Her and Wu proposed a two-step recording mechanism that in the laser spot the Cu layer first reacts with Si to form crystalline Cu_3Si nanoparticles, which subsequently serve as nucleation sites for crystallization of the remaining amorphous Si. They also pointed out that the rapid solid-phase crystallization of a-Si with a thin Cu layer under pulsed laser annealing is different from the melt crystallization of a-Si under high-power pulsed laser irradiation because the laser energy density applied in their study is only high enough for the crystallization of the a-Si/Cu bilayer but not enough for the crystallization or melting of pure a-Si.

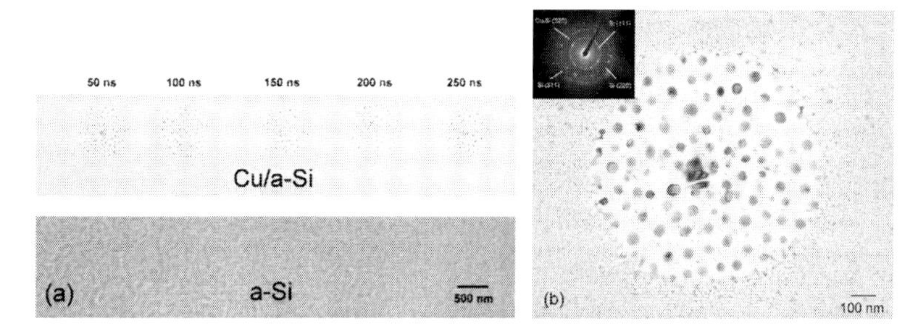

Figure 7.1 (a) TEM images of an a-Si/Cu bilayer and an a-Si single layer after irradiation by a 405 nm blue pulsed laser with a power of 6 mW for 50, 100, 150, 200, and 250 ns. (b) TEM image and diffraction patterns from the recording mark in the a-Si/Cu bilayer. Reproduced from Ref. [9], Copyright 2004, American Institute of Physics.

To further investigate the crystallization kinetics of the a-Si/Cu bilayer under pulsed laser annealing, Her and Wu monitored the reflectivity variations of the a-Si/Cu bilayer in real time when blue pulsed lasers with powers of 6 mW, 8 mW, and 10 mW were applied for 40 ns, as shown in Fig. 7.2a. Typical S-shaped transformation curves were observed. When irradiated by a pulsed laser, the a-Si/Cu bilayer exhibited a rapid increase in reflectivity once the crystallization of a-Si was triggered, and the reflectivity of the recording film reached a saturated value after irradiation for 90 ns, when the crystallization process was complete. The saturated value was found to increase with the laser power because the temperature profile of the laser spot with a higher laser power is higher than that with a lower laser power so that the crystallization area in the laser spot with a higher laser power will be larger than that with a lower laser power, leading to a higher saturated reflectivity. The crystallization characteristics of the a-Si/Cu bilayer under pulsed laser irradiation can be described by the Johnson–Mehl–Avrami (JMA) equation [39–41]. In the JMA equation, the fraction of crystallization X as a function of time t is given by

$$X(t) = 1 - \exp[-K(t - t_0)^m] \quad (7.1)$$

Here, t_0 is the incubation time, m is the reaction exponent, and K is the rate constant. The reaction exponent m, which can be determined from the $\ln[-\ln(1 - X)]$ versus $\ln(t - t_0)$ plots, as shown in Fig. 7.2b, was found to vary from 1.2 to 1.4, indicating the Cu-induced crystallization of a-Si under pulsed laser annealing was a grain-growth-controlled process associated with instantaneous nucleation [42].

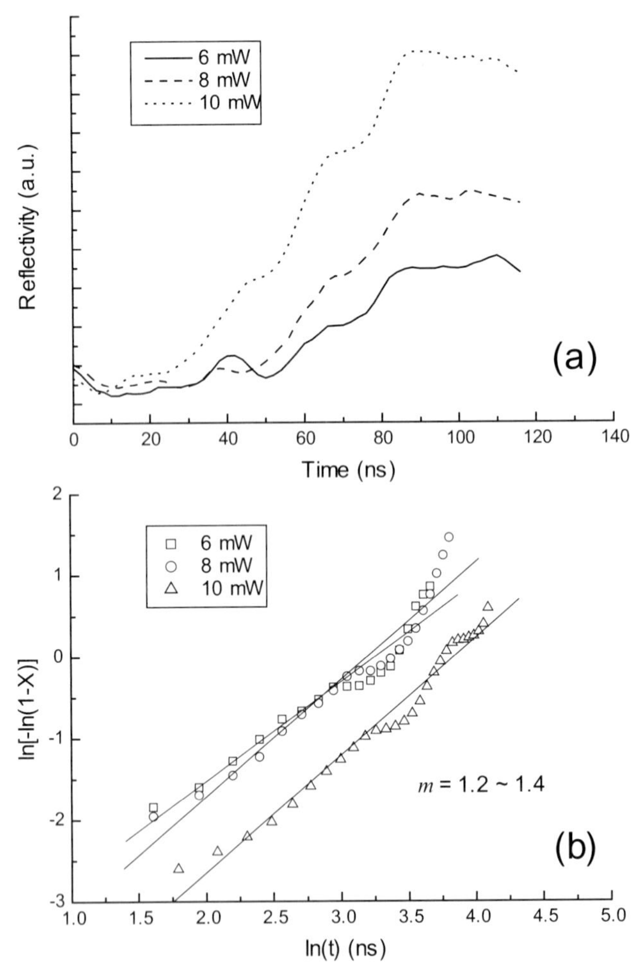

Figure 7.2 (a) Reflectivity changes with time, (b) $\ln[-\ln(1 - X)]$ vs. $\ln(t - t_0)$ plots for the a-Si/Cu bilayers irradiated by blue laser pulses with powers of 6, 8, and 10 mW for 40 ns. Reproduced from Ref. [9], Copyright 2004, American Institute of Physics.

To estimate the activation energy for crystallization of the a-Si/Cu bilayer under pulsed laser annealing, temperature distribution at the recording layer is calculated using a two-dimensional finite-difference method to solve the heat transfer equation, which can be expressed as

$$C_{\mathrm{P}}\frac{\partial T}{\partial t}=k\nabla^2 T+Q \tag{7.2}$$

Here C_c is the specific heat, k is the material's thermal conductivity, and Q is the heat source. Light absorption by the recording film is assumed to be the main source. The heat source is obtained from the results of the optical characteristic simulation by considering multiple reflections. The laser beam intensity is described by the Gaussian distribution

$$I(x,y)=[P/(\pi r_0{}^2)]\exp[-(x^2+y^2)/r_0{}^2] \tag{7.3}$$

Here P is the instantaneous laser beam power, and r_0 is the $1/e^2$ radius of the Gaussian beam. Thus the heat source term Q can be calculated as the summation of the product of energy absorbance and light intensity at each mesh. Table 7.1 lists the optical and thermal properties of the materials used in our simulation [43–46]. Here the thermal conductivity of a Cu thin film is assumed to be one-tenth of the value for bulk material. The simulated temperature profiles show that the a-Si layer will be heated to a saturation temperature immediately after the laser pulse is turned on. The saturation temperatures at the center of the laser spot on a-Si layer are calculated to be 543°C, 724°C, and 905°C, respectively, as the blue laser pulses with powers of 6 mW, 8 mW, and 10 mW are applied for 40 ns. Considering the slope in the acceleration regime of the transformation curve in Fig. 7.2a as the crystallization velocity, the Arrhenius plot of crystallization velocity versus reciprocal temperature for the a-Si/Cu bilayer under blue pulsed laser annealing, as shown in Fig. 7.3, yields an activation energy of ~0.22 eV. Comparing with the activation energy for crystallization under thermal annealing, the activation energy for crystallization of the a-Si/Cu bilayer under pulsed laser annealing is nearly 1 order of magnitude lower. Her and Wu ascribed to the EC of a-Si by mechanical impact with a high-power pulsed laser. As the a-Si/Cu bilayer is irradiated by a high-power pulsed laser, the covalent Si–Si bonds in the irradiated region would be broken and the temperature of the a-Si/Cu bilayer would have risen instantaneously above the threshold temperature for solid-phase EC. The formation of the Cu_3Si phase prior to the crystallization of a-Si can lower the system free energy and provide nucleation sites for the subsequent EC of a-Si, resulting in much lower activation energy.

Table 7.1 Optical and thermal properties of a PC substrate and Cu and a-Si thin films

Materials	Refractive index	Extinction coefficient	C_p (J/cm^3/K)	k (J/cmKs)
PC substrate	1.46	0	1.7	0.0022
Cu thin film	1.205	2.14	4.436	0.4
a-Si thin film	5.185	1.956	1.733	0.0243

Source: Reproduced from Ref. [9], Copyright 2004, American Institute of Physics.

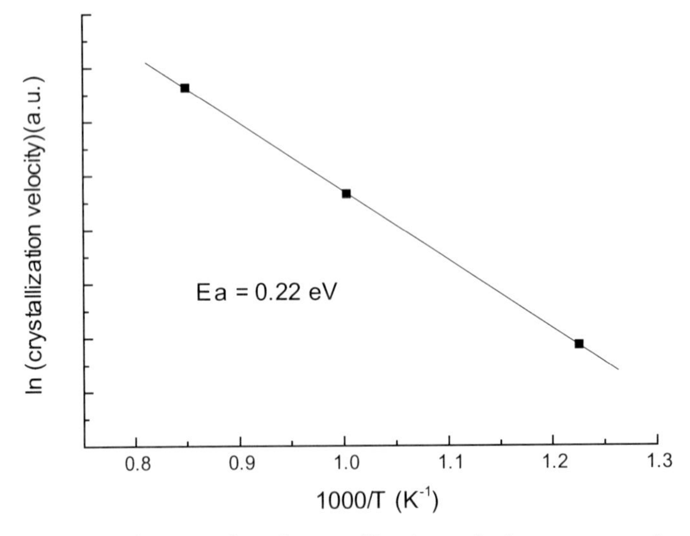

Figure 7.3 Arrhenius plot of crystallization velocity versus reciprocal temperature for the Cu/a-Si bilayer under blue pulsed laser annealing. Reproduced from Ref. [9], Copyright 2004, American Institute of Physics.

Kuiper et al. performed Auger electron spectroscopy (AES) measurements on optical disks, both before and after laser-induced writing [47]. To mimic the actual situation in an optical disk, they prepared recording tracks on a PC disk with the following general structure: PC/50 nm Ag/15 nm ZnS:SiO_2/X nm Cu/Y nm Si/60 nm ZnS:SiO_2. Part of the disk surface was homogeneously written with a laser in an initiator (l = 810 nm, laser power 400–500 mW, linear rotation velocity 5 m/s, pitch per rotation 6 μm). Figures 7.4a and 7.4b show the AES depth profiles of recording stacks with 12 nm Cu/5 nm Si and 6 nm Cu/10 nm Si, respectively, before and after initiation. The depth profiles reveal that Si and Cu have mixed upon initialization when the Cu layer is much thicker than the Si layer, while there seems to be a maximum Si concentration in the CuSi mixture upon initialization, leaving unreacted Si on top. To reveal the mixing process of the Cu and Si layers in more detail, the AES profiles of a recording stack with 6 nm Cu/6 nm Si on a glass substrate after annealing at different temperatures were measured, as shown in Fig. 7.5. It was found that there is hardly any diffusion of Cu into the overlying Si up to 150°C. In contrast, Si is diffusing into the Cu layer already at the lowest temperature. At 200°C, Cu is penetrating into the Si layer, but Si has diffused into the entire Cu layer. On the basis of the AES results, they concluded that the recording process under laser irradiation consists primarily of the diffusion of Si into Cu, leading to the formation of a CuSi alloy containing 25–30 at.% Si. Any excess of Si is left behind as unreacted film.

Mai et al. added a small amount of Al in the Cu layer to modulate the crystallization temperature of the Cu/Si system and investigated the evolution of the microstructure and composition in the Cu-Al/Si BD by TEM [48]. They prepared a BD-R with a multilayer structure of a PC substrate (1.1 mm)/Ag reflection layer (80 nm)/ZnS–SiO_2 (30 nm)/CuAl (5 nm)/Si (5 nm)/ZnS–SiO_2 (30 nm)/cover layer (0.1 mm) Figures 7.6a and 7.6b show the representative TEM images for disk samples subjected to 2X and 4X recording, respectively. Signal marks with various recording lengths residing in the grooves of disks can be readily seen. Figures 7.7a and 7.7b depict the locations subjected to energy dispersive spectroscopy (EDS) analysis and the chemical compositions corresponding to these locations. No obvious composition fluctuation was observed when the detection shifts from the nonmark area to the mark area, implying the absence of silicide phases such as Cu_3Si in the mark area. They ascribed to the ultrafast heating/cooling feature of laser writing, which inhibits the long-range atomic diffusion for silicide

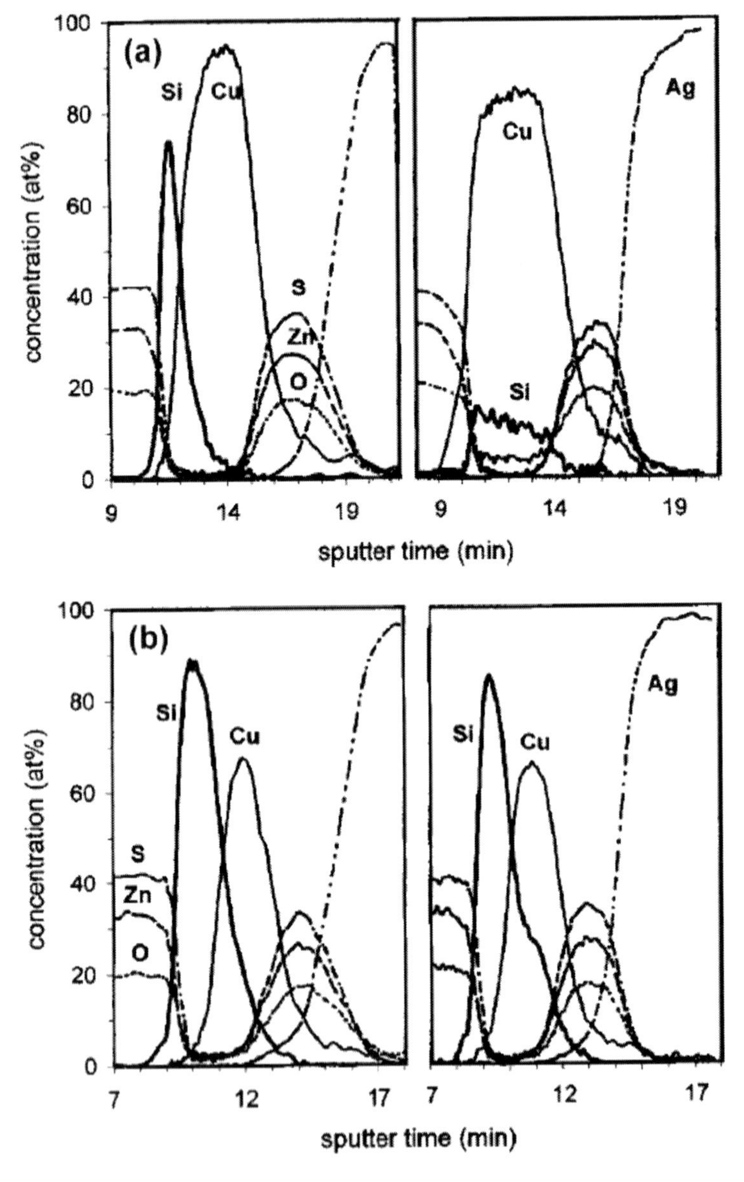

Figure 7.4 AES depth profiles of recording stacks with (a) 12 nm Cu/5 nm Si and (b) 6 nm Cu/10 nm Si, respectively, before and after initiation. Reproduced from Ref. [47], Copyright 2005, American Institute of Physics.

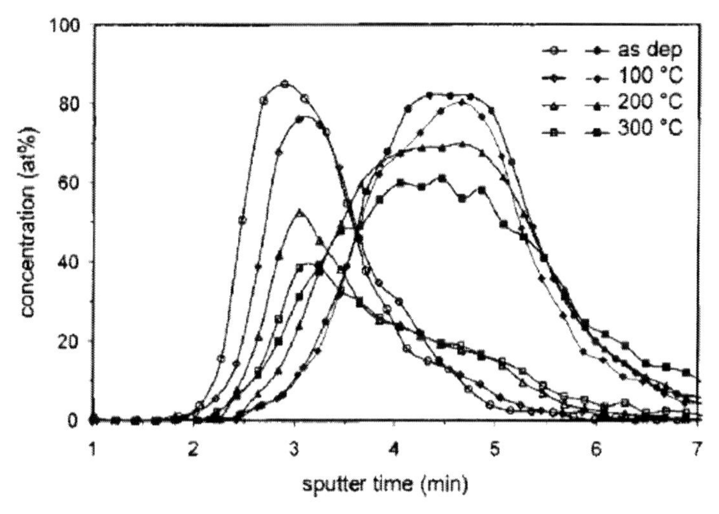

Figure 7.5 AES profiles of a recording stack with 6 nm Cu/6 nm Si on a glass substrate after annealing at different temperatures. Reproduced from Ref. [47], Copyright 2005, American Institute of Physics.

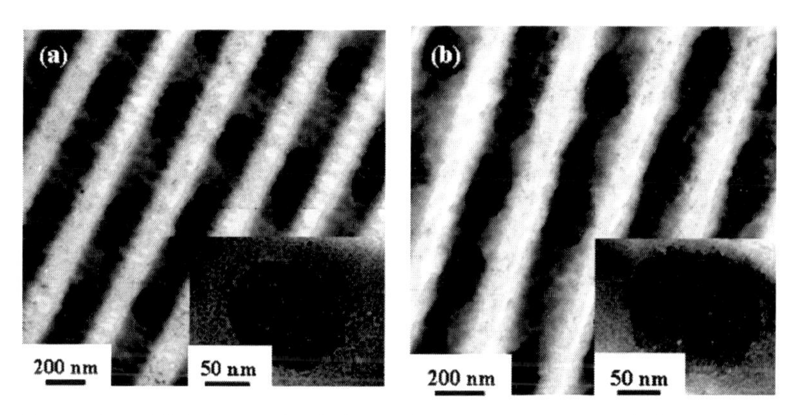

Figure 7.6 TEM images for disk samples subjected to 2X and 4X recording. Reproduced from Ref. [48], Copyright 2011, American Institute of Physics.

formation and recrystallization during signal recording. Figure 7.8a shows the TEM image and selected area electron diffraction (SAED) patterns taken from the mark and nonmark areas. The broad concentric diffraction rings in both SAED patterns indicate the recording layer comprise nanoscale crystallites regardless of the signal recording. As they closely examined the enlarged SAED patterns taken from the mark area, as shown in Fig. 7.8b, blur, bright spots decorated in the broad diffraction rings were observed, implying a moderate crystallinity improvement by laser heating. Since the disk sample is unlikely heated over 800°C in their work and the writing process was completed in an extremely short time span, Mai et al. suggested that the preliminary stages of annealing, for example, recovery through subgrain motion and polygonization dominated the structure rearrangement. Again, no match could be found for the Cu_3Si phase. Instead, Cu- and Si-rich solid-solution phases were formed due to the element mixing in the mark area caused by laser heating.

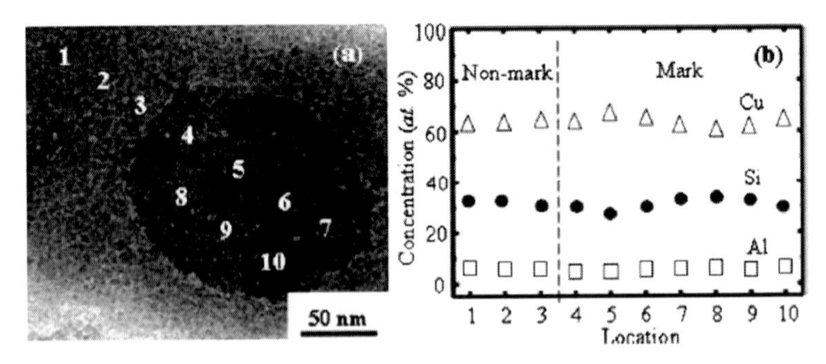

Figure 7.7 (a) Locations of EDS analysis of a single mark and its vicinity and (b) variations in element compositions deduced from EDS analysis on the nonmark and mark areas. Reproduced from Ref. [48], Copyright 2011, American Institute of Physics.

Ou et al. proposed the Si/CuSi bilayer as a recording film of the BD-R and investigated its crystallization mechanism by thermal analysis [49]. The Si (3 nm)/CuSi (16 nm) bilayer was deposited on a nature-oxidized Si wafer. The composition of CuSi was controlled to $Cu_{70}Si_{30}$. The microstructures of as-deposited and annealed samples were characterized by TEM. Figures 7.9a–c show the TEM bright-field images and electron diffraction patterns of the as-deposited, 270°C and 500°C annealed Si/CuSi bilayer films, respectively. It was found that supersaturated Cu_3Si nanocrystal grains with a grain size of 3–5 nm appeared in the CuSi film of the as-deposited Si/CuSi bilayer (Fig. 7.9a). After the Si/CuSi bilayer was annealed at 270°C for 15 minutes, the supersaturated Si atoms would obtain enough energy to diffuse out of the Cu_3Si phase and then segregate and crystallize into the cubic Si phase so that small c-Si grains (gray) with sizes of about 10–20 nm and large Cu_3Si grains (dark) with sizes of about 50–200 nm were observed (Fig. 7.9b). After the Si/CuSi bilayer was annealed at 500°C for 15 minutes, due to the diffusion of Si atoms in the Si layer into the CuSi film, the grain size of the cubic Si phase (dark grains) grew to about 200–300 nm and a new hexagonal Si phase was observed (Fig. 7.9c). The results indicate that the crystallization temperature of a-Si can be significantly reduced to 172°C to 206°C by the Si/CuSi bilayer structure.

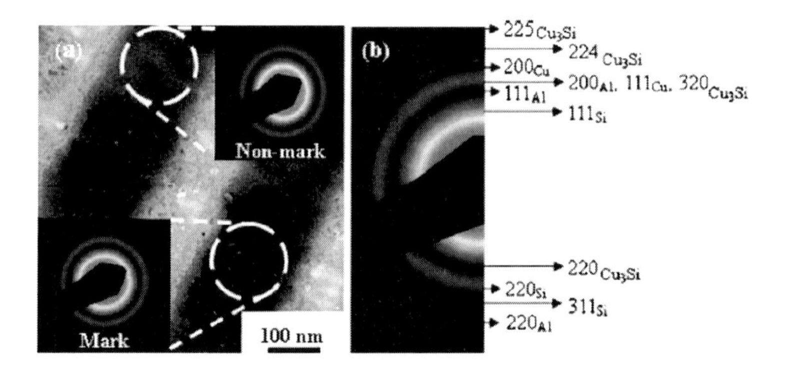

Figure 7.8 (a) SAED patterns taken from the nonmark (upper right-hand corner) and mark (lower left-hand corner) areas of the disk sample and (b) the indexing of SAED patterns taken from marked areas. Reproduced from Ref. [48], Copyright 2011, American Institute of Physics.

7.2.2 The a-Si/Ni System

Using a Ni layer to promote the crystallization of a-Si is another effective method to manufacture large-scale and high-performance polycrystalline Si [50]. During thermal annealing, the Ni reacts with Si to form octahedral $NiSi_2$ precipitates, which have the same lattice structure as crystalline Si (c-Si) with a very small mismatch of 0.4%. Then, the $NiSi_2$ precipitates, acting as nuclei, can promote the subsequent crystallization of a-Si at ~500°C, where needle-like Si crystallites form as a result of migration of the silicide precipitates through the a-Si network. Since the lattice mismatch between $NiSi_2$ and c-Si is only 0.4%, Ni is expected to work better than Cu for the

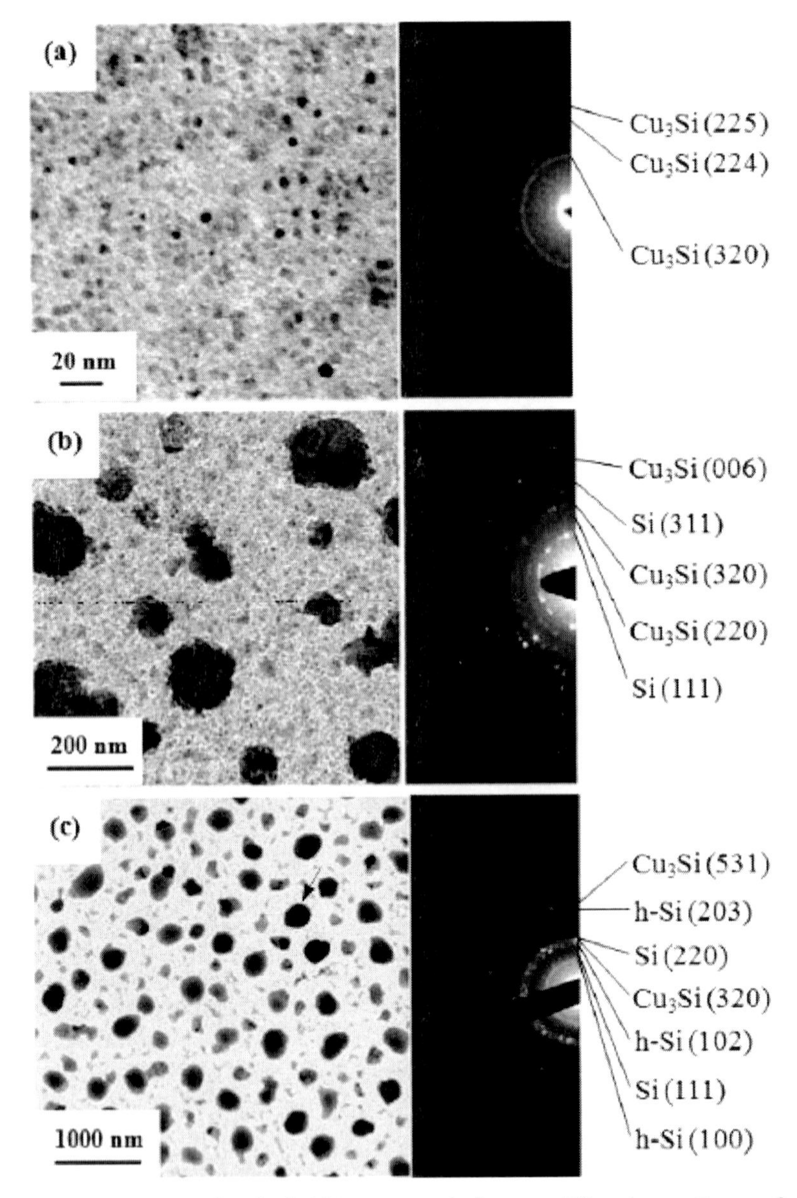

Figure 7.9 TEM bright-field images and electron diffraction patterns of the (a) as-deposited and (b) 270°C and (c) 500°C annealed Si/CuSi bilayer films. Reproduced from Ref. [49], Copyright 2011, American Institute of Physics.

MIC of a-Si. In addition, as the amount of Ni atoms consumed for $NiSi_2$ formation is much less than that for Cu_3Si formation, a thinner Ni layer is required for mediating the crystallization of a-Si, which will be suitable for multilayer recording. Therefore, the a-Si/Ni bilayer is expected to be a better recording film than the a-Si/Cu bilayer for the BD-R, especially for multilayer recording.

Her et al. prepared a-Si/Ni bilayers on pregrooved PC substrates and studied their crystallization mechanism under pulsed laser irradiation [20]. The thicknesses of a-Si and Ni layers were controlled at 20 nm and 1 nm, respectively, giving overall atomic concentration ratios of approximately 11 Si atoms to 1 Ni atom. The crystallization behavior of the a-Si/Ni bilayer under high-power pulsed laser irradiation was also performed by a static tester from Tueoptics, where samples were irradiated by a blue laser with powers ranging from 3 mW to 8 mW and pulse durations varying from 200 ns to 400 ns. The crystalline structural transition of the a-Si/Ni bilayer after irradiation by blue laser pulses was characterized by TEM to identify the crystallization mechanism of the a-Si/Ni bilayer under pulsed laser irradiation. Figures 7.10a and 7.10b show the TEM images of the a-Si/Ni bilayer after irradiation by a pulsed laser with a power of 6 mW for various durations and by a pulsed laser with different powers for 300 ns. Obviously, the recording marks could be formed after irradiation by laser pulses with very low powers for very short durations. However, the microstructure in the recording mark would depend on the pulsed laser power and duration under irradiation. When the a-Si/Ni bilayer was irradiated by a pulsed laser with a low power for a short duration, needle-like precipitates formed radically in the recording mark and the contrast between the recording mark and the background was obscured. As the laser power and duration were increased, granular precipitates would form in the central region of the recording mark and be surrounded by the needle-like precipitates and the contrast between the recording mark and the background became clear. To investigate the crystallization behavior of the a-Si/Ni bilayer under high-power pulsed laser irradiation, the bilayer recording film was irradiated by a pulsed laser with a power of 8 mW for 200 ns, 300 ns, and 400 ns and then examined by TEM, respectively, as shown in Fig. 7.11a–c. For the pulsed duration of 200 ns, needle-like precipitates were found to form radically in the recording mark, and the constituent phases were identified to be $NiSi_2$ and a-Si from the SAD patterns, indicating that the energy provided by the laser pulse can only activate the reaction between Si and Ni atoms to form the $NiSi_2$ phase but is not high enough to trigger the subsequent crystallization of unreacted a-Si. As the pulsed duration was increased to 300 ns, a bigger recording mark with equiaxed grains in the central region surrounded by needle-like precipitates would form, and the constituent phases were identified to be $NiSi_2$ and c-Si, indicating that the energy provided by the laser pulse is high enough to activate the formation of the $NiSi_2$ phase and the subsequent crystallization of unreacted a-Si. As the pulsed duration was further increased to 400 ns, both the recording mark and the area occupied by the equiaxed grains expanded, while the constituent phases remained as $NiSi_2$ and c-Si. It was found that the crystallization behavior of the a-Si (20 nm)/Ni (1 nm) bilayer under pulsed laser irradiation is similar to that under thermal annealing. During the recording process, the $NiSi_2$ phase will precipitate first and serve as the nucleation site for the following crystallization of the remaining a-Si.

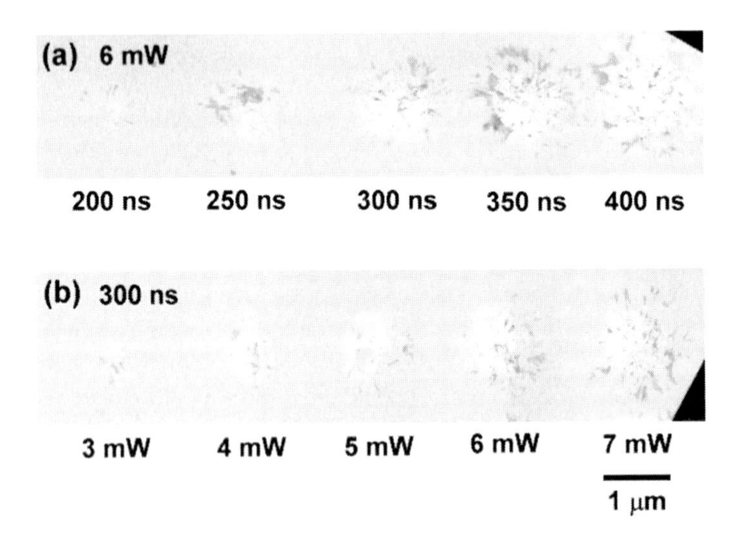

Figure 7.10 TEM images of the a-Si/Ni bilayer after irradiation (a) by a pulsed laser with a power of 6 mW for various durations and (b) by a pulsed laser with different powers for 300 ns. Reproduced from Ref. [20], Copyright 2007, American Institute of Physics.

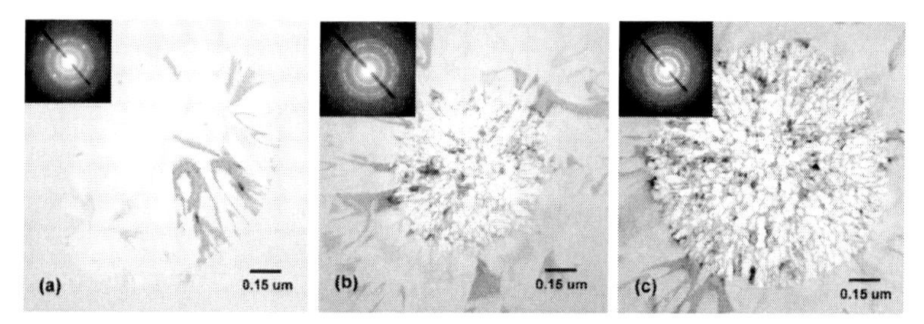

Figure 7.11 TEM bright-field images and SAED patterns of the a-Si (20 nm)/Ni (1 nm) bilayer irradiated by a pulsed laser with a power of 8 mW for (a) 200 ns, (b) 300 ns, and (c) 400 ns. Reproduced from Ref. [20], Copyright 2007, American Institute of Physics.

7.2.3 The a-Si/Al System

It has been reported that a-Si in contact with Al can be crystallized at 250°C or 300°C [51, 52], which is much lower than the crystallization temperature of 485°C for a-Si in contact with Cu. As a result, the a-Si/Al bilayer may be a better recording material for high data-transfer rate, write-once Blu-ray recording. Al-induced crystallization of a-Si under the eutectics temperature of the AlSi system is believed to be a solid-state process [32]. The crystallization of a-Si has been suggested to start within the metallic layer [27, 28], while the crystallization initiated by the diffusion of Al into a-Si has also been proposed [29]. Some researchers also reported the intermixing of Al with Si and the formation of an alloyed thin layer of high metal concentration at the a-Si/Al interface before crystallization [27, 30, 31]. It appears that the mechanism of Al-induced crystallization of a-Si is still not clear. Besides, most studies related with Al-induced crystallization of a-Si were conducted under thermal annealing, and the thickness of the a-Si film was several hundred nanometers. The crystallization behavior of ultrathin a-Si induced by Al under pulsed laser irradiation has not been studied yet, which is important when the a-Si/Al bilayer is used for write-once Blu-ray recording.

Her and Chen studied the crystallization kinetics of the a-Si (10 nm)/Al (5 nm) bilayer on pregrooved PC substrates under high-power pulsed laser irradiation by a static tester [53]. The samples were irradiated by a blue laser with powers ranging from 6 mW to 10 mW and pulse durations varying from 20 ns to 100 ns. The activation energies for crystallization, reaction exponents, and crystallization mechanisms of the a-Si/Al bilayer were determined and discussed. Figure 7.12a shows the TEM image of the a-Si/Al bilayer after irradiation by a pulsed laser with a power of 6 mW for various durations. It is seen that recording marks were clearly formed in the a-Si/Al bilayer, even at a duration of 50 ns. As the microstructures of the recording marks formed in the a-Si/Al bilayer were closely examined, as shown in Fig. 7.12b, a-Si surrounded by radial crystalline Al grains was found in the a-Si/Al bilayer. It has been reported that the crystalline phase of a pulsed-laser-irradiated Si thin film strongly depends on the laser energy density and Si film thickness [54]. When a laser energy density beyond the crystallization threshold was applied, the phase transformation to crystalline occurred. Once the laser energy density exceeded the melting threshold, the amorphization of Si can be observed in a-Si thin films thinner than about 30 nm, whereas the crystallization of Si can be observed in a-Si thin films thicker than 30 nm. Since the laser energy density applied in this study is not high enough to melt pure a-Si, it is believed that the reamorphization of a-Si in contact with an Al metal layer can be attributed to the melting of a-Si/Al initiated at the interface, due to the low melting temperature of SiAl alloy and the rapid solidification that followed. However, the crystallization of a-Si occurring before the amorphization of a-Si in the a-Si/Al bilayer cannot yet be ruled out.

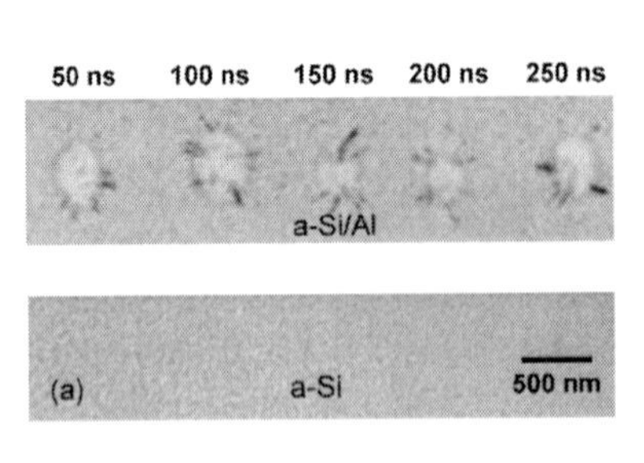

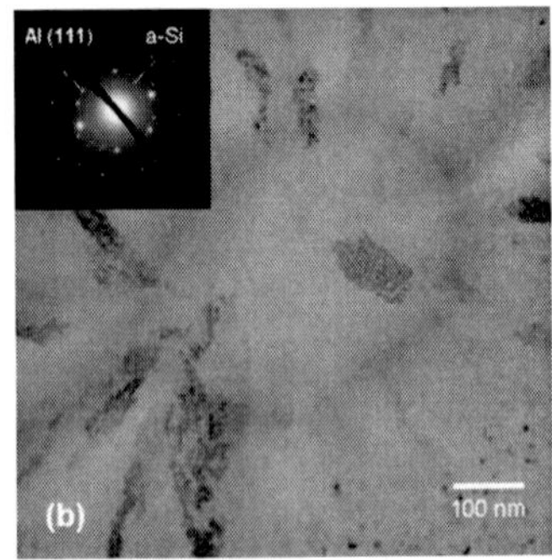

Figure 7.12 TEM images of (a) a-Si/Al bilayer and a-Si single layer after irradiation by a blue pulsed laser with a power of 6 mW for various durations and (b) the recording mark formed in a-Si/Al. Reproduced from Ref. [53], Copyright 2007, American Institute of Physics.

Figure 7.13a shows the reflectivity variations with time of the as-deposited a-Si/Al bilayer irradiated by blue pulsed lasers with powers of 6 mW, 8 mW, and 10 mW for 50 ns. Similarly, S-shaped transformation curves were found in the a-Si/Al bilayers at the early stage of transformation under pulsed laser irradiation, indicating that the crystallization of a-Si also took place in the a-Si/Al bilayer. In the incubation regime, the temperature in the a-Si layer was below the crystallization temperature of a-Si, so the reflectivity change with time was not appreciable. Once the temperature in the a-Si layer reached the crystallization temperature of a-Si, the crystallization of a-Si would be triggered immediately, leading to a rapid increase in reflectivity in the acceleration regime. After the acceleration regime, the crystallization process went into the saturation regime, where the reflectivity of the bilayer reached a saturated value with minimal variation. However, the saturation regime would last only for a short period of time and was followed by an abrupt drop of reflectivity. This was caused by pulsed-laser-induced amorphization of the a-Si/Al bilayer through melting and rapid solidification of the a-Si-Al alloy, which was consistent with the resulting microstructures observed by TEM. Considering only the crystallization process, the reaction exponents m for the a-Si/Al bilayers under pulsed laser irradiation, determined from the $\ln[-\ln(1 - X)]$ versus $\ln(t - t_0)$ plots in the acceleration regimes of the transformation curves, as shown in Fig. 7.13b, were found to vary from 1.9 to 2.2, which was slightly higher that under thermal annealing. This means that the crystallization of a-Si induced by Al under pulsed laser irradiation at an ultrafast heating rate is similar to that under thermal annealing at a much lower heating rate. However, the decrease of the nucleation rate with the progress of grain growth under pulsed laser irradiation is slower than that under thermal annealing. This may be because the diffusion of Si into Al was enhanced by mechanical impact with a high-power pulsed laser, leading to extensive reaction of Al atoms with Si atoms and formation of Al silicides.

To estimate the activation energy for crystallization of a-Si induced by Al under pulsed laser irradiation, the temperature distribution at the a-Si layer was also calculated by using a two-dimensional finite-difference method to solve the heat transfer equation. The simulated temperature profiles show that the a-Si layer would be heated to a saturation temperature immediately after the laser pulse

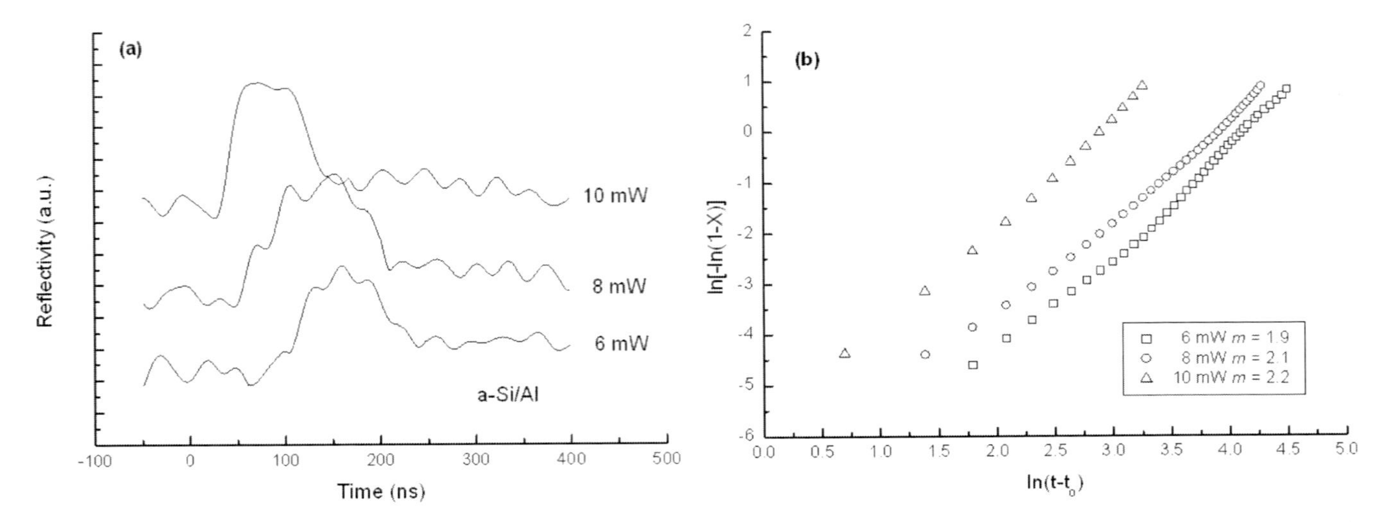

Figure 7.13 (a) Reflectivity variations with time for a-Si/Al irradiated by blue laser pulses with powers of 6, 8, and 10 mW for 50 ns. (b) ln[−ln(1 − X)] vs. ln($t - t_0$) for a-Si/Al under pulsed laser irradiation. Reproduced from Ref. [53], Copyright 2007, American Institute of Physics.

is turned on. The saturation temperatures at the center of the laser spot on the a-Si layer are calculated to be 903 K, 1204 K, and 1505 K, respectively, when blue laser pulses with powers of 6 mW, 8 mW, and 10 mW were applied for 50 ns. Considering the reciprocal of time required for the completion of crystallization in the acceleration regime as the crystallization velocity, the Arrhenius plot of the logarithm of the crystallization velocity versus the reciprocal of the absolute temperature for the a-Si/Al bilayer under pulsed laser irradiation, as shown in Fig. 7.14, yields an activation energy of 0.22 ± 0.03 eV, which is about an order of magnitude lower than that under thermal annealing. Similar to that obtained in the a-Si/Cu bilayer, this may be explained by the EC of a-Si by mechanical impact with a high-power pulsed laser.

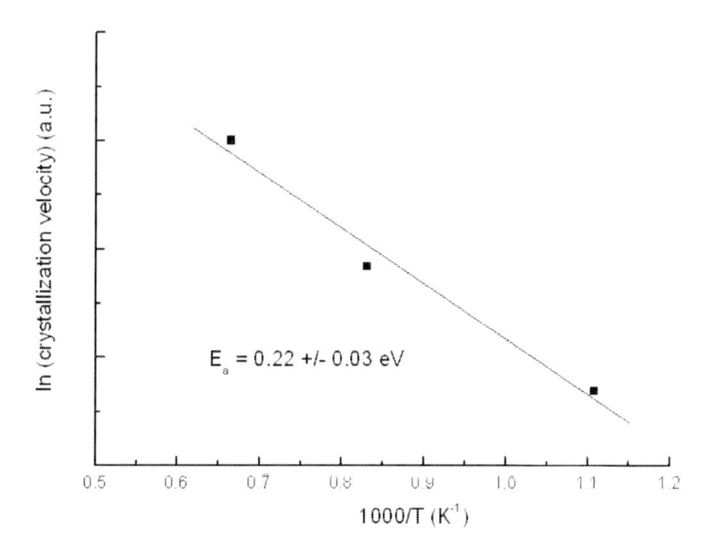

Figure 7.14 Arrhenius plot of the logarithm of the crystallization velocity vs. reciprocal of the absolute temperature for a-Si/Al under pulsed laser irradiation. Reproduced from Ref. [53], Copyright 2007, American Institute of Physics.

7.3 Laser-Assisted Metal-Induced Crystallization of Amorphous Germanium

Germanium has the same crystal structure and similar physical and chemical properties as Si. However, the crystallization temperature of a-Ge is only ~450°C, which is much lower that of a-Si [22]. Accordingly, lower recording power and higher recording speed will be expected for the a-Ge/metal bilayer recording film used in the BD-R. The feasibilities of a-Ge/Au [23, 55], a-Ge/Ni [24], and a-Ge/Cu [25] bilayer recording films for use in BD-Rs have been demonstrated, and the crystallization characteristics of these bilayer systems under thermal annealing and pulsed laser irradiation have also been investigated to reveal the possible recording mechanism.

7.3.1 The a-Ge/Au System

On the basis of the phase diagram of GeAu [56], the eutectic temperature of GeAu is 361°C, which is a suitable character for optical recording media. Wu et al. proposed Ge/Au bilayer media for a write-once optical disk and reported the microstructure, diffusion, and crystallization (recording) mechanism of the Ge/Au bilayer [23, 55]. The lower dielectric layer, Ge (8 nm)/Au (8 nm) bilayer, upper dielectric layer, and Ag reflective layer were deposited on nature-oxidized (110) silicon substrates, and the microstructure changes before and after heating were discussed in accordance with the nanoscale TEM cross-sectional image of the Ge/Au bilayer. The bright-field images and electron diffraction patterns of the Ge/Au bilayer in the as-deposited state and after 220°C and 320°C annealing, as shown in Fig. 7.15a–c, reveal that the as-deposited Ge/Au bilayer comprises Ge_2Au_3 and Au crystalline phases. After 220°C annealing, the formation of dendrites was found, while the constituent phases remained as Ge_2Au_3 and Au. As the annealing temperature was increased to 320°C, the dendrites are mixed and the crystalline Ge (c-Ge) appears. The cross-sectional views of the as-deposited and 220°C annealed ZnS–SiO_2/Ge (8 nm)/Au (8 nm)/Si substrate, as shown in Figs. 7.16a and 7.16b, respectively, demonstrate that the Ge_2Au_3 interface layer of 1 nm thickness is formed between a-Ge and the crystalline Au layer. The Ge_2Au_3 crystallites act as nucleation sites during recording. After annealing at 220°C, interdiffusion of Au and Ge atoms begins at the interface layer and forms the metastable phase of Ge_2Au_3 (321) because the *d* spacing of Ge_2Au_3 (321) is about equivalent to the closed-packed plane of Ge (111). After annealing at 310°C–320°C, the formation of Ge (111) grains was found. Clearly, the MIC mechanism can be confirmed.

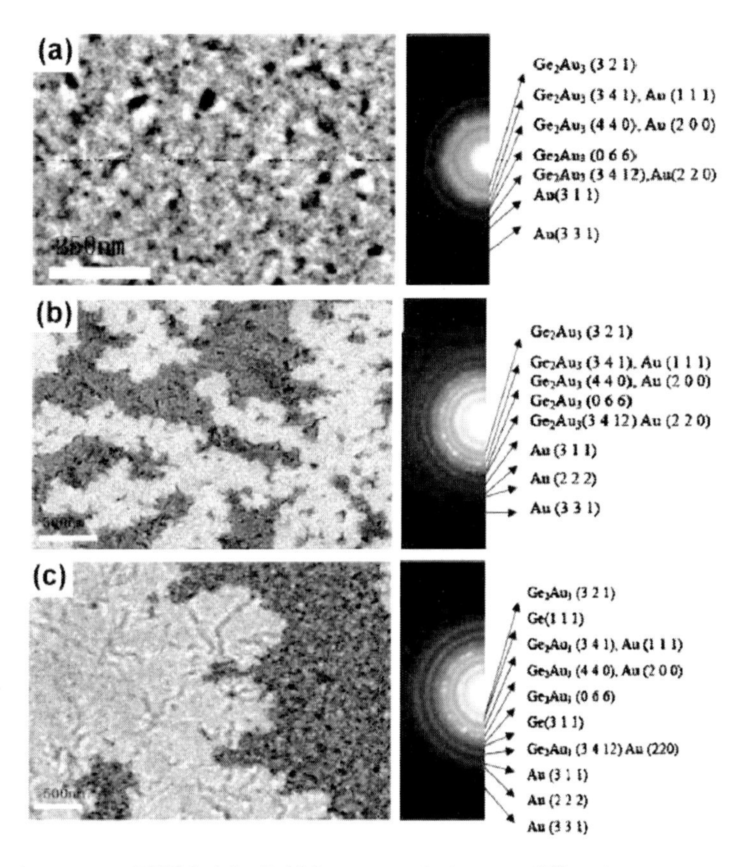

Figure 7.15 TEM bright-field images and electron diffraction patterns of the Ge/Au bilayers (a) in the as-deposited state and after (b) 220°C and (c) 320°C, 30-minute annealing. Reproduced from Ref. [23], Copyright 2007, American Institute of Physics.

To identify the diffusion path and dominant diffusion species, the element concentration–depth profiles of the as-deposited and recording regions were also measured by AES. Figure 7.17a shows the depth profile of the Ge/Au bilayer. A three-layered structure composed of an Au layer, a Ge/Au interface layer, and a Ge layer can be clearly observed in the bilayer. The Ge and Au concentration–depth profiles of as-deposited and laser-initialized Ge/Au bilayers, as shown in Figs. 7.17b and 7.17c, respectively, reveal that during the first 30 seconds of the sputtering (about 0–6 nm depth), the Ge atoms did not diffuse to such a depth after being laser initialized.

Between 30 seconds and 40 seconds (about 6–8 nm depth), the Ge concentration in the laser-recorded sample is slightly larger than that of the as-deposited sample. After that, the Ge concentration of the laser-initialized sample is lower than that of the as-deposited sample. On the other hand, the Au atoms diffused into the Ge layer as the depth is deeper than 16 nm (sputtering time is longer than 80 seconds). The Au concentration in the laser-recorded sample is slightly smaller but is higher than that of the as-deposited sample at a sputtering time of 30–40 seconds and 40–80 seconds, respectively. The results indicate that the dominant diffusion element is Au and the diffusion path is from the Au layer to the Ge layer during recording.

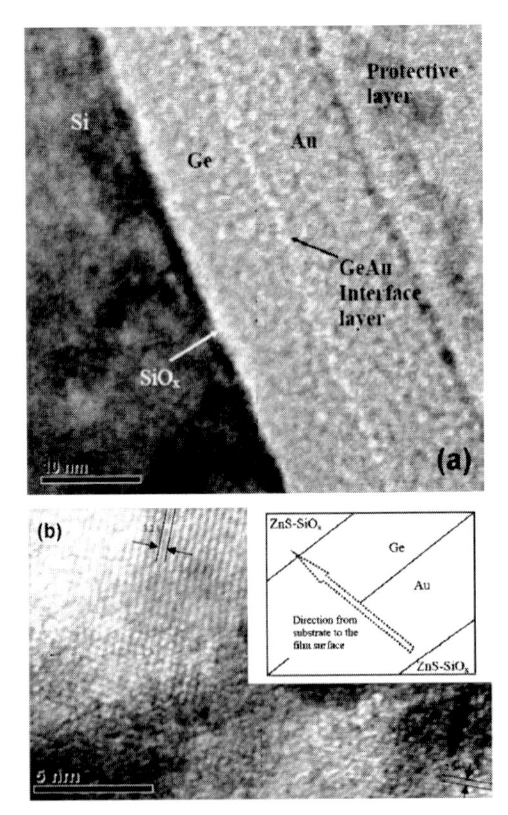

Figure 7.16 Cross-sectional views of the (a) as-deposited and (b) 220°C annealed ZnS–SiO_2/Ge (8 nm)/Au (8 nm)/Si substrate. Reproduced from Ref. [23], Copyright 2007, American Institute of Physics.

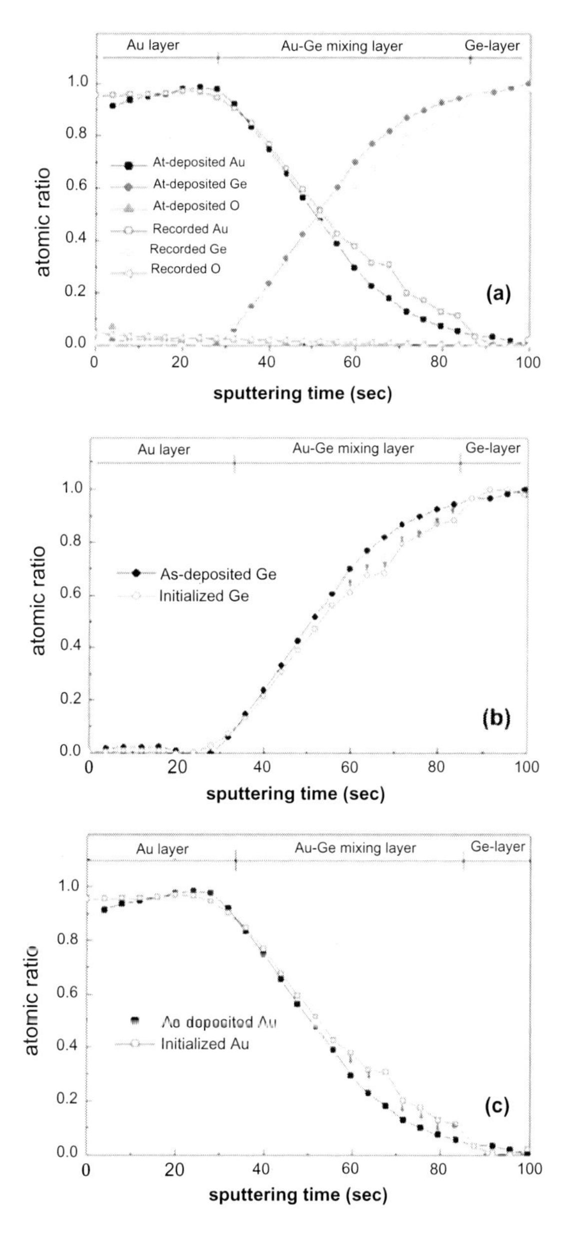

Figure 7.17 (a) The concentration–depth profile of as-deposited and laser-initialized Ge/Au bilayer samples and the concentration–depth profile of as-deposited and laser initialized (b) Ge and (c) Au atoms in the Ge/Au bilayer samples. Reproduced from Ref. [55], Copyright 2008, American Institute of Physics.

7.3.2 The a-Ge/Ni System

It has been reported that the presence of a Ni layer on an a-Ge film can lower the crystallization temperature of a-Ge by as much as 180°C [57]. Accordingly, the a-Ge/Ni bilayer is expected to exhibit a higher recording sensitivity for use in the BD-R. In the past decade, an extensive amount of renewed interest in germanide thin films has occurred for their use in microelectronic devices as Ohmic contacts [58]. As the monogermanide NiGe, exhibiting high thermal stability and low resistivity, could be formed at low temperature, the solid-state reactions of a thin Ni film with Ge have been investigated to understand the mechanism of formation of NiGe [57, 59, 60]. For the Ni germanide reactions, the sequential phase formation of Ni_5Ge_3 followed by NiGe has been observed in the a-Ge/Ni bilayer subjected to ramp anneals. Recently, the simultaneous growths of Ni_5Ge_3 and NiGe have also been observed during 150°C–190°C isothermal heat treatments of Ni thin films on amorphous and polycrystalline Ge [61]. Although the formation mechanism of Ni germanide during Ge metallization is important for the integration, the crystallization of a-Ge induced by Ni under thermal and pulsed laser annealing is a key issue when the a-Ge/Ni bilayer is used for write-once Blu-ray recording.

Her et al. deposited an a-Ge/Ni bilayer on a pregrooved PC substrate and investigated its crystalline structural transitions under pulsed laser irradiation by a static tester to understand the possible recording mechanism when used for the BD-R [24]. The thicknesses of the a-Ge thin film and the Ni ultrathin metal layer were controlled at 20 nm and 2 nm, respectively, giving an overall atomic concentration ratio of approximately 4.8 Ge atoms to 1 Ni atom. The crystalline structural transitions of the a-Ge/Ni bilayer, irradiated by a blue laser with powers ranging from 3 mW to 6 mW and pulse durations varying from 30 ns to 300 ns, were characterized by TEM. Figure 7.18a shows the TEM images of the a-Ge/Ni bilayer after irradiation by a 399 nm blue pulsed laser with powers of 3 mW, 4 mW, 5 mW, and 6 mW for durations of 30 ns to 300 ns. Obviously, the recording marks could be readily formed in the a-Ge/Ni bilayer recording film by laser pulses with very low powers at very short durations. As they closely examined the microstructure of the recording mark formed in the a-Ge/Ni bilayer recording film irradiated by a laser power of 5 mW for 50 ns, as shown in Fig. 7.18b, granular precipitates were found to form in the central region of the recording mark, surrounded by the needle-like precipitates and plenty of nanovoids (bright dots). From the SAED patterns, the constituent phases were identified to be a mixture of NiGe and c-Ge for the granular precipitates in the central region and Ni_5Ge_3 and NiGe for the needle-like precipitates in the outer region. The formation of nanovoids might be attributed to the volume shrinkage caused by the formation of germanides (Ni_5Ge_3 and NiGe) and/or the crystallization of a-Ge. Normally, the laser beam intensity is considered as the Gaussian distribution so that the temperature on the recording layer will decrease exponentially with the distance from the center of the laser spot. When the a-Ge/Ni bilayer was irradiated by a pulsed laser of 5 mW for 50 ns, the temperature in the central region was high enough to activate the formation of the NiGe phase and the subsequent crystallization of a-Ge, while the temperature in the outer region could only activate the reaction between Ni and Ge to form Ni_5Ge_3 and NiGe phases but could not trigger the subsequent crystallization of a-Ge. The crystallization behavior of the a-Ge/Ni bilayer under pulsed laser irradiation was found to be similar to that under thermal annealing. During the recording process, the Ni_5Ge_3 and NiGe phases will precipitate in sequence and the NiGe precipitates will serve as nucleation sites for the following crystallization of the remaining a-Ge.

7.3.3 The a-Ge/Cu System

Compared to the Au and Ni metal layers, a thin Cu layer can effectively reduce the crystallization temperature of a-Ge to a lower temperature [62]. It has also been reported that the Cu_3Ge phase can be formed at about 200°C and remains stable up to ~420°C prior to the crystallization of a-Ge [63]. Since the Cu_3Ge phase may be able to provide enough optical contrast at the blue wavelength of 405 nm against the a-Ge matrix, we can also employ the Cu_3Ge phase to make a successful recording at even lower powers in the BD. As a result, it is expected that the a-Ge/Cu bilayer will exhibit a much higher recording speed and sensitivity than those reported bilayers for use in the BD-R.

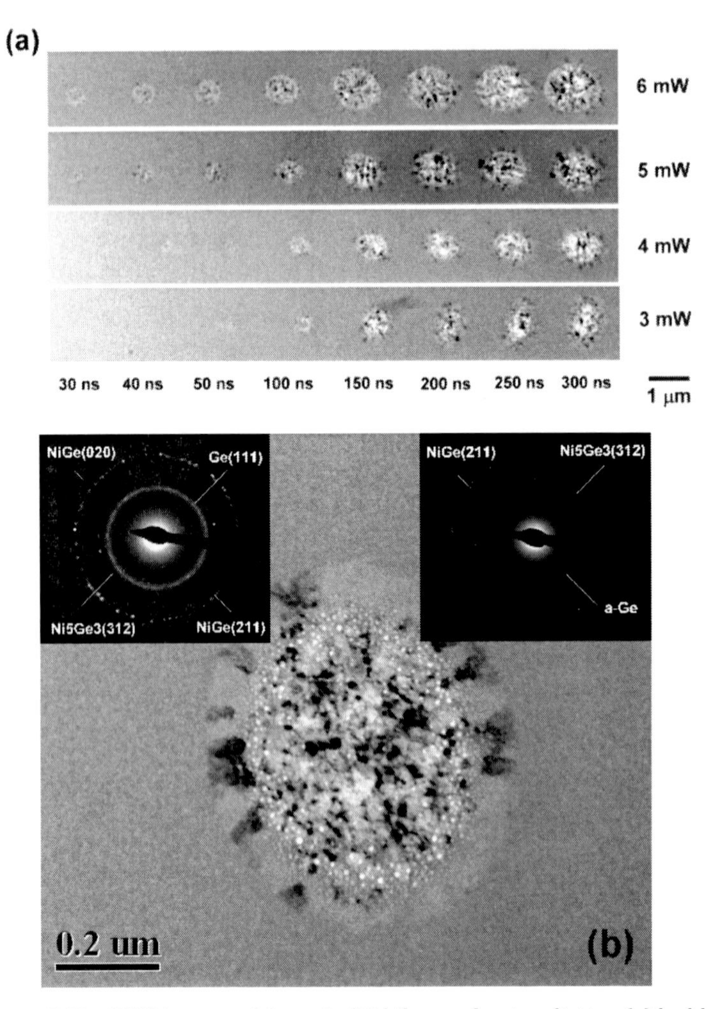

Figure 7.18 TEM images of the a-Ge/Ni bilayer after irradiation (a) by blue pulsed lasers with various powers and durations and (b) by a pulsed laser with a power of 5 mW for 50 ns. Reproduced from Ref. [24], Copyright 2009, American Institute of Physics.

Her et al. also investigated the structural transitions of the a-Ge (15 nm)/Cu (5 nm) bilayer under pulsed laser irradiation to understand the possible recording mechanism [25]. The thicknesses of the a-Ge thin film and the Cu metal layer were controlled at 15 nm and 5 nm, respectively, giving an overall atomic concentration ratio of approximately 2 Ge atoms to 1 Cu atom. The phase formation as well as the crystallization behavior of the a-Ge/Cu bilayer under pulsed laser irradiation were also performed by a static tester, where samples were irradiated by a blue laser with various powers and duration. Figure 7.19a shows the TEM images of the a-Ge/Cu bilayer after irradiation by a 399 nm blue pulsed laser with powers of 3 mW to 6 mW for durations of 20 ns to 250 ns. Obviously, the recording marks could be formed in the a-Ge/Cu bilayer by laser pulses with very low powers at very short durations. At the recording power of 3 mW, the minimum duration to form a recording mark is around 50 ns. As the recording powers were increased to 4 mW, 5 mW, and 6 mW, the minimum durations would reduce to around 40 ns, 30 ns, and 20 ns, respectively. Apparently, the a-Ge/Cu bilayer is a promising candidate for high-speed write-once Blu-ray recording. To investigate the recording mechanism of the a-Ge/Cu bilayer, the crystalline structures of the recording marks formed in the a-Ge/Cu bilayer irradiated by a pulsed laser with a power of 5 mW for various durations were closely examined by TEM, as shown in Fig. 7.19b–d. It can be seen that only needle-like grains formed radically in the recording marks at pulsed durations of 30 ns and 50 ns, and the constituent phase was identified to be Cu_3Ge by SAED patterns. When the pulsed duration was increased to 100 ns, equiaxed grains surrounded by radically needle-like grains would form in the recording mark. From the SAED patterns, the constituent phases were identified to be a mixture of Cu_3Ge and c-Ge for the equiaxed grains in the central region and Cu_3Ge for the needle-like grains in the outer region. Similar to the a-Ge/Ni bilayer, numerous nanovoids (bright dots) were formed in the a-Ge/Cu bilayer after pulsed laser irradiation, which can also be ascribed to the volume shrinkage due to the formation of Cu_3Ge and/or the crystallization of a-Ge. Likewise, the laser beam intensity can be considered as the Gaussian distribution so that the temperature on the recording layer will decrease exponentially with the distance from the center of the laser spot. When the a-Ge/Cu bilayer was irradiated by a pulsed laser of 5 mW for durations shorter than 50 ns, the temperature in the central region was higher than the Cu_3Ge phase formation temperature but lower than the a-Ge crystallization temperature, resulting in the formation of a recording mark composed of only the Cu_3Ge phase. When the pulsed duration was longer than 100 ns, the temperature in the central region was high enough to activate the formation of the

Cu_3Ge phase and the subsequent crystallization of a-Ge, while the temperature in the outer region could only trigger the formation of the Cu_3Ge phase, leading to the formation of a recording mark with a central region consisting of a mixture of Cu_3Ge and c-Ge, surrounded by an annular Cu_3Ge phase. It is concluded that the Cu_3Ge phase will precipitate first and then serve as the nucleation site for the following crystallization of the remaining a-Ge during the recording process, which is similar to the recording mechanisms observed in the a-Ge/Ni bilayer as well as in the a-Si/Cu and a-Si/Ni bilayers.

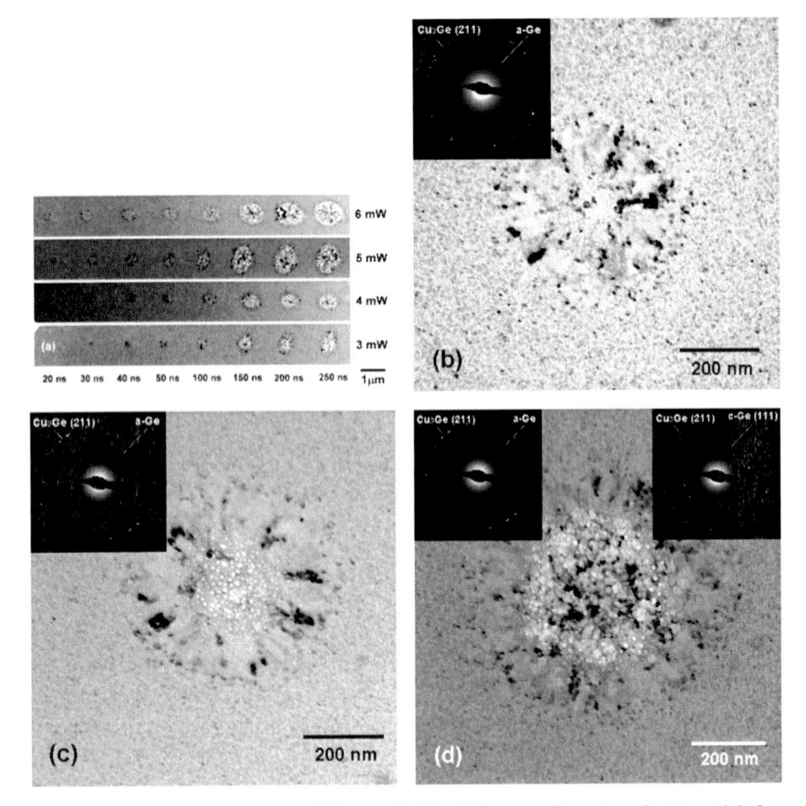

Figure 7.19 TEM images of the a-Ge/Cu bilayer after irradiation (a) by blue pulsed lasers with various powers and durations and by a pulsed laser with a power of 5 mW for (b) 30 ns, (c) 50 ns, and (d) 100 ns. Reproduced from Ref. [25], Copyright 2012, American Institute of Physics.

7.4 Recording Characteristics of Amorphous Silicon/Metal and Amorphous Germanium/Metal Bilayers

7.4.1 The a-Si/Metal Bilayer System

An inorganic BD-R was first developed by Inoue et al. in 2003 [16]. Figure 7.20a shows the cross section of this disk. A reflective layer (Ag alloy), a protective layer (ZnS–SiO_2), a recording stack (Cu alloy layer and Si layer), and a protective layer (ZnS–SiO_2) were deposited in this order. Finally, a transparent cover layer of 0.1 mm thickness was spin-coated on the top of the disk. The write-once disk was written by a modulated laser with a new write strategy. The write strategy of the 4T signal is shown in Fig. 7.20b, where they controlled the recording power by adjusting the bias power level. The bias power is increased along with high-speed recording. Figure 7.20c,d shows the dependence of jitter on the recording power at the recording rates of 36 Mbps (1X), 72 Mbps (2X), and 144 Mbps (4X). Jitter values of less than 8% with a recording peak power of 5 mW can be achieved in the range of 36 Mbps to 144 Mbps.

Kuiper's group continued to improve the recording properties and increase the recording speed by adjusting the ratio of Cu/Si layers [47] and optimizing the write strategy [64]. Figure 7.21 shows the multitrack jitter and modulation for various Cu/Si combinations: Cu/Si = 1:1, x nm Cu/6 nm Si, and 6 nm Cu/x nm Si. The best results (low jitter, high modulation) are consistently obtained for equally thick Cu/Si layers. Further optimizing such bilayers yielded a lowest jitter value of 4% for Cu/Si = 4 nm/4 nm. To improve the lifetime of the disk, a CuAl alloy was used instead of pure Cu. A so-called castle write strategy, as shown in Fig. 7.22a, was chosen to improve the performance at high-speed recording tests. For 2T and 3T marks, a single pulse with power P_w is used, while for 4T to 9T marks an intermediate power level (P_2) is added. By adding extra large pulses at the start and at the end, the temperature gradient will be steeper, hence sharpening the transition between marks and spaces. The jitter data of a 25 GB disk at different recording speeds is shown in Fig. 7.22b. Even at a 7X recording speed, good jitter values can be obtained using a castle-type strategy.

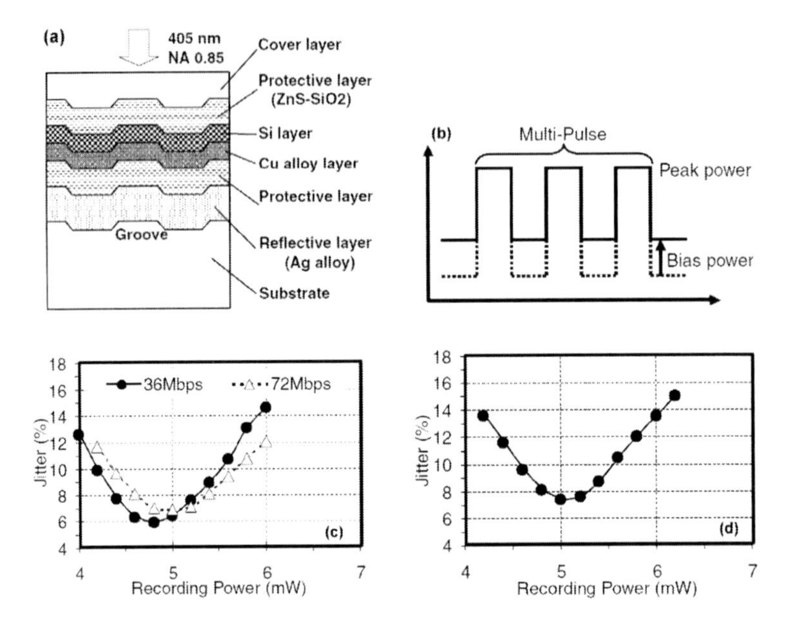

Figure 7.20 (a) Cross section of a write-once disk, (b) write strategy of a 4T signal, and dependence of jitter on the recording power at the recording rates of (c) 36 Mbps (1X), 72 Mbps (2X), and (d) 144 Mbps (4X). Reproduced from Ref. [16], Copyright 2003, Japan Society of Applied Physics.

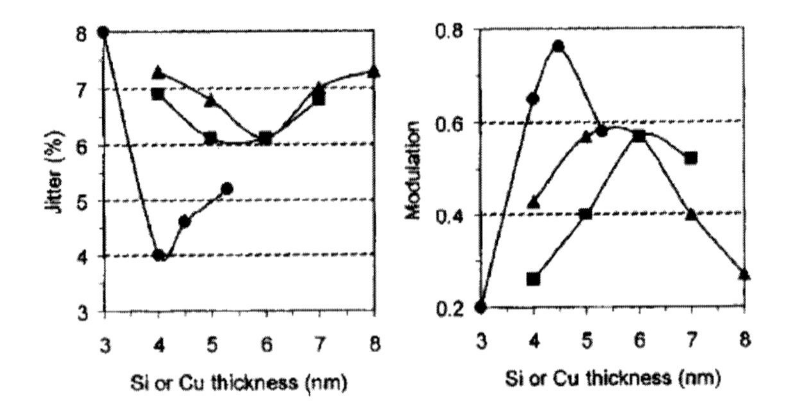

Figure 7.21 Multitrack jitter and modulation for various Cu/Si combinations: Cu/Si = 1:1 (circles), x nm Cu/6 nm Si (triangles), and 6 nm Cu/x nm Si (squares). Reproduced from Ref. [47], Copyright 2005, American Institute of Physics.

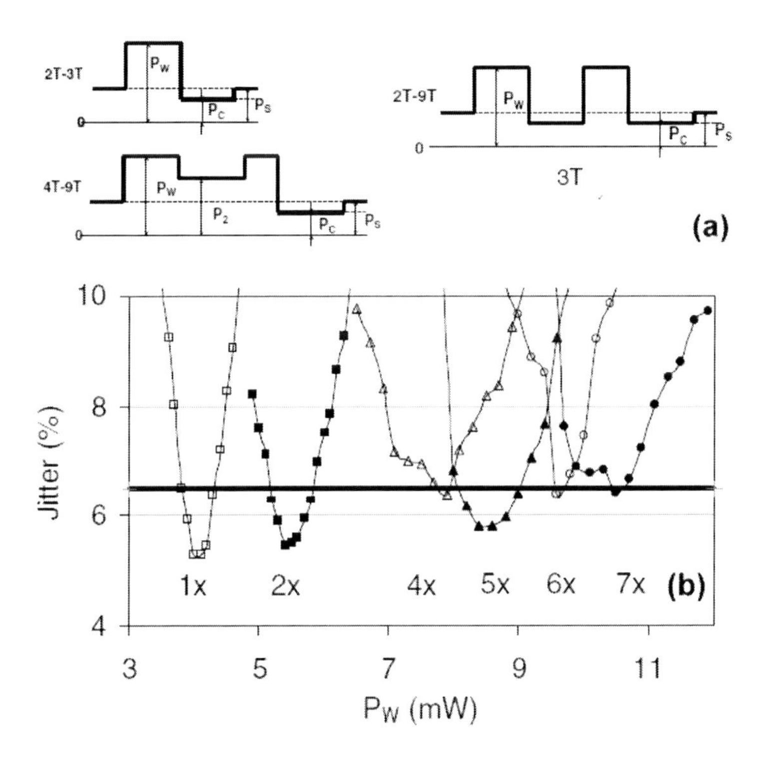

Figure 7.22 (a) The castle-type write strategy (left) and the conventional (n – 1) strategy (right) and (b) the jitter data of a 25 GB disk at different recording speeds. Reproduced from Ref. [64], Copyright 2006, Japan Society of Applied Physics.

On the basis of the mechanical properties of the standard 12 cm diameter PC disks, and because of safety arguments, the disc revolution speed is maximized at 10,000 rpm. For BDs, the speed race for conventional recording would stop at 12X. To reach this ultimate recording speed, Schleipen et al. used a power-controlled transition (PCT) addressing scheme for precise and well-controlled timing of the optical laser output [18]. The applied PCT scheme allows for a time resolution in the 10 ps range and does not require special ultrahigh speed electronics, making it a very suitable candidate for future drive implementation. For their recording experiments, a standard 25 GB BD with CuSi as a recording layer and a castle-type write strategy were used. Figure 7.23 shows the readout performance of disks written at 8X recording speed, with and without applying the PCT scheme. Apparently, the PCT scheme is required to achieve good recording performance with bottom jitters down to 6% for 8X recording. For 10X recording, the bottom jitter was found to slightly increase to 7.2%. At higher speeds > 10X, the recording/readout performance clearly starts to degrade, as illustrated in Fig. 7.24. Although the disks have not been optimized yet for high-speed recording, good jitter levels can still be obtained for single-track recording, illustrating the feasibility of CuSi as a recording material for high speed (up to 12X) BD recording.

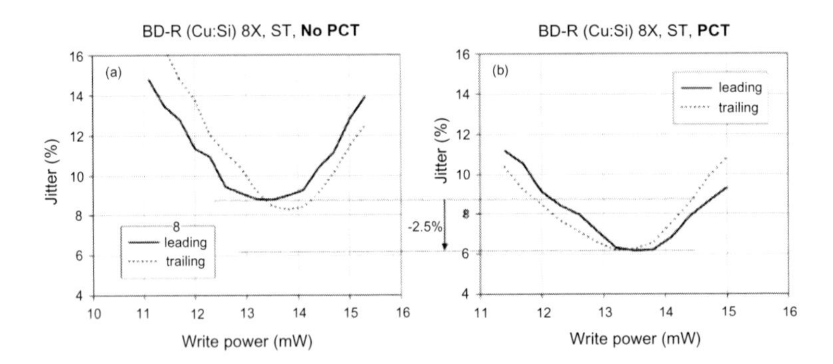

Figure 7.23 Readout performance of disks written at 8X recording speed, with and without applying the PCT scheme. Reproduced from Ref. [18], Copyright 2006, Japan Society of Applied Physics.

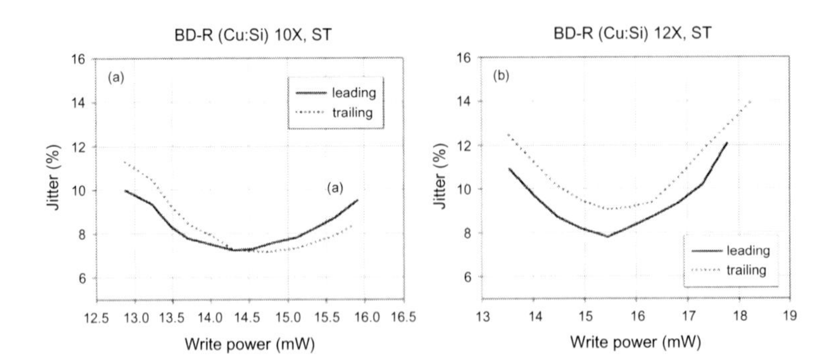

Figure 7.24 Readout performance of disks written at (a) 10X and (b) 12X recording speeds, with applying the PCT scheme. Reproduced from Ref. [18], Copyright 2006, Japan Society of Applied Physics.

Her et al. estimated the maximum data transfer rates that can be achieved by the a-Si/Cu bilayer at different blue laser powers based upon the static testing results [17]. Figure 7.25a–c shows the reflectivity variations with time for a Cu/a-Si bilayer being irradiated by blue laser pulses with various powers for 50 ns, 20 ns, and 10 ns. It was found that the Cu/a-Si bilayer showed a significant increase in reflectivity, which corresponds to the crystallization of a-Si, once the laser power was above a certain level. Here the laser power required to trigger the crystallization of the Cu/a-Si bilayer was defined as the threshold power. As the durations of irradiation were set at 50 ns, 20 ns, and 10 ns, the threshold powers to trigger the crystallization of the Cu/a-Si bilayer were found to be 8 mW, 10 mW, and 10 mW, respectively. In other words, the minimum durations to achieve successful recording for laser powers of 6 mW, 8 mW, and 10 mW were about 100 ns, 50 ns, and 10 ns, respectively. For a BD with a wavelength of 405 nm and a numerical aperture of 0.85, the effective laser spot size (d) is about 290 nm and the data bit length (l_b) is 130 nm. Since the effective recording time (t_{eff}) is equal to the ratio of the effective laser spot size to linear scan velocity (V), that is, $t_{eff} = d/V$, the maximum linear scan velocity and maximum data transfer rate can be estimated if the minimum duration for successful recording is known. Since the minimum durations to achieve successful recording for laser powers of 6 mW, 8 mW, and 10 mW were about 100 ns, 50 ns, and 10 ns, respectively, the maximum linear scan velocities and data transfer rates for the BD using a Cu/a-Si bilayer as a recording film could be estimated to be approximately 3 m/s, 6 m/s, and 29 m/s, and 23 Mb/s, 46 Mb/s, and 223 Mb/s, respectively. Apparently, high data transfer rates can be readily achieved by the Cu/a-Si bilayer.

The dynamic tests were also performed by Her et al. to evaluate the feasibility of the a-Si/Ni bilayer for write-once Blu-ray recording [20]. Figure 7.26 shows the dependence of the carrier-to-noise ratio (CNR) for a 3T mark on the recording power at linear velocities of 4.4 m/s and 6.6 m/s. It is evident that as the linear velocities were adjusted at 4.4 m/s and 6.6 m/s, the ideal recording powers were located at 6 mW and 7 mW and the maximum values of CNR for 3T could reach 41 dB and 43 dB, respectively, demonstrating high potentiality for practical use.

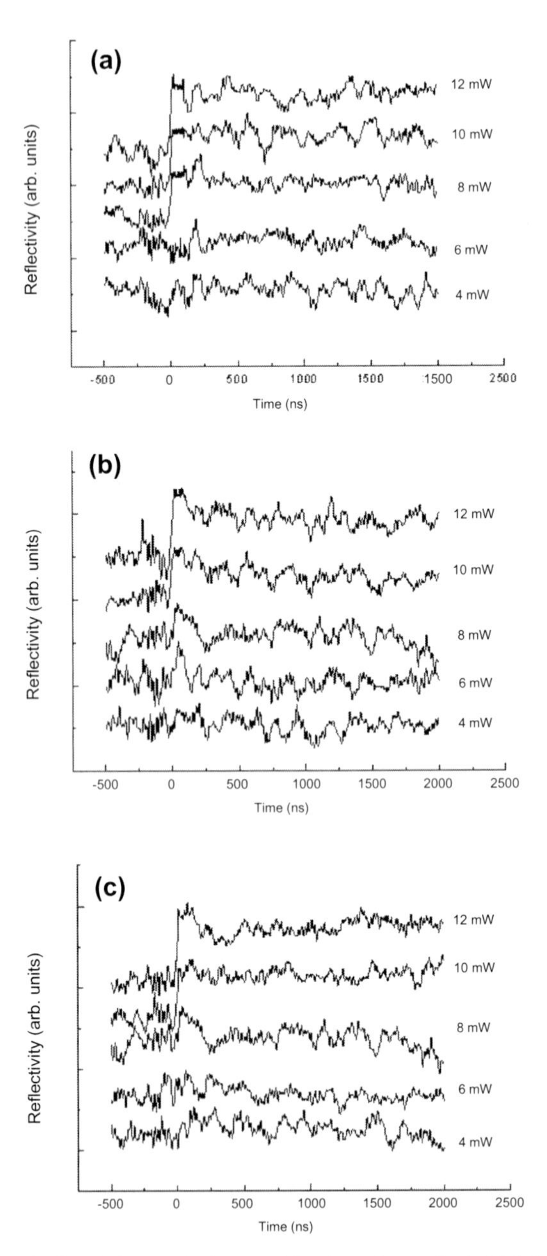

Figure 7.25 Reflectivity variations with time for a Cu/a-Si bilayer being irradiated by blue laser pulses with various powers for (a) 50 ns, (b) 20 ns, and (c) 10 ns. Reproduced from Ref. [17], Copyright 2004, Japan Society of Applied Physics.

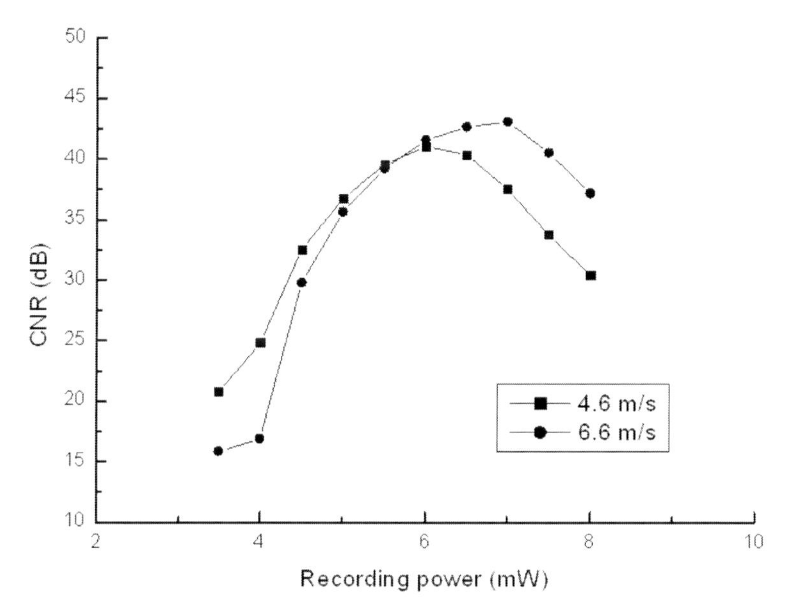

Figure 7.26 Dependence of the CNR for a 3T mark on the recording power at the linear velocities of 4.4 and 6.6 m/s. Reproduced from Ref. [20], Copyright 2004, Japan Society of Applied Physics.

Recently, Ou et al. demonstrated the feasibility of the Si/CuSi bilayer for a 1X–4X BD-R [49]. The Ag reflective layer (95 nm), the ZnS–SiO_2 upper dielectric layer (35 nm), the Si (3 nm)/CuSi (16 nm) recording layer, and the ZnS–SiO_2 lower dielectric layer (24 nm) were sequentially deposited on the 1.1 mm thick PC substrate, and then a PC transparent cover layer was spin-coated on the top of these layers. The dynamic test results of the BD-R with various writing powers at 1X and 4X recording speeds was shown in Fig. 7.27. The optimum jitter values were found to be 5.2% at 4.75 mW and 6.1% at 7.9 mW, respectively, for 1X and 4X recording speeds. The Si/CuSi BD also exhibited a wider recording power margin than that in the Si/Cu BD. This proved that the Si/CuSi bilayer is suitable for a BD-R.

7.4.2 The a-Ge/Metal Bilayer System

Wu et al. reported the feasibility of Ge/Au bilayer for a high-density digital versatile disk (DVD) write-once disk [23]. A lower dielectric

Ge/Au bilayer, an upper dielectric layer, and a Ag reflective layer were sequentially deposited on a 0.6 mm thick PC substrate, and then the sample was bonded with a dummy substrate 0.6 mm thick. The partial response signal-to-noise ratio (PRSNR) and the simulated bit error rate (SbER) were evaluated by using a dynamic tester (ODU 1000, Pulstec), as shown in Fig. 7.28. For a high-density DVD, the suggested PRSNR and SbER need to be larger than 15 dB and lower than 5×10^{-5}, respectively. Accordingly, a suitable power for 1X recording speed is 7.8–8.2 mW and 8.8–9.1 mW for 2X recording speed. The contrast was between 70% and 78% at a wavelength of 405 nm.

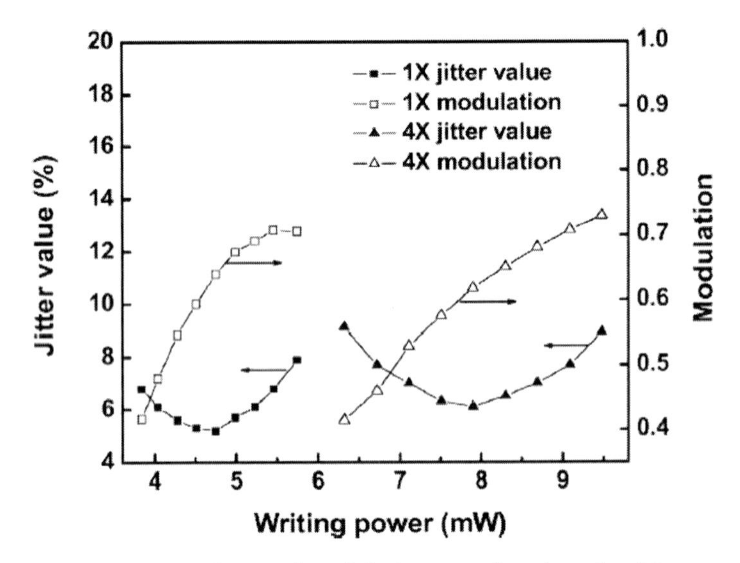

Figure 7.27 Jitter values and modulations as a function of writing power at 1X and 4X recording speeds. Reproduced from Ref. [49], Copyright 2011, American Institute of Physics.

Her et al. estimated the maximum data transfer rates that can be achieved by the a-Ge/Ni [24] and a-Ge/Cu [25] bilayers at different blue laser powers by using static tests. Figures 7.29a and 7.29b show TEM images of the a-Ge/Ni and the a-Ge/Cu bilayer, respectively, after irradiation by a 399 nm blue pulsed laser with various powers for different durations. For the a-Ge/Ni bilayer, at the recording powers of 3 mW, 4 mW, 5 mW, and 6 mW, the minimum durations to form a clear recording mark were found to be 100 ns, 40 ns, 30 ns, and 20 ns, respectively, corresponding to the maximum data transfer rates of 22 Mbits/s, 56 Mbits/s, 74 Mbits/s, and 112 Mbits/s. For the a-Ge/Cu bilayer, the maximum data transfer rates of 44 Mbits/s, 56 Mbits/s, 74 Mbits/s, and 112 Mbits/s, respectively, could be achieved at recording powers of 3 mW, 4 mW, 5 mW, and 6 mW, respectively.

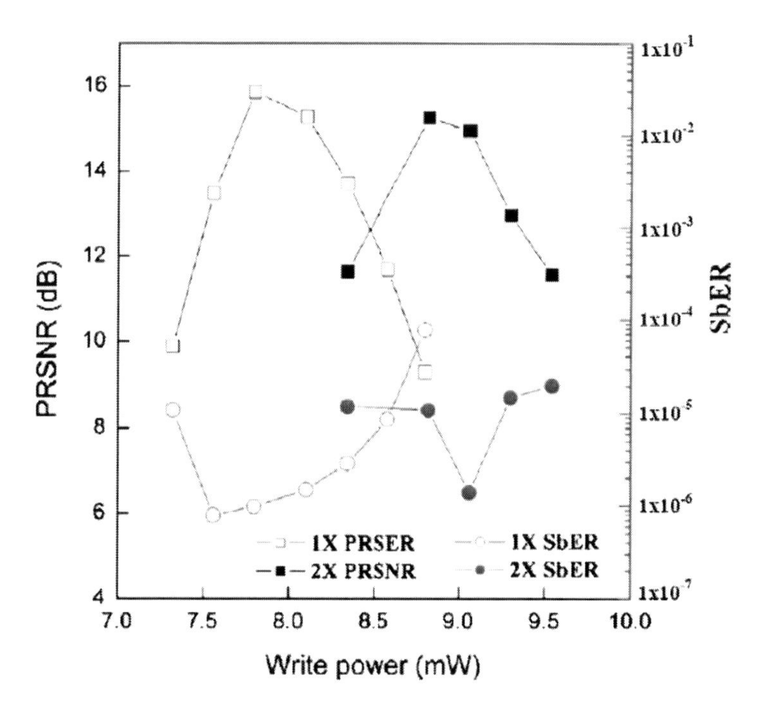

Figure 7.28 PRSNR and SbER as a function of the peak power at 1X and 2X recording speeds. Reproduced from Ref. [23], Copyright 2007, American Institute of Physics.

7.5 Conclusion

The crystallization characteristics of various a-Si/metal and a-Ge/metal bilayers with the assistance of laser irradiation and their recording performances for use in BD-Rs have been thoroughly reviewed. As the a-Si/Cu bilayer was first adopted by TDK Corporation for use in the BD-R, the laser-assisted crystallization characteristics of a-Si induced by Cu metal and Cu alloys (CuAl and CuSi) have been extensively studied. It is generally accepted that

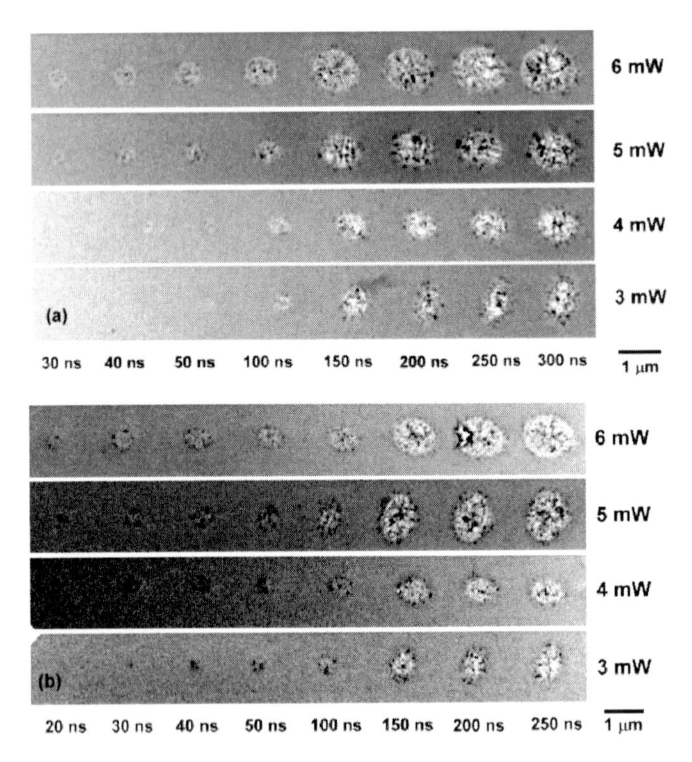

Figure 7.29 TEM images of the (a) a-Ge/Ni and the (b) a-Ge/Cu bilayer, respectively, after irradiation by a 399 nm blue pulsed laser with various powers for different durations. Reproduced from Ref. [24, 25], Copyright 2007, American Institute of Physics.

a crystalline Cu_3Si compound or CuSi alloy nanoparticles will be formed in the initial stage due to the reaction between Cu and Si and then serve as nucleation sites for the crystallization of the unreacted a-Si when the a-Si/(Cu or CuSi) bilayer is subjected to high-power pulsed laser irradiation. However, no Cu_3Si phase is found in the a-Si/Cu-Al bilayer. Instead, Cu- and Si-rich solid-solution phases are formed due to the element mixing caused by laser heating. The laser-assisted crystallization characteristics of a-Si induced by Ni and Al metal thin films are similar to those of a-Si induced by pure Cu metal and CuAl alloy thin films, respectively. On the basis of kinetics studies, it is also found that the activation energies for crystallization of a-Si induced by Cu and Al metal layers under pulsed laser annealing are about 1 order of magnitude lower than those under thermal annealing, which can be ascribed to the formation of a Cu_3Si phase and the subsequent EC of a-Si by mechanical impact with a high-power pulsed laser. For a-Ge/metal bilayer systems, the laser-assisted crystallization characteristics of a-Ge induced by Au, Cu, and Ni metal layers are found to be analogous to those of a-Si induced by Cu and Ni, where Ge_2Au_3 crystallites or germanides (Cu_3Ge, Ni_5Ge_3, and NiGe) will precipitate first and then serve as nucleation sites for the crystallization of the remaining a-Ge.

Regarding the practical recording performances for use in BD-Rs, the a-Si/Cu bilayer recording film achieves jitter values of less than 8%, with a recording peak power of 5 mW at recording rates ranging from 36 Mbps (2X) to 144 Mbps (4X). By adjusting the ratio of Cu/Si layers and optimizing the write strategy, a lowest jitter value of 4% can be achieved in the Cu (4 nm)/Si (4nm) bilayer and good jitter values can be obtained at 7X recording speed using a castle-type strategy. The recording speed of the a-Si/Cu bilayer can be further increased to 12X by applying a PCT addressing scheme. Besides the a-Si/Cu bilayer, the a-Si/Ni bilayer can achieve 41 dB and 43 dB of CNR for 3T at recording powers of 6 mW and 7 mW, respectively, and the Si/Cu–Si bilayer can obtain jitter values of 5.2% and 6.1% at 4.75 mW and 7.9 mW, respectively, for 1X and 4X recording speeds, demonstrating their high potentiality for practical use. The a-Ge/Au, a-Ge/Ni, and a-Ge/Cu bilayers also show their feasibility for use in BD-Rs. The a-Ge/Au bilayer can be recorded at 2X recording speed at a power of 8.8–9.1 mW and can obtain optical contrast between 70% and 78% at a wavelength of 405 nm. The a-Ge/Ni and a-Ge/Cu bilayers demonstrate that they can be successfully recorded at a recording power as low as 3 mW. However, they may also suffer from poor thermal stability, which might be the reason why a-Ge/metal bilayers have not been practically adopted for use in BD-Rs.

Acknowledgments

This work was sponsored mainly by the National Science Council of the Republic of China under grant no. NSC98-2221-E005-033-MY2 and supported in part by the Ministry of Education under the ATU plan.

References

1. Kasami, Y., Kuroda, Y., Seo, K., Kawakubo, O., Takagawa, S., Ono, M., Yamada, M. (2000) Large capacity and high-data-rate phase-change disks. *Japanese Journal of Applied Physics*, **39**, 756–761.
2. Tieke, B., Dekker, M., Pfeffer, N., van Wouderberg, R., Zhou, G.-F., Ubbens, I. P. D. (2000) High data-rate phase-change media for the digital video recording system. *Japanese Journal of Applied Physics*, **39**, 762–765.
3. Inoue, H., Hirata, H., Kato, T., Shingai, H., Utsunomyia, H. (2001) Phase change disc for high data rate recording. *Japanese Journal of Applied Physics*, **40**, 1641–1642.
4. Schep, K., Stek, B., Woudenerg, R. von, Blum, M., Kobyashi, S., Narahara, T., Yamagami, T., Ogawa, H. (2001) Format description and evaluation of the 22.5 GB digital-video-recording disc. *Japanese Journal of Applied Physics*, **40**, 1813–1816.
5. Kato, T., Hirata, H., Komaki, T., Inoue, H., Shingai, H., Hayashida, N., Utsunomiya, H. (2001) The phase change optical disc with the data recording rate of 140 Mbps. *Japanese Journal of Applied Physics*, **41**, 1664–1667.
6. Ohkubo, S, Ide, T., Okada, M. (2001) Basic study of write-once media for blue-violet laser. *Technical Digest Optical Data Storage*, 34–36.
7. Kawazu, Y., Kudo, H., Onari, S., Arai, T. (1990) Low-temperature crystallization of hydrogenated amorphous silicon induced by nickel silicide formation. *Japanese Journal of Applied Physics*, **29**, 2698–2700.
8. Lee, J.-B., Lee, C.-J., Choi, D.-K. (2001) Influences of various metal elements on field aided lateral crystallization of amorphous silicon film. *Japanese Journal of Applied Physics*, **40**, 6177–6181.
9. Her, Y. C., Wu, C. L. (2004) Crystallization kinetics of Cu/a-Si bilayer recording film under thermal and pulsed laser annealing. *Journal of Applied Physics*, **96**, 5563–5568.
10. Lee, S. W., Jeon, Y. C., Joo, S. K. (1995) Pd induced lateral crystallization of amorphous Si thin films. *Applied Physics Letters*, **66**, 1671–1673.
11. Kawazu, Y., Kudo, H., Onari, S., Arai, T. (1990) Initial stage of the interfacial reaction between nickel and hydrogenated amorphous silicon. *Japanese Journal of Applied Physics*, **29**, 729–738.
12. Hultman, L., Robertsson, A., Hentzell, H. T. G., Engstrom, I., Psaras, P. A. (1987) Crystallization of amorphous silicon during thin-film gold reaction. *Journal of Applied Physics*, **62**, 3647–3655.
13. Bian, B., Yie, J., Li, B., Wu, Z. (1993) Fractal formation in a-Si:H/Ag/a-Si:H films after annealing. *Journal of Applied Physics*, **73**, 7402–7406.

14. Haque, M. S., Naseem, H. A., Brown, W. D. (1994) Interaction of aluminum with hydrogenated amorphous silicon at low temperatures. *Journal of Applied Physics*, **75**, 3928–3935.

15. Yoon, S. Y., Park, S. J., Kim, K. H., Jang, J. (2001) Metal-induced crystallization of amorphous silicon. *Thin Solid Films*, **383**, 34–38.

16. Inoue, H., Mishima, K., Aoshima, M., Hirata, H., Kato, T., Utsunomiya, H. (2003) Inorganic write-once disc for high speed recording. *Japanese Journal of Applied Physics*, **42**, 1059–1061.

17. Her, Y. C., Wu, C. L. (2004) Feasibility of Cu/a-Si bilayer for high data-transfer-rate write-once blue-ray recording. *Japanese Journal of Applied Physics*, **43**, 1013–1017.

18. Schleipen, J., Hesen, R., Jaegers, J., Janssen, A., Rijpers, J., Sonneville, P. (2007) High speed laser modulation for up to 12X Blu-ray disc CuSi recording. *Japanese Journal of Applied Physics*, **46**, 4007–4010.

19. Chen, B. M., Chen, H. F., Yeh, R. L., Chung, J. M. (2004) Inorganic writ once media. *Japanese Journal of Applied Physics*, **43**, 5018–5019.

20. Her, Y. C., Jean, S. T., Wu, J. L. (2007) Crystallization kinetics and recording mechanism of a-Si/Ni bilayer for write-once blue-ray recording. *Journal of Applied Physics*, **102**, 093503-1–093503-7.

21. Hu, X., Shi, L. P., Chua, E. K., Wang, W. W., Lee, M. L., Chong, T. C., Miao, X. S. (2007) Dual-speed inorganic writ once disk with low-to-high polarity. *Japanese Journal of Applied Physics*, **46**, 3922–3925.

22. Germain, P., Squelard, S., Bourgoin, J., Gheorghiu, A. (1977) Crystallization kinetics of amorphous germanium. *Journal of Applied Physics*, **48**, 1909–1913.

23. Wu, T. H., Kuo, P. C., Fang, Y. H., Chen, J. P., Yen, J. P., Jeng, T. R., Wu, C. Y., Huang, D. R. (2007) Microstructure and recording mechanism of Ge/Au bilayer media for write-once optical disc. *Applied Physics Letters*, **90**, 151111-1–151111-3.

24. Her, Y. C., Chen, J. H., Tsai, M. H., Tu, W. T. (2009) Nickel-induced crystallization of amorphous Ge film for blue-ray recording under thermal annealing and pulsed laser irradiation. *Journal of Applied Physics*, **106**, 023530-1–023530-5.

25. Her, Y. C., Tu, W. T., Tsai, M. H. (2012) Phase transformation and crystallization kinetics of a-Ge/Cu bilayer for blue-ray recording under thermal annealing and pulsed laser irradiation. *Journal of Applied Physics*, **111**, 043503-1–043503-6.

26. Herd, S. R., Chaudhari, P., Brodsky, M. H. (1972) Metal contact induced crystallization in films of amorphous silicon and germanium. *Journal of Non-Crystalline Solids*, **7**, 309–327.

27. Nast, O., Wenham, S. R. (2000) Elucidation of the layer exchange mechanism in the formation of polycrystalline silicon by aluminum-induced crystallization. *Journal of Applied Physics*, **88**, 124–132.

28. Widenborg, P. I., Aberle, A. G. (2002) Surface morphology of poly-Si films made by aluminium-induced crystallisation on glass substrates. *Journal of Crystal Growth*, **242**, 270–282.

29. Masaki, Y., Ogata, T., Ogawa, H., Jones, D. I. (1994) Kinetics of solid phase interaction between Al and a-Si:H. *Journal of Applied Physics*, **76**, 5225–5231.

30. Radnoczi, G., Robertsson, A., Hentzell, H. T. G., Gong, S. F., Hasan, M. A. (1991) Al induced crystallization of a-Si. *Journal of Applied Physics*, **69**, 6394–6399.

31. Ashtikar, M. S., Sharma, G. L. (1995) Silicide mediated low temperature crystallization of hydrogenated amorphous silicon in contact with aluminum. *Journal of Applied Physics*, **78**, 913–918.

32. Konno, T. J., Sinclair, R. (1992) Crystallization of silicon in aluminium/amorphous-silicon multilayers. *Philosophical Magazine B: Physics of Condensed Matter Statistical Mechanics Electronic Optical and Magnetic Properties*, **66**, 749–765.

33. Gore, G. (1855) On a peculiar phenomenon in the electro-deposition of antimony. *Philosophical Magazine*, **9**, 73–74.

34. Thompson, M. O., Galvin, G. J., Mayer. J. W., Peercy, P. S., Poate, J. M., Jacobson, D. C., Cullis, A. G., Chew, N. G. (1984) Melting temperature and explosive crystallization of amorphous silicon during pulsed laser irradiation. *Physical Review Letters*, **52**, 2360–2363.

35. Wickersham, C. E., Poole, J. E. (1988) Explosive crystallization in zirconium/ silicon multilayers. *Journal of Vacuum Science & Technology A: Vacuum. Surfaces, and Films*, **6**, 1699–1702.

36. Mineo, A., Matsuda, A., Kurosu, T., Kikuchi, M. (1973) Observations of propagation in shock-crystallization of sputtered amorphous germanium films. *Solid State Communications*, **13**, 1307–1310.

37. Floro, J. A. (1986) Propagation of expolsive crystallization in thin Rh/ Si multilayer films. *Journal of Vacuum Science & Technology A: Vacuum. Surfaces, and Films*, **4**, 631–636.

38. Koba, R., Wickersham, C. E. (1982) Temperature and thickness effects on the explosive crystallization of amorphous germanium films. *Applied Physics Letters*, **40**, 672–675.

39. Avrami, M. (1939) Kinetics of phase change. I general theory. *Journal of Chemical Physics*, **7**, 1103–1112.

40. Avrami, M. (1940) Kinetics of Phase Change. II transformation–time relations for random distribution of nuclei. *Journal of Chemical Physics*, **8**, 212–224.

41. Avrami, M. (1941) Granulation. Phase change, and microstructure kinetics of phase change. III. *Journal of Chemical Physics*, **9**, 177–184.

42. Ohshima, N. (1996) Crystallization of germanium–antimony–tellurium amorphous thin film sandwiched between various dielectric protective films. *Journal of Applied Physics*, **79**, 8357–8363.

43. Peng, C., Cheng, L., Mansuripur, M. (1997) Experimental and theoretical investigations of laser-induced crystallization and amorphization in phase-change optical recording media. *Journal of Applied Physics*, **82**, 4183–4191.

44. Miyamoto, M., Hirotsune, A., Miyauchi, Y., Ando K., Terao, M., Tokusyuku, N., Tamura, R. (1998) Analysis of mark-formation process for phase-change media. *IEEE Journal of Selected Topics in Quantum Electronics*, **4**, 826–831.

45. Adach, S. (1999) *Optical Constants of Crystalline and Amorphous Semiconductors* (Kluwer Academic Publishers, MA, USA).

46. Moon, S., Hatano, M., Lee, M., Grigoropoulos, C. P. (2002) Thermal conductivity of amorphous silicon thin films. *International Journal of Heat Mass Transfer*, **45**, 2439–2447.

47. Kuiper, A. E. T., Vullers, R. J. M., Pasquariello, D., Naburgh, E. P. (2005) Cu-Si bilayers as storage medium in optical recording. *Applied Physics Letters*, **86**, 221921-1–221921-3.

48. Mai, H.-C., Hsieh, T.-E., Jeng, S.-Y. (2011) Characterization of write-once Blu-ray disc containing Cu-Al/Si recording layer using transmission electron microscopy. *Applied Physics Letters*, **98**, 094103-1–094103-3.

49. Ou, S.-L., Kuo, P.-C., Chen, S.-C., Tsai, T.-L., Yeh, C.-Y., Chang, H.-F., Lee, C.-T., Chiang, D. (2011) Crystallization mechanisms and recording characteristics of Si/CuSi bilayer for write-once Blu-ray disc. *Applied Physics Letters*, **99**, 121908-1–121908-3.

50. Lee, S. W., Joo, S. K. (1996) Low temperature poly-Si thin-film transistor fabrication by metal-induced lateral crystallization, *IEEE Electron Device Letters*, **17**, 160–162.

51. Haque, M. S., Naseem, H. A., Brown, W. D. (1994) Interaction of aluminum with hydrogenated amorphous silicon at low temperatures. *Journal of Applied Physics*, **75**, 3928–3935.

52. Ishihara, S., Kitagawa, M., Hirao, T. (1987) Low-temperature crystallization of hydrogenated amorphous silicon films in contact

with evaporated aluminum electrodes. *Journal of Applied Physics*, **62**, 837–840.

53. Her, Y. C., Chen, C. W. (2007) Crystallization kinetics of ultrathin amorphous Si film induced by Al metal layer under thermal annealing and pulsed laser irradiation. *Journal of Applied Physics*, **101**, 043518-1–043518-7.

54. Higashi, S., Sameshima, T. (2001) Pulsed-laser-induced microcrystallization and amorphization of silicon thin films. *Japanese Journal of Applied Physics*, **40**, 480–485.

55. Wu, T. H., Kuo, P. C., Ou, S. L., Chen, J. P., Yen, J. P., Jeng, T. R., Wu, C. Y., Huang, D. R. (2008) Diffusion and crystallization mechanism of Ge/Au bilayer media for write-once optical disc. *Applied Physics Letters*, **92**, 011126-1–011126-3.

56. Thaddeus, B., Murray, J. L., Bennett, L. H., Baker, H. (1986) *Binary Alloys Phase Diagrams*, 2nd ed., Vol. 1 (Amercian Society for Metals, Metals Park, OH, USA), 373.

57. Gaudet, S., Detavernier, C., Lavoie, C., Desjardins, P. (2006) Reaction of thin Ni films with Ge: phase formation and texture. *Journal of Applied Physics*, **100**, 034306-1–034306-10.

58. Zhang, S.-L. (2003) Nickel-based contact metallization for SiGe MOSFETs: progress and challenges. *Microelectronic Engineering*, **70**, 174–185.

59. Jin, L. J., Pey, K. L., Choi, W. K., Fitzgerald, E. A., Antoniadis, D. A., Piters, A. J., Lee, M. L., Chi, D. Z., Tung, C. H. (2004) The interfacial reaction of Ni with (111)Ge, (100)$Si_{0.75}Ge_{0.25}$ and (100)Si at 400 °C. *Thin Solid Films*, **462–463**, 151–155.

60. Hsu, S. L., Chien, C. H., Yang, M. J., Huang, R. H., Leu, C. C., Shen, S. W., Yang, T. H. (2005) Study of thermal stability of nickel monogermanide on single- and polycrystalline germanium substrates. *Applied Physics Letters*, **86**, 251906-1–251906-3.

61. Nemouchi, F., Mangelinck, D., Bergman, C., Clugnet, G., Gas, P., Labar, J. L. (2007) Simultaneous growth of Ni_5Ge_3 and NiGe by reaction of Ni film with Ge. *Applied Physics Letters*, **89**, 131920-1–131920-3.

62. Doyle, J. P., Svensson, B. G., Johansson, S. (1995) Morphological instability of bilayers of copper germanide films and amorphous germanium. *Applied Physics Letters*, **67**, 2804–2806.

63. Hong, S. Q., Comrie, C. M., Russell, S. W., Mayer, J. W. (1991) Phase formation in Cu-Si and Cu-Ge. *Journal of Applied Physics*, **70**, 3655–3660.

64. Vullers, R. J. M., Kuiper, A. E. T., Pasquariello, D. (2006) High-speed 7X CuSi-based write-once Blu-ray disc. *Japanese Journal of Applied Physics*, **45**, 1219–1222.

Index